HIGH–ENERGY CHEMISTRY

Ellis Horwood and Prentice Hall
are pleased to announce their collaboration in a new imprint whose list will encompass outstandi
works by world-class chemists, aimed at professionals in research, industry and academia. It is
intended that the list will become a by-word for quality, and the range of disciplines in chemical
science to be covered is:

ANALYTICAL CHEMISTRY
ORGANIC CHEMISTRY
INORGANIC CHEMISTRY
PHYSICAL CHEMISTRY
POLYMER SCIENCE & TECHNOLOGY
ENVIRONMENTAL CHEMISTRY
CHEMICAL COMPUTING & INFORMATION SYSTEMS
BIOCHEMISTRY
BIOTECHNOLOGY

Ellis Horwood   PTR Prentice Hall
PHYSICAL CHEMISTRY SERIES

Series Editors:
Ellis Horwood, M.B.E.
Professor T. J. Kemp, University of Warwick

Current titles in the
Ellis Horwood   PTR Prentice Hall
Physical Chemistry Series

Blandamer	**CHEMICAL EQUILIBRIA IN SOLUTION:** **Dependence of Rate and Equilibrium Constants on Temperature and Pressure**
Bugaenko	**HIGH-ENERGY CHEMISTRY**
Horspool & Armesto	**ORGANIC PHOTOCHEMISTRY:** **A Comprehensive Treatment**
Navratil	**NUCLEAR CHEMISTRY**

HIGH–ENERGY CHEMISTRY

L.T. BUGAENKO, Dr. Sc. (Chem.),
M.G. KUZMIN, Dr. Sc. (Chem.),
both Professors of Moscow University, Russia

L.S. POLAK, Dr. Sc. (Phys. Math.),
Professor of the Institute of Petrochemical Synthesis of the
Academy of Sciences, Moscow, Russia

Translation Editor:
Professor T.J. KEMP
Department of Chemistry, University of Warwick

ELLIS HORWOOD   PTR PRENTICE HALL
NEW YORK   LONDON   TORONTO   SYDNEY   TOKYO   SINGAPORE

English Edition first published in 1993
in coedition between
ELLIS HORWOOD LIMITED
Market Cross House, Cooper Street,
Chichester, West Sussex, PO19 1EB, England

A division of
Simon & Schuster International Group
A Paramount Communications Company

and

KHIMIYA PUBLISHERS
13 Stromynka Street, Moscow

© 1993 Translation: © Ellis Horwood Limited, 1993
Translated by Dr S.V. Zatonsky

*Distributed in Albania, Bulgaria, Chinese People's Republic, Cuba, Czechoslovakia, Germany
Hungary, Korean People's Democratic Republic, Mongolia, Poland, Rumania, Vietnam,
the C.I.S., Yugoslavia*

Printed and bound in Great Britain
by Bookcraft Ltd, Midsomer Norton, Avon

British Library Cataloguing in Publication Data

A catalogue record for this book is available from the British Library

ISBN 0–13–388646–8

Library of Congress Cataloging-in-Publication Data

Available from the publisher

Table of contents

Introduction

This book[†] represents the first attempt undertaken to combine under the common heading 'high-energy chemistry' (HEC) three historically distinct subjects, namely plasma chemistry, radiation chemistry, and photochemistry, the motivation being that they have strong common roots and aim to solve identical problems. In trying to define the term 'high-energy chemistry', which perhaps does not seem clear to the novice, one has to formulate an idea as to what 'high' (or large) energy means. In science, the word 'large' makes sense only when something is compared with something else, choosing a particular scale and/or coordinate system. Here, 'high' refers to energy introduced into a system which is larger ('hyperthermal' energy $> kT$) than that needed for most energy-consuming molecular processes and the resulting reactions. This energy is supplied not in the form of heat but through the impact either of an electron, ion, atom, molecule, or photon or by application of electric, magnetic, etc. fields. Under these circumstances a thermodynamically non-equilibrium concentration of highly reactive species, viz. electrons, ions, excited molecules, radicals, and molecular fragments is created which causes processes to occur in an essentially different way compared with normal or 'thermal' chemistry. The specific conditions generated in the reacting system by this energy induce, in turn, new types of process unlike normal (thermal) ones. The investigation of these processes is the mission of high-energy chemistry. We cannot, at the present, suggest a more correct or precise definition of the subject. Let this be the task of our successors in this field of science, who might choose to write an entirely different sort of book.

In the first place it should be noted that the reacting systems studied in high-energy chemistry are mostly open systems, that is, they interact with their surroundings through movements of mass, energy, etc. For closed system, the

---

† The Introduction was written by L. S. Polak; Chapters 1, 2, 3, 7, and 8, by L. T. Bugaenko, M. G. Kuzmin, and L. S. Polak; Chapter 4 by M. G. Kuzmin; Chapter 5 by L. T. Bugaenko; and Chapter 6 by L. S. Polak.

the second law of thermodynamics is strictly fulfilled, and the concept of one and only one temperature has a clear physical meaning as the key parameter of the Maxwell-Boltzmann distribution. The Arrhenius law can be applied to closed systems. However, in the case of open systems, the entropy does not necessarily increase at any rate as demanded by thermodynamics, namely an entropy increment dS can have both positive and negative values:

$$dS = dS_i + dS_e$$

where dS_i is the entropy flow due to exchange with the surrounding and dS_e is the entropy production inside the system due to irreversible processes (chemical reactions, diffusion, and heat conductance).

Whereas dS_i never can be negative, dS_e has no definite sign. For an isolated system

$$dS_e = 0; \qquad dS_i \geq 0$$

In the stationary state $dS = 0$ and $dS_e = -dS_i < 0$ for a closed system.

In open systems, which are not at equilibrium, at least with $dS < 0$, it may turn out that their total or local order increases (entropy is a measure of disorder, $dS = k \ln W$). Exchanges involving mass and energy in these systems is a necessary factor of their existence. Chemical reactions in systems of this kind are studied by nonequilibrium chemical kinetics. The evolution of systems where the chemical reactions occurring are conjugate, to diffusion-like processes, is described by using nonlinear partial differential equations.

Thus, high-energy chemistry, aims to study the specific chemical and physical processes developing when a large amount of energy is supplied to unit volume of a reacting system during unit time. This energy is great, not only with respect to the value of kT of the surroundings, but exceeds the energy needed to realized essentially new physical and chemical processes. Here, the mutual influence of these processes (channels) on each other and on the surroundings plays an important part. In introducing large amounts of energy into a system, what are the features of the system's behaviour that we should be talking about? These features must be identical for every branch of HEC. Naturally, in addition to the common characteristics specifying HEC as a whole, separate fields of high-energy chemistry have their own distinguishing features. For instance, plasma chemistry is characterized by the reactions of metastable species and by Maxwellization of the distribution functions; radiation chemistry by the formation and dissipation of tracks and by thermalization; and photochemistry by the reactions of excited states, by thermalization of the excitation energy, and by radiationless transitions to other electronic states, etc. Nevertheless, the basic characteristics have to be rather similar, if not identical. Let us consider them briefly.

First of all, there must be some physical source for stimulation of the chemical reactions.[†] The stimulation means not only (if at all) a change (an increase) in rate of the chemical reactions, but the realization of novel, mutually interacting pathways (channels) of physicochemical and chemical processes, the creation of highly reactive species, and a change in the nature of the relation between the reacting system and its surroundings. The physical sources of stimulation render a reacting system multichannel, nonstationary, and unstable to a certain extent. Depending on the energy introduced into the system, the degree of 'openness' of the different channels varies, as does their interaction with each other and with their surroundings.

Accordingly, here are the typical features of processes related to high-energy chemistry:

(a) a high-energy density (volume density or surface density);

(b) nonequilibrium (non-Maxwell and non-Boltzmann distributions of translational energy and quantum level populations for particles of a system);

(c) approximation and convergence of the characteristic times of physical, physicochemical, and chemical processes;

(d) the dominant role played in chemical processes by electrons, excited molecules, ions, and radicals, that is, by those highly-reactive species which are met in conventional chemistry but have a lesser role;

(e) the multichannel character of processes in reacting systems.

In addition, one important circumstance should be noted since it is often forgotten. Being introduced by different physical means, the energy can affect the state and behaviour of the reacting species either directly (electron impact, for example) as observed in most cases, or via an intermediate stage of 'thermal' energy (in dissipation or relaxation processes, for example). This also leaves a certain imprint on the processes of high-energy chemistry.

In Chapters 4 to 6, the reader will see how the general concepts presented above are realized under the specific conditions of the three largest branches of high-energy chemistry. In Table I.1 we have tried to compare the basic characteristics of photochemistry, plasma chemistry, and radiation chemistry. It shows unambiguously how close are the fundamental features of these fields of HEC.

Let us dwell briefly on some of the essential characteristics of chemical reactions taking place under the high energy densities introduced into the reacting system. In addition to the features discussed above, the fraction of the species of energy exceeding the threshold value for a given chemical reaction varies under these conditions, while the high energy ' tail' of the energy distribution is not recovered by energy-exchange collisions since the number of such

---

† Some authors consider the features of high-energy chemistry to be necessarily related to the 'physical methods of induction of chemical reactions' excluding the action of heat on a reacting system. However, this is incorrect. The most important property of chemical systems experiencing high-energy flow, the nonequilibrium character, can also emerge in reactions occurring at high temperatures.

Table I.1. Characteristics of the three branches of high-energy chemistry

Quality	Branch		
	Plasma chemistry	Radiation chemistry	Photochemistry
Nature of system	Open	Open	Open
Nature of process and reaction	Nonequilibrium (quasi-equilibrium in the limit)	Nonequilibrium	Nonequilibrium
Energy distribution function	Non-Maxwell	Non-Maxwell	Non-Maxwell
Population – of electronic levels	Non-Boltzmann	Non-Boltzmann	Non-Boltzmann (up to inversion in the limit)
– of vibrational levels	Non-Boltzmann (shifted Boltzmann in the limit)	Boltzmann (in most cases)	Boltzmann (in most cases)
Reactive transient species	e^-, vibrationally excited molecules, excited atoms and ions	e^-, ions, electroncially excited particles	Electronically excited and vibrationally excited molecules; ions and electrons
Major primary processes	$e^- + M \to M^* + e^-$ $(M^{\pm}) + e^- \to$ Products	$e^-, h\nu + M \to M^+, e^-,$ $(M^+)^*, M^* \to$ Products	$h\nu + M \to M^* \to$ Products
Time	$\tau_e \approx \tau_t$	$\tau_c \approx \tau_d$	$\tau_v < \tau_c \lesssim \tau_e$
Nature of transitions	Diabatic and adiabatic	Diabatic and adiabatic	Diabatic and adiabatic

collisions is quite small but the reaction rates are high. The conventional kinetic theory of chemical reactions and the absolute reaction rate theory turn out to be inappropriate to describe a situation like this.

There is still no general solution to the problem of how strongly a chemical reaction affects the distribution functions depending on the nature of the system and whether this effect, in turn, influences the rate and direction (selectivity) of chemical reactions. In many cases, translational, vibrational, and rotational temperatures differ from each other significantly. In addition, various components of the system (electrons, ions, or heavy particles) may also have different temperatures or can be ascribed no temperature at all. Obviously, classical chemical kinetics are inappropriate for cases like this; the expression deduced for the reaction rate constant is meaningless, and the concentrations of some of the components do not correspond to thermodynamically-equilibrium concentrations.

In conventional chemical kinetics, it is legitimately assumed that reactions have a threshold nature and that there is only one energy barrier from where reaction begins, while the number of particles with a super-barrier energy is usually small. Reactions in high-energy chemistry involve particles in various quantum states including those with a super-barrier energy. The role played by this latter type of species in the mechanism of chemical process may be significant.

These characteristics of high-energy chemistry are typical for any kind of nonequilibrium process.[†] Since the chemical reactions of high energy chemical concern are generally nonequilibrium and, moreover, strongly nonequilibrium in many cases, including the most important, thermodynamic approaches to their analysis give insufficient information or are simply inapplicable. Nonequilibrium chemical kinetics are mainly used to describe processes of this kind.

This discipline studies the evolution in time and space of systems where reactions of excited (and unexcited), charged, and neutral species take place, the translational energy distribution functions and the populations of electronic, vibrational, and rotational levels of these species being usually non-Maxwell and non-Boltzmann. The above reactions occur both in systems in stationary equilibrium and in relaxing systems, and deviations from the equilibrium state cannot be considered as weak.

Assuming the dynamics of collisions (for example, binary collisions) are known, the kinetic behaviour of an ensemble of particles can be described by using the populations of the quantum levels of the molecules, with translational energy distribution functions for the particles, and by the ratio of flux probabilities for different channels.

---

† Of these, heterogeneous chemical processes are of particular significance as playing an important role in plasma chemistry and radiation chemistry.

This situation is not found in explicit form in classical chemical kinetics.[†] The point is that the Maxwell distribution is assumed to exist at zero time, which is not violated by chemical reaction (or is disturbed only slightly) and the population is considered to be Boltzmann-like, with predominant (by many orders of magnitude) occupation of the ground levels; all reacting particles possess the same mean energy (temperature) of the unified Maxwell distribution.

In nonequilibrium chemical kinetics all (or some) of the typical features of classical kinetics are either perturbed or absent. The task of nonequilibrium kinetics is to develop a theory that will be able to predict the experimental kinetic attributes of a chemical reaction, namely the dependence of the reaction rate coefficients on different types of energy (temperature) and pressure, on the initial composition of reactants, on the nature of the reaction products, on the type of 'pumping', and on the cross-section of formation of some products in certain quantum states.

A nonequilibrium system left to itself relaxes to its equilibrium (most probable) state via a sequence of intermediate states whose probability increases monotonically in a closed system. However, most reacting systems in high-energy chemistry are open systems interacting with their surroundings through mass and energy fluxes directed from the surroundings to the system and vice versa. Because of these flows, the overall entropy (or probability) of the system may both increase and decrease, as has been already mentioned. This means the possibility of a decrease in the amount of disorder and the emergence of a new structure with novel properties (self-organization). Fluctuations in an open non-equilibrium system of this sort can result in new stable states. In a system like this, gas-phase reactions (heterogeneous reactions included) may be accompanied by formation of a solid phase with unique and unforeseen structure and properties (from the standpoint of thermodynamics).

We have now presented the major properties of HEC processes as revealed for radiation chemistry, photochemistry, and plasma chemistry. However, these features (or some of them) are typical of a number of other branches of chemistry which should be also assigned to HEC. The following comprise a perhaps incomplete list: photoradiation chemistry, laser chemistry, mechano-chemistry, chemistry of ultrasonic and infrasonic waves (sonochemistry), chemistry of processes induced by IR and microwave radiation, chemistry of extraterrestrial atmospheres, cosmochemistry, chemistry of processes induced by nuclear transformations (Mössbauer effect, recoil atom chemistry, etc.), chemistry of 'new' atoms (positronium, muonium), chemistry of dissolution of activated solids, electrochemistry of electrode processes, photoelectrochemistry,

---

† Strictly speaking, chemical kinetics is still a semi-quantitative branch of science. The cross-sections are quite rarely known with an accuracy better than 50%. Activation energies for chemical reactions are determined mostly from measurements of reaction rate constants $k(T)$, and the Born–Oppenheimer potential barrier even in the best cases is determined with an accuracy of only 4 to 8 kJ mol^{-1}.

molecular beam chemistry, electric discharge chemistry of solids and liquids, exoelectronic and exoionic emission chemistry, cryochemistry, chemistry of processes in ultrasonic jets, flame chemistry, and chemistry of processes in plasma accelerators and MHD-generators.

To work out the correct strategy for controlling chemical reactions and their space-and-time scale, it is necessary to obtain sufficient information on the dynamics and kinetics of the reactions, on the energy and space-time distributions of substances participating in reactions, on the effect of hydrodynamic conditions upon chemical processes, on the reverse effects of these processes, etc.

The study of the behaviour of open, nonequilibrium, unstable, nonstationary, and nonlinear reacting systems, using a realistic model of a molecule (a molecule in a given quantal, electronic, vibrational, and rotational state with a defined energy of translational motion, for example N_2 ($X'\Sigma_g^+$, $v = 15$, ε_{kin}), makes it possible to solve the above problems, thus providing the development of a scientific basis for new advanced technologies.

Chapters 4 (photochemistry), 5 (radiation chemistry), and 6 (plasma chemistry) disclose to the reader implementation of the general ideas developed in the Introduction and in the first three chapters of this book to each of these fields. The way in which we present the material, for example in Chapter 4 on photochemistry, is intended primarily for radiation chemists, plasma chemists, and research workers engaged in other fields of high-energy chemistry, but not specifically for professionals active in photochemistry. The same approach is also adopted in Chapters 5 and 6.

We are pleased to express our gratitude to I. A. Abramenkova, V. M. Byakov, E. P. Kalyazin, A. A. Levitskii, and V. V. Saraeva for helpful discussions, to V. A. Goldin for some of the figures, and to N. M. Rytova for her help in preparing the typesecript.

1

Absorption and transfer of energy in processes of high-energy chemistry

As mentioned in the Introduction, the different branches of HEC use their own types of physical stimulation of processes. From the viewpoint of interaction with matter, we will consider those radiations or flows of particles which concern the subjects outlined. They are electromagnetic radiation with an energy less than or nearly equal to the first ionization potential of a substance for photochemistry and laser chemistry, fast electrons and ions or neutrons for radiation chemistry, and fast electrons for plasma chemistry.

Owing to the variety of types of action and the wide energy range of particles and radiations used in HEC, the nature of their interactions with matter is quite different. The mode of transfer of energy and its spatial distribution determine the features of subsequent chemical processes.

The interaction of radiations with matter can be described with different quantities, such as effective cross-sections, molar, linear, or mass absorption coefficients or scattering coefficients, absorption half-layer, etc. The most general characteristic for the different types of interaction is the effective cross-section of the interaction (or simply the cross-section, σ). When a particle of energy E passes through a layer l of matter containing n absorbing atoms or molecules per unit volume, so that it can interact with these species in a certain way, then the product $\sigma(E)nl$ represents the probability of this interaction. The quantity $\sigma(E)$, having the dimensions of area and depending on the particle energy, is defined as the overall cross-section of the interaction. It is numerically equal to the probability of interaction for an incident particle passing through a target where a unit area contains one molecule of matter. If the area is 1 cm^2, the cross-section is measured in 10^{-28}m^2; an off-system unit, the barn, equal to 10^{-24} cm^2, is also used. The incident particle may lose part of its energy and change its direction of motion upon interaction, which is why a differential cross-section is specified. This characterizes only one side of an interaction, namely the fraction of energy transferred within an interval ε, $\varepsilon + d\varepsilon$ or an angle of alteration of direction θ, $\theta + d\theta$ and φ, $\varphi + d\varphi$ in polar coordinates. The overall cross-section $\sigma(E)$ is defined through the differential cross-sections as follows:

$$\sigma(E) = \int_{\varepsilon=0}^{\varepsilon_{max}} \int_{\theta=0}^{\theta=\pi} \int_{\varphi=0}^{\varphi=2\pi} \sigma(E, r, \theta, \varphi) \sin \theta \ d\varepsilon \ d\theta \ d\varphi$$

The processes of interaction of radiation with matter can also be described with other quantities, for example by using mass or molar coefficients of absorption and scattering, stopping power, half-layer of absorption, etc.

There is a simple relationship between the different quantities characterizing the interaction of radiation (particle) with matter. The linear absorption coefficient μ (denoted α in optics) is defined as the natural logarithm of the ratio of the radiant power of the incident radiation (P_0) to the radiant power of the transmitted radiation (P) divided by the pathlength l:

$$\mu = (1/l) \ln (P_0/P).$$

The linear and mass (μ/ρ) absorption coefficients are related to the cross-sections of the corresponding processes via the equations

$$\mu = \sigma \rho N_A / A$$

and

$$\mu/\rho = \sigma \rho n_A$$

where A is the atomic (molecular) mass of the absorbing substance.

The molar (decadic) coefficient of light absorption ε is defined as

$$\varepsilon = (1/cl) \log (P_0/P)$$

where c is the concentration of absorbing compound, in mol dm^{-3}, related to the cross-section of absorption as

$$\varepsilon = \sigma N_A / \ln 10$$

To describe a reduction in radiation flux, a free path length $1/\mu$ is frequently used, which corresponds to the thickness of the layer of material reducing the flux e-fold.

All types of interaction of radiation with matter may be subdivided into two groups: as processes of absorption or scattering. In absorption, which is typical mainly of electromagnetic and neutron radiations, an incident particle 'vanishes', that is, it completely loses its energy in the excitation of the atoms and molecules of matter (light absorption, neutron capture), or, in addition, transfers the energy to secondary particles (photoeffect, pair production). In scattering processes the incident particle also transfers its energy to matter, but a simultaneous change in its direction of motion takes place, which is important from the viewpoint of the spatial distribution of interaction events in the medium. Elastic and inelastic scatterings need to be distinguished. In elastic collisions, the kinetic energy of a system consisting of an incident particle (electron, photon, etc.) interacting with a stationary particle of matter (molecule, atom, or atomic nucleus) does not change as a whole. In inelastic scattering, the kinetic energy of the system decreases. In the process of absorption or inelastic scattering, the atoms and molecules of a gaseous medium undergo a transition from their ground state to a state with higher energy (excited rotational, vibrational, electronic, or nuclear levels) or are ionized. In addition, collective excited states (phonons, excitons, and plasmons) as well as delocalized charged species (holes and conduction electrons) can be formed in condensed media. A detailed list of all of the transient species mentioned here will be discussed in Chapter 2. At this point we shall consider the general characteristics of their interaction with matter of different types of radiation, and their dependence on the energy of the radiation and on the composition of the medium.

1.1 INTERACTION OF ELECTROMAGNETIC RADIATION WITH MATTER

Electromagnetic radiation is usually considered to be of different types according to its energy. A classification of electromagnetic waves is given in Table 1.1.

Table 1.1 Electromagnetic spectrum

Radiation	Wavelength region	Energy range/eV
Gamma- and X-rays	< 100 nm	> 12.4
soft	0.2 to 100 nm	12.4 to 6200
hard	< 0.2 nm	> 6200
Ultraviolet	10 to 400 nm	3.1 to 124
near	200 to 400 nm	3.1 to 6.2
far (vacuum)	10 to 200 nm	6.2 to 124
Visible light	400 to 740 nm	1.67 to 3.1
Infrared	0.74 to 2000 μm	6.02×10^{-4} to 1.67
near	0.74 to 2.5 μm	0.50 to 1.67
medium	2.5 to 50 μm	0.025 to 0.50
far	50 to 2000 μm	0.000 62 to 0.025
Radio waves	> 500 μm	< 0.025
SHF	0.3 to 60 cm	2.6×10^{-6} to 4×10^{-4}
HF	0.6 to 100 m	1.24×10^{-8} to 2×0^{-6}

On passing through a layer of substance, a beam of photons interacts with particles of the material, transferring to them its energy, which results in a reduction of the radiant power. For a parallel beam of monochromatic radiation of moderate intensity, the total reduction of radiant power caused by photon absorption is exponential, and it is described by a simple equation, the Lambert's law

$$P = P_0 e^{-\mu l} \qquad (1.1)$$

where P_0 is the incident radiatant power, P the power after passing through a layer of substance l cm thick, and μ the linear coefficient of reduction of radiant power (flux), in cm^{-1}.

Absorption spectra are usually presented in the form of a dependence of the molar extinction coefficient ε on the frequency or the wavelength of radiation. According to the Beer–Lambert law, the absorbance A of a flat layer of solution is expressed as follows

$$A = \log(P/P_0) = \varepsilon c l. \qquad (1.2)$$

Provided the concentration c of the light-absorbing substance is expressed in mol dm^{-3} and the optical pathlength l in cm, then the dimensions of ε are M^{-1} cm^{-1} which almost equals 0.1 m^2 mol^{-1}. Simple molecules have line spectra, but complex molecules usually have continuous absorption spectra mainly due to vibronic and solvent interaction (inhomogeneous broadening). The area under an absorption spectrum is related to the probability of the electronic transition.

The coefficient of reduction of radiant power (absorption coefficient) may also be expressed by other means, that is, the mass absorption coefficient μ/ρ in cm^{2}g^{-1} referred to the unit mass and electronic, atomic, or molecular absorption coefficients (μ_e, μ_a, or μ_{mol}), respectively.

The exponential law of absorption is not fulfilled for non-parallel beams, and so some corrections are required, depending on the linear size of the source, the thickness of the irradiated target, etc. If a target is sufficiently large compared with the distance to the source, it is necessary to take into account the reduction in photon flux to take place proportionally to the square of the distance.

The process of absorption of radiation involves the transition of the target molecules from their ground state to a state of higher energy under the action of the electromagnetic field [1]. Depending on the energy of the absorbed photon, the molecules are stimulated to their excited rotational, vibrational, electronic, or nuclear states, or they eliminate an electron (photoionization). In a condensed phase, associated excited states are also formed (excitons, plasmons, phonons, etc.).

The power P_a of absorption of radiation of frequency v, which is the energy absorbed per unit time per unit area, is given by

$$P_a = \rho(N_1 - N_2)B_{12}h\nu \tag{1.3}$$

where ρ is the radiation flux density (the quantity of energy passing through unit volume in unit time), N_1 and N_2 the number of molecules of substance in the ground and excited states, respectively (per unit volume), and B_{12} the so-called Einstein coefficient describing the probability of transition between these states induced by a photon with energy $h\nu$.

The Einstein coefficient is related to quantities used in spectroscopy, namely, to the oscillator strength f_{12} and to the transition moment M_{12} as follows:

$$B_{12} = \frac{\pi e^2}{h\nu m_e} f_{12} = \frac{8\pi^3}{3h^2}|M_{12}|^2 \tag{1.4}$$

The transition dipole moment M_{nm} can be calculated from the integral taken over the molar absorption coefficient in the range of the particular electronic transition (in the wave number scale):

$$|M_{12}|^2 = \frac{3 \times 2303hc}{8\pi^3 N_A} \int \frac{\varepsilon(\tilde{\nu})}{\tilde{\nu}}\,d\tilde{\nu} = 1.022 \times 10^{-57} \int \frac{\varepsilon(\tilde{\nu})}{\tilde{\nu}}\,d\tilde{\nu} \tag{1.5}$$

where the numerical factor 2303 is due to the conversion from the absorption coefficient of a gas at atmospheric pressure and 0°C to the molar absorption coefficient.

The processes of absorption of electromagnetic radiation are usually treated in wave mechanics by means of nonstationary perturbation theory, where the Schrödinger equation includes a perturbation operator $\mathbf{H}$ account for the effect of the radiation field, in addition to the static Hamiltonian $\mathbf{H}_0$:

$$(\mathbf{H} + \mathbf{H}_0)\,\psi(x, t) = E\psi(x, t) \tag{1.6}$$

The time-dependent wave functions $\psi(x, t)$ may be represented as the sum of the wave functions for an unperturbed system taken with time-dependent factors $a_k(t)$:

$$\psi(x,t) \approx \sum_k \alpha_k(t)\,\psi_k(x). \tag{1.7}$$

Thus, the transition of molecules from one state to another induced by the absorption of radiation appears to be possible. The transition moment M_{12} is equal to

$$M_{12} = \int \psi_1 \mu \psi_2\,d\tau \tag{1.8}$$

where ψ_1 and ψ_2 are the wave functions for the initial and final states, respectively.

The transition operator μ is a vector coinciding with the direction of the electric vector of the electromagnetic field. For molecules possessing same symmetry elements, it is convenient to decompose the operator into its components in terms of the orthogonal axes of the molecule

$$(M_{12})_x = \int \psi_1 \mu_x \psi_2 \, d\tau. \tag{1.9}$$

Electromagnetic radiations of different energy interact with matter in different ways, so we shall consider them separately in terms of their basic types.

For high intensities of incident radiation (especially on laser irradiation), multiphoton transitions, which involve the simultaneous absorption of two or more photons either of identical or different energies, become significant [1]. Under these conditions the total energy of the absorbed photons is equal to the difference in energy of the levels between which the quantum transition takes place, and the existence of any intermediate levels is unimportant. Multiphoton processes are the subject matter of nonlinear optics. The probability of a multiphoton transition is proportional to the product of the flux densities of the absorbed photons.

The selection rules for multiphoton transitions are different from those of one-photon processes. Thus, for molecules possessing a centre of symmetry, electric dipolar transitions involving an even number of photons are allowed only between states with identical parity, while transitions involving an odd number of photons take place only between states of opposite parity. This enables such electronic states to be populated by high-intensity radiation which cannot be accessed directly on one-photon excitation; moreover, photons of much lower energy than those required for one-photon excitation may be used.

Multiphoton transitions should be distinguished from successive photon absorptions, where relatively stable intermediate electronic levels or transient products are involved [2].

1.1.1 Visible light and UV-radiation

The major processes of interaction of visible and UV radiation with matter were summarized in Table 1.1.

The absorption of photons of visible light and UV radiation is associated with the transition of molecules to their electronically excited states. Far-ultraviolet radiation, with photon energies (usually > 6 eV) exceeding the ionization potential of a molecule, also causes photoionization. By analysing the absorption spectra of molecules it is possible to determine the nature of the excited states produced on absorption of photons.

The states of molecules are described in quantum mechanics by a wave function Ψ representing their electronic state, their nuclear motion, and their magnetic properties. To simplify this approach, the Born–Oppenheimer approximation is frequently used. It assumes the electronic (φ_e), vibrational (χ), and spin (s) wave functions to be mutually independent and the total wave function to be therefore the product of these three functions

$$\Psi = \varphi_e \chi s. \tag{1.10}$$

One should note that this approximation often appears to be too approximate for excited electronic states. Many radiationless processes in molecules occur which are due just to vibronic interaction, and intersystem crossing accompanied by spin inversion is caused by spin–orbit coupling. Nevertheless, the Born–Oppenheimer approximation is widely used, mainly because of the lack of anything better, and then these and other relevant interactions are considered as perturbations.

The simplest types of molecular orbital (MO) are as follows: σ-orbitals, having axial symmetry with respect to the bond and a maximum electron density along this axis, and π-orbitals possessing a nodal plane crossing the bond axis (the electron density along the bond axis is zero while its maxima are on both sides of the plane). The interaction of two atomic orbitals (AOs) results in the formation of two MOs, one having an energy less than the original AO (bonding orbital) with the other having a higher energy (antibonding orbital). These MOs are denoted as σ and σ^* or π and π^*, respectively. Coplanar with π-and π^*-orbitals are atomic p- and d-orbitals, either occupied (l) or vacant (v), which can interact (couple) with these π-orbitals. Orthogonal to the π-system, p- and d-orbitals or n-orbitals (n from nonbonding) do not couple but can participate in the formation of excited states donating or withdrawing the electrons. Such a scheme of MOs is certainly oversimplified since it neglects electron–electron repulsion and many other effects. Nevertheless, this representation is widely used. The real electronic wave functions for a molecule can be obtained by taking into account all possible electronic configurations (the so-called configuration interaction method), constructed on the basis of orbital approximation.

In denoting excited electronic states, it is customary to indicate only the orbitals with unpaired electrons, for example $\sigma\sigma^*$, $\pi\pi^*$, or $n\pi^*$.

If a molecule displays any type of symmetry, its electronic wave function may either conserve its sign, or change it to the opposite, after a particular symmetry operation; thus the wave function is termed *even* (and labelled with subscript g from the German *gerade*) in the first case or *odd* (with subscript u from *ungerade*) in the second. Different excited states, especially in spectroscopy, are frequently designated in terms of their group symmetry. This can be done, however, only for molecules which have been reasonably well-studied. In dealing with complex molecules, one may do no more than indicate their MOs or just the sequential number of a state (with respect to energy, beginning with the ground state which is taken as the zeroth state).

The vibrational wave functions of a molecule are expressed to a first approximation in the form of the wave functions for a harmonic or anharmonic oscillator.

The spin wave functions of molecule are considered to a first approximation as the product of the spin wave functions of unpaired electrons. The nuclear spins are usually neglected. Molecules with an even number of electrons must have a total spin equal to an integer (n = 0, 1, 3, etc.) The quantity $2n + 1$ is known as the multiplicity, and it is indicated by the left superscript in the state symbol (^1M, ^3M, $^1M^*$, $^3(n, \pi^*)$, or $^1(\pi, \pi^*)$). States with zero total spin are called singlet states, those with a spin equal to unity are triplet states, while those with a spin equal to two are quintet states. In most molecules with an even number of electrons, the electrons are paired in the ground state and the total spin is zero. However, in molecules having degenerate nonbonding orbitals of identical energy, the two electrons of highest energy can occupy these orbitals and have the same spin, so that the total spin is 1. If a molecule with an even number of electrons is excited, at least two electrons must occupy different orbitals, and they may possess either identical or different spins. Therefore, these molecules have several sets of electronic states: singlet, triplet, etc.

Molecules with an odd number of electrons have a total spin equal to $n = m + 1/2$, namely 1/2, 3/2, etc. Such states are referred to as doublet, quartet, etc.

The transition moment can sometimes be estimated qualitatively without detailed calculation if based on the symmetry of the electronic, vibrational, and spin wave functions of the state participating in the transition. In this case we use the so-called selection rules which define the conditions when the transition moment is non-zero; that is, (the transition is not 'forbidden'). If both of the states involved in the radiative transition are of the same parity (g or u), then the integral $\int \varphi_e \mu \varphi_e^* \, d\tau$ is equal to zero since the operator μ is odd and this type of transition is forbidden. Thus radiative transitions are allowed to occur only between states of opposite parity ($g \leftrightarrow u$). As far as radiationless transitions are concerned, the transition operator is even, and the selection rules are the opposite ($g \leftrightarrow g$, $u \leftrightarrow u$).

The requirement for the overlap integral of the vibrational wave functions $\int \chi_1 \chi_2 d\tau_v$ to be different from zero is known as the Franck–Condon principle. In its classical statement, the principle means that the internuclear distance in a molecule does not really change during a radiative electronic transition, and that radiationless transitions occur at the point of intersection of or quasi-intersection of the potential surfaces of electronic states. Since the equilibrium distances between the atoms in a molecule are not the same for the ground and excited states, this leads to the situation that the state formed on absorption (or emission) of electromagnetic radiation also appears in its vibrationally excited form (the so-called vibronic states as a combination of 'vibrational + electronic').

The spin selection rule requires conservation of the state multiplicity. The absorption of photons results in populating excited states mostly of the same multiplicity as the initial state, because the probability of radiative transition to states of different multiplicity is much less (by several orders of magnitude).

An important role is played by fast radiationless transitions between excited states of identical multiplicity (internal conversion), when molecules gaining access on excitation to higher excited states return rapidly (10^{-12} s) to a lower excited state of longer lifetime. Thus most photochemical processes take place from the lowest excited state (Kasha's rule) regardless of the initial excited state accessed through photon absorption.

States of multiplicity differing from the ground state are populated mainly via radiationless transitions accompanied by spin inversion (intersystem crossing). The probability of such transitions depends strongly on the molecular structure (the symmetry and type of excited states, the presence of heavy atoms, etc.).

1.1.2 IR-radiation

The absorption of infrared radiation is related to the excitation of vibrations and rotations of molecules. It also obeys the Beer–Lambert law. In addition to the molar extinction coefficient, the cross-section of absorption $\sigma = 10^3 \ \varepsilon/N_A$ is frequently used. Typical values of the molar extinction coefficients and cross-sections for allowed transitions in the IR-region are *ca.* $10^4 \ M^{-1}cm^{-1}$ and 10^{17} cm^2, respectively. The absorption bands in rotational–vibrational spectra are usually narrower than in electronic spectra, so that the selective excitation of levels is possible even in a mixture of identical, but isotopically nonuniform molecules.

An inherent property of the IR spectrum for many types of molecule, especially for organic species, is the presence of so-called characteristic frequencies, which are related to the vibrations of bonds or groups of a particular type.

A molecule being in its excited electronic, vibrational, or rotational states can absorb electromagnetic radiation and hence multiquantum processes are rather likely. The basic condition for the latter process is for the lifetime of the excited state to be large enough to enable another photon to hit the excited species at a given light intensity.

1.1.3 Ionizing electromagnetic radiation (γ- and X-rays, far-UV)

As already defined, ionizing radiation is that type the energy of which exceeds the first ionization potential of a substance exposed to the radiation. Thus, γ-rays, X-rays and far-UV light are referred to as ionizing electromagnetic radiations.

Energy quanta of electromagnetic radiation exceeding the ionization potential of matter can be produced in the following physical processes.

(1) On the transition of an electron from an outermost orbital to a vacancy in an inner orbital created by a charged particle or quantum, or in K-capture. This

process is known as X-ray fluorescence and its photons as characteristic X-ray emission. A spectral line set of this emission is characteristic for each element (no isotope effect is apparent here). The energy of characteristic X-rays may reach 140 keV for elements from the bottom of the periodic table.

(2) On the slowing down of electrons or other charged particles by the Coulombic field of an atomic nucleus (X-rays with a continuum spectrum, or *Bremsstrahlung*). There is no upper limit for the energy.

(3) In nuclear transitions and nuclear reactions (γ-radiation). The maximum energy is determined by the nuclear energy levels and does not exceed 10 MeV.

(4) On the slowing down of electrons in a cyclic electromagnetic field (that is, synchroton radiation). This process generates electromagnetic quanta with an energy from several eV to several keV and with a continuous energy spectrum.

There are some other processes which can produce electromagnetic quanta of high energy, namely annihilation radiation and radiation resulting from meson decay or from the decay of other elementary particles, but these radiation sources are not employed in HEC, because of their low intensity.

Highly-energetic quanta can interact with atoms in different ways, as shown in Table 1.2. In this table we emphasize the interactions important for HEC processes which will be considered below. Each of these processes is described by its cross-section and its own absorption coefficient. The overall linear absorption coefficient comprises the coefficients for all types of interaction

$$\mu = \tau + \sigma_{cl} + \sigma + \tau_{pn} + \tau_{pe} \tag{1.11}$$

where τ, σ_{cl}, σ, τ_{pn}, and τ_{pe} are the linear absorption coefficients for the photoelectric effect, for classical and Compton scattering, and for pair production at the nuclei and electrons respectively.

The individual coefficients and, consequently, the total absorption coefficient, depend on the photon energy and on the atomic number of the material.

Photoeffect. On photoelectric absorption (photoeffect) a quantum of energy equal to or exceeding the ionization potential of an electronic level is completely absorbed by an atom. The excess energy is transformed to the kinetic energy of the photoelectron, which can be expelled from any AO or MO. The linear coefficient of the photoeffect decreases approximately in proportion to the cube of the quantum energy, but as the apparent atomic number of the medium increases, it increases in proportion to Z^4 for elements with small atomic number and *ca.* Z^5 for elements with large Z. If the quantum energy exceeds the maximum ionization potential for a given atom, the photoeffect occurs from the K-shell to an 80 to 90% extent.

Classical (Rayleigh) scattering. This kind of interaction changes only the direction of motion of the photon. The cross-section for classical scattering decreases with increasing quantum energy and increases with increase in the apparent atomic number of the material. It does not exceed 10% of the total cross-section of interaction.

Table 1.2. Principal processes in the interaction of radiation with matter

Type if radiation	Affected object	Type of interaction		
		Absorption	Scattering	
			Elastic	Inelastic
Fast electrons [4]	Atomic electrons	–	–	**Ionization losses** Radiation losses
	Atomic nuclei	–	+	Radiation losses
	External electro-magnetic field applied to matter	–	–	Synchrotron radiation
	Isotropic medium	–	–	Cerenkov emission
Ion beams [4]	Atomic electrons	–	–	**Ionization losses** Charge transfer Ionization by capture of atomic electron
	Atomic nuclei	Nuclear reactions	+	Radiation losses
	Isotropic medium	–	–	Cerenkov emission

Table 1.2. (continued)

Type of radiation	Affected object	Type of interaction		
			Scattering	
		Absorption	Elastic	Inelastic
Neutrons [5, 6]	Atomic electrons	–	–	Ionization on passing nearby
	Atomic nuclei	**Nuclear reactions**	**Potential scatting** **Resonance scattering**	+
	Atoms and molecules of matter	–	+	+
Electromagnetic radiation [3]	Atomic electrons	**Electronic, vibrational, and rotational excitation; photoeffect**	**Classical (Rayleigh) scattering**	**Compton scattering** two-photon Compton scattering, Raman scattering
	Atomic nuclei	Nuclear photoreactions	+	+
	Electric field of nucleus	**Electron–positron pair production**	Thomson scattering, Delbruck scattering	–
	Electric field of electron	**Electron–positron pair production**	–	–
	Mesonic field of nucleus	Nucleon-antinucleon pair production	+	–
	Isotropic medium	–	Refraction	–

Note: Items in bold refer to most important interactions in HEC. The 'minus' sign stands for no interaction, while the 'plus' indicates the presence of an interaction.

Compton scattering. In Compton scattering, an incident quantum expels an electron, and a new (secondary) quantum of lower energy is produced. Since the binding energy of an atomic electric is, to a first approximation, low compared with the energy of the incident quantum for materials with not very large Z and quantum energies above 100 keV, all the electrons in an atom turn out to be energetically identical for this process. Consequently, the total efficiency of Compton scattering is proportional to the number of electrons in an atom.

In considering Compton scattering, one usually distinguishes the overall (total) atomic linear extinction coefficient σ, which is composed of the linear coefficient of scattering of electromagnetic radiation σ_s and the linear absorption coefficient σ_a. Since the Compton electrons produce virtually no secondary electromagnetic quanta, σ_a is called the linear coefficient of true absorption, while σ_s is the linear coefficient of true scattering. Values for these coefficients calculated per electron are given in Table 1.3.

Pair-production effect. When a quantum of electromagnetic radiation is of energy exceeding 1.022 Mev, it can undergo conversion to the 'electron + positron' pair in the field of an atomic nucleus (1.022 Mev = $2m_0c^2$, where m_0 is the rest mass of the electron). However, if an electromagnetic quantum has an energy above 2.044 MeV, it can also transform into an electron–positron pair in the field of an atomic electron (the expulsion of the atomic electron occurs

Table 1.3. Coefficients of Compton scattering σ_{se} and of absorption σ_{ae}, per electron, and mean energies of secondary quantum and secondary electron [3]

Quantum energy/MeV	$\sigma/10^{-28}\ m^2$			E/MeV	
	σ_e	σ_{ae}	σ_{se}	quantum	electron
0.01	0.640	0.012	0.628	0.0098	0.0002
0.05	0.561	0.047	0.514	0.046	0.004
0.10	0.493	0.0649	0.428	0.0862	0.0138
0.25	0.379	0.0947	0.284	0.190	0.060
0.50	0.289	0.0996	0.189	0.329	0.1712
0.75	0.250	0.1016	0.148	0.450	0.300
1.0	0.211	0.0925	0.118	0.560	0.440
1.25	0.191	0.0905	0.1005	0.665	0.585
1.50	0.172	0.0860	0.0860	0.758	0.742
1.75	0.159	0.0824	0.0766	0.85	0.89
2.0	0.146	0.0772	0.0687	0.939	1.061
2.5	0.135	0.0754	0.0596	1.113	1.397
5.0	0.0828	0.0514	0.0314	1.86	3.14
10.0	0.051	0.0348	0.0161	3.16	6.83
25.0	0.0261	0.0204	0.0057	6.53	18.3
100.0	0.0082	0.0065	0.0017	20.6	79.4

simultaneously), although this process is less probable than pair production near an atomic nucleus. The distribution of excess energy (above the threshold) between the electron and the positron is random, but it is most probable for these particles to acquire similar energies. The cross-section of pair production in a nuclear field is nearly proportional to the square of the atomic number, and it increases as the energy of the quantum increases above the threshold. The cross-section of pair production at electrons is proportional to the number of electrons, and it increases with increasing excess energy of the radiation.

The total cross-sections for photon interaction with matter. There exist the mass extinction coefficient

$$(\mu/\rho)_{ext} = (\sigma_{cl} + \sigma_s + \sigma_a + \tau + \tau_{pn} + \tau_{pe})/\rho \tag{1.12}$$

and the mass absorption coefficient

$$(\mu/\rho)_{abs} = (\sigma_{cl} + \tau + \tau_{pn} + \tau_{pe})/\rho \tag{1.13}$$

which evidently differ from one another by the terms responsible for scattering. The total mass absorption coefficient is used in dosimetry (see section 7.7.2), and the total mass extinction coefficient is used in calculations of the spatial distribution of the electromagnetic radiation field in a medium and in calculations of radiation protective shielding.

Fig. 1.1 shows the dependence of the mass extinction and mass absorption coefficients on the energy of γ-radiation for water. Similar plots can be drawn for any chemical compound. Values of these coefficients for some elements are listed in Table 1.4. If lacking data for a particular chemical element, the relevant values may be found as the mean of those for its neighbouring elements.

A convenient characteristic for the extinction (but not absorption) of a flow of electromagnetic radiation is the half-layer of extinction $\Delta_{1/2}$, which decreases two-fold the flux density of γ-radiation. For example, water reduces the directed flux of ^{60}Co γ-radiation two-fold with a 10.8 cm-thick layer.

1.2 INTERACTION OF ELECTRONS WITH MATTER

One should distinguish between ionizing and nonionizing electrons depending on their energy. 'Ionizing electron' is a term for a particle having an energy exceeding the first ionization potential of the target material (usually, one does not differentiate between the processes of ionization and electronic excitation). The nonionizing electrons, correspondingly, possess an energy below the first excitation potential (subexcitation electrons).

1.2.1 Ionizing electrons. Slowing down of electrons
The course of energy loss by ionizing electrons is known as the slowing down of electrons, and the processes responsible were shown in Table 1.2. Let us consider the major characteristics of the most import HEC processes such as elastic

$\log(\mu \, \rho^{-1}/\text{cm}^2\text{g}^{-1})$

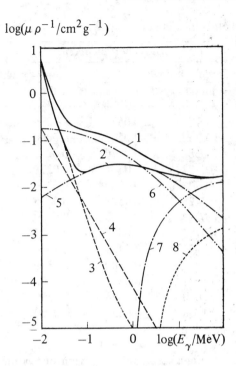

$\log(E_\gamma/\text{MeV})$

Fig. 1.1. Dependence of total mass extinction (1) and mass absorption coefficients (2), and of the mass coefficients of constituent interaction modes on energy for water as an absorbing medium [3]. 3—photoeffect; 4—classical scattering; 5—Compton absorption; 6—Compton scattering; 7—pair-production at nuclei; 8—pair-production at electrons.

Table 1.4. Mass extinction and mass absorption coefficients for ^{60}Co and ^{137}Cs γ-radiations [3]

Element	Mass coefficient/cm² g⁻¹			
	Extinction		Absorption	
	^{137}Cs	^{60}Co	^{137}Cs	^{60}Co
H	0.154	0.114	0.058	0.054
Be	0.0689	0.0512	0.0262	0.0243
C	0.0707	0.0577	0.0268	0.0273
N	0.0776	0.0576	0.0295	0.0273
O	0.0777	0.0577	0.0295	0.0273
Na	0.0742	0.0552	0.0282	0.0266
Mg	0.0766	0.0569	0.0291	0.0270
Al	0.0749	0.0557	0.0285	0.0264
Si	0.0773	0.0576	0.0294	0.0273
P	0.0751	0.0559	0.0286	0.0265

Table 1.4. (continued)

Element	Mass coefficient/cm^2 g^{-1}			
	Extinction		Absorption	
	^{137}Cs	^{60}Co	^{137}Cs	^{60}Co
S	0.0776	0.0577	0.0295	0.0273
K	0.0757	0.0561	0.0288	0.0266
Ca	0.0588	0.0576	0.0296	0.0273
Fe	0.0733	0.0540	0.0278	0.0256
Cu	0.0722	0.0530	0.0274	0.0251
Mo	0.0727	0.0521	0.0276	0.0246
Sn	0.0737	0.0513	0.0280	0.0243
I	0.0723	0.0525	0.0275	0.0244
W	0.0936	0.0464	0.0356	0.0220
Pt	0.0898	0.0580	0.0376	0.0275
Tl	0.1031	0.0591	0.0392	0.0280
Pb	0.1049	0.0598	0.0398	0.0283
U	0.1238	0.0652	0.0470	0.0309

collisions, and ionization and radiation losses. (Ionizing positrons interact with matter essentially in the same way as electrons of identical energy).

Electrons undergo elastic collisions only when encountering atomic nuclei. The cross-section of elastic scattering is directly proportional to the atomic number of the material and nearly inversely proportional to the square of the energy of the electron, and it decreases dramatically with increasing scattering angle. The pathlengths for electrons calculated for some scattering angles are shown in Table 1.5. These data are referred to single scattering for the condition $l \ll A/\sigma N_A \rho$ (where l is the target thickness). If $l \gg A/\sigma N_A \rho$, then all electrons from an incident flow suffer multiple scattering.

When undergoing elastic scattering, electrons become polarized, the degree of polarization depending both on the electron energy and the atomic number of the nucleus encountered. The larger the scattering angle, the stronger the polarization.

Two types of inelastic collision are important for HEC. These are (a) the ionization and electronic excitation of molecules and atoms of matter (ionization losses) and (b) the emission of electromagnetic radiation upon slowing down, or 'Bremsstrahlung' (radiation losses). The former process predominates at electron energies up to a few tens of MeV.

Ionization losses. On interacting with atoms and molecules of matter, incident electrons can transfer part of their kinetic energy to atomic electrons. The fraction of energy transferred depends on the duration of the interaction, which

Table 1.5. Mean free path of electrons in elastic collisions in water and heptane

Substance	Electron energy/ MeV	Free path, in cm, for scattering angle of		
		5°	15°	90°
Heptane	0.0020	1.9×10^{-7}	4.0×10^{-6}	1.5×10^{-4}
Water	0.0020	1.0×10^{-7}	2.0×10^{-6}	6.7×10^{-5}
Heptane	0.0105	5.9×10^{-6}	7.7×10^{-5}	6.7×10^{-3}
Water	0.0105	3.0×10^{-6}	3.7×10^{-5}	3.0×10^{-3}
Heptane	0.046	0.000 16	0.0016	0.15
Water	0.046	7.7×10^{-5}	0.000 48	0.067
Heptane	0.128	0.0010	0.0097	1.3
Water	0.128	0.000 53	0.0045	0.55
Heptane	0.664	0.011	0.17	42
Water	0.664	0.006	0.08	17

is determined mainly by the velocity (energy) of the incident electron. An atomic electron may receive an energy either lower or higher than its ionization potential, and the molecule will be either excited or ionized, respectively. Both processes are considered together under the common name 'ionization losses'. The electron generated in the ionization is called a secondary electron. If the secondary electrons possess energy sufficient to cause ionization of other atoms and molecules of the material, they are referred as δ-electrons: in fact the δ-electrons transfer to matter most of the energy carried by incident electrons.

The dependence of the cross-section of ionization losses on electron energy is shown in Fig. 1.2. It is seen that the probability of excitation is higher than that of ionization for the gas phase. The opposite is observed for the condensed phase. The maximum cross-section of ionization losses corresponds to an electron energy exceeding by 3–10 times the first excitation and ionization potentials.

The theory describing ionization losses is known to be valid for incident electron energies > 1 keV. All theoretical approaches made to explain the experimental dasta for lower energies seem to be rather approximate. In practice, the ionization losses are characterized not by the cross-section but by the stopping power $S_M = (dE/dx)_{ion}$ which can be determined from the slowing down of electrons in different substances. The stopping power is related to the properties of the incident electron and the atoms of the material via the Bethe equation:

σ/relative units

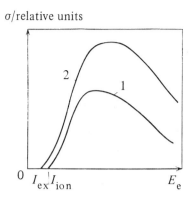

Fig. 1.2. Cross-section of inelastic collisions as a function of incident electron energy
for ionization (1) and excitation (2) in gas phase.

$$-\left(\frac{dE}{dx}\right)_{ion} = \frac{2\pi ne^4}{m_e v_e^2}\left[\begin{array}{l} \ln \dfrac{m_e v_e^2 E}{2(1-\beta^2)I(Z)^2} + (1-\beta^2) \\[2mm] -\left(\sqrt{1-\beta^2}-(1-\beta^2)\right)\ln 2 + \dfrac{1}{8}\left(1-\sqrt{(1-\beta^2)}\right) \end{array}\right]$$

(1.14)

where $n = N_A \sigma Z/A$ is the number of electrons per cm^3 of matter, Z the atomic number, and $I(Z)$ the mean ionization potential.

The stopping power is usually measured in g cm^{-3}. It should be noted that the ionization losses are, to a first approximation, proportional to the electron density of the material.

The Bethe equation involves the so-called mean ionization potential of the material $I(Z)$ which accounts for the fact that ionization occurs from any orbital, but its efficiency decreases as the ionization potential of the atomic electron decreases. The value of $I(Z)$ is found experimentally. The dependence of $I(Z)$ on the atomic number of the material is described satisfactorily with the relationship $I(Z) = 9.3 + 11.3\,Z - 0.011\,Z^2$.

Radiation losses of electrons, that is, the energy consumed producing Bremsstrahlung, occur mainly because of the interaction of the electrons with the Coulombic field of the atomic nucleus. The intensity of the Bremsstrahlung is proportional to the square of the atomic number, and it increases with increasing electron energy. The radiation losses for 1 MeV-electrons are quite small compared with the ionization losses.

Energy transfer on slowing down. The process of energy loss by an ionizing electron may be considered as occurring in two stages, namely deceleration, when the electron energy becomes less than the lowest potential of electronic excitation of the material, and thermalization, which is a process of further decrease in energy down to thermal energies. In fact, an ionizing electron can transfer to an atomic electron any amount of its energy up to the whole of it (electron exchange). However, the larger the fraction of energy to be transferred, the less the probability of the transfer. There are two extreme cases of interaction which are usually considered, namely, a grazing collision, when the energy transferred is close to the ionization potential, and a head-on collision, when a molecule acquires an energy exceeding its ionization potential many times.

An important feature of the interaction of accelerated electrons with matter is the rate of energy transfer to matter, — dE/dx. For electron energies above 1 keV, the energy loss is described by Eq. (1.14), and the rate of energy loss for a condensed phase is equal to 10^{17} eV s^{-1} (Fig. 1.3). The rate of energy loss in the above region can be easily determined with the equation

$$- dE/dx = [LET]v_e \tag{1.15}$$

where LET is the linear energy transfer.

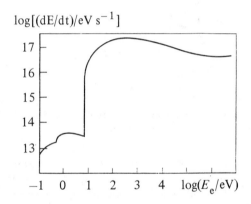

Fig. 1.3. Energy dependence of rate of energy loss in liquid water by an incident electron [7].

If the energy of the incident electrons is less than 1 keV, one must take into account the difference in cross-sections of their interaction with various electronic levels of the molecules of the material, that is, the cross-sections themselves must be known. The latter are related to the oscillator strength f_i. A set of oscillator strengths over the electromagnetic spectrum (from the IR to the X-ray region) forms an optical absorption spectrum $df/dE = f(E)$ (a discrete band and a continuum). An example of such an optical absorption spectrum is shown in Fig. 1.4 for ethanol. According to Bethe, the probability of a particular transition is proportional to f_i/E_i, where E_i is the energy of the corresponding electronic transition. Thus the idea of an excitation (ionization) spectrum $(Ry/E) \, df/dE = f(E)$ may be proposed, where Ry is the Rydberg constant (see

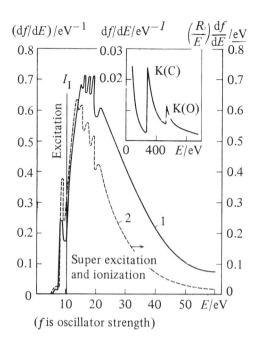

Fig. 1.4. Optical absorption spectrum (1) and excitation spectrum (2) for ethanol
vapour [7].

curve 2 in Fig. 1.4). The excitation spectrum appears to be similar to the optical
spectrum, although in the region of higher energies the probability of excitation
decreases faster than the probability of absorption of electromagnetic radiation.
Meanwhile, the intensity of the excitation spectrum for any of the energy bands
is significant in spite of the fact that the probability of interaction of the incident
electron with inner-shell electrons is lower by several orders of magnitude compared with the outermost electrons. It is typical of the excitation spectrum, as
well as of the absorption spectrum, that their integral lies mostly in the region of
comparatively high E values. The excitation spectrum can also be used for
calculating the rate of energy loss by an incident electron carrying an energy
ranging from the first ionization potential to 1 keV. The rate of energy loss for
this range is shown in Fig. 1.2 as calculated from the excitation spectrum.

When slowing down in a medium, an electron travels some distance. The tree
path R (or range) of the electron is defined as the distance from the point of its
entering the medium (for a δ-electron it is the site of its origin) to the point
where its energy is decreased so that it cannot further ionize the material. This
range is often called the deceleration pathlength. The range of the electron is
considerably shorter than its true path (multiple scatterings). The less the

Table 1.6. Ranges R and LET-values for electrons in water
[10]

E_e/keV	LET/eV nm^{-1}	R/μm
0.010	1.0	0.000 40
0.020	13.3	0.000 82
0.0501	23.2	0.002 85
0.100	30.3	0.044 8
0.200	22.0	0.008 00
0.501	19.5	0.022 0
1.0	13.0	0.053 4
2.0	7.75	0.164
5.01	3.92	0.770
10.00	2.32	2.50
20.00	1.35	8.33
50.1	0.675	42.2
100	0.420	140
200	0.284	440
501	0.206	1 740
1 000	0.188	4 300
2 000	0.186	9 610
5 010	0.193	25 000
10 000	0.200	48 800
20 000	0.206	91 800
50 000	0.2124	197 000
100 000	0.220	325 000

electron energy, the larger is the difference between the path and the range of the electron. In fact, the range can be calculated by integrating the cross-sections for all types of electron interaction as a function of electron energy, but it is usually determined experimentally by stopping electrons in films prepared from different materials. Under these conditions, different empirical equations are used to describe the relation between the energy of the electron and its range. Some of the ranges are listed in Table 1.6.

When passing through a thick layer of substance, a parallel beam of monoenergetic electrons is scattered, the electron energy falls, and electrons of different energy emerge, that is, the homogeneity of the energy is broken. One of the widely used quantities in describing the properties of electrons and all of the other types of ionizing radiation is the linear energy transfer (LET) which is the energy lost by an ionizing particle per unit pathlength. The LET is measured in eV nm^{-1}. An example of the LET dependence on electron energy is Table 1.6. By definition, the LET function represents the spatial loss of energy in the form of a linear loss.

1.2.2 Nonionizing electrons. Thermalization of electrons

The energy of nonionizing electrons can be consumed via the following channels: (i) polarization losses in the perturbation of dipoles in polar media (Debye losses), (ii) the vibrational excitation of molecules (the energy of a vibration quantum is *ca.* 0.05 to 0.5 eV), (iii) the rotational excitation of molecules (the energy of a rotational quantum is *ca.* 0.001 to 0.1 eV), (iv) the excitation of intermolecular vibrational transitions (intermolecular vibrational levels are *ca.* 0.0001 to 0.001 eV), (v) elastic collisions (the energy transferred in one event is ≤ 0.0001 eV), (vi) overcoming the Coulombic field of the parent ion (for secondary electrons), and (vii) resonance capture of an electron by a molecule to form an unstable ion as well as possible solvation (trapping) of an electron.

Resonance capture does not occur in every type of material. Energy losses in elastic collisions are quite small, so they may be neglected to a first approximation. As far as the energy consumption to overcome the Coulombic field of the parent ion is concerned, it has yet to be estimated.

Meanwhile, the first four types of energy loss are associated with certain energy levels appearing in the optical spectrum, so the rate of energy loss for the energy region below the first excitation potential can be described with the use of the optical approximation, similar to that used for the region above the excitation potential. The rates of energy losses for low-energy electrons shown in Fig. 1.3 were calculated simply by using the optical approximation.

It turns out that the mechanism of energy loss by nonionizing electrons in water and ice consists mainly of the vibrational excitation of water molecules; the thermalization time for these electrons is *ca.* 10^{-13} s, which is temperature-independent. As regards alcohols, Debye losses take place, so the time for thermalization, which is *ca.* 10^{-12} to 110^{-11} s, is temperature-dependent [7].

It should be noted that the optical approximation is related to the optical spectrum, that is, to allowed transitions between energy levels in a molecule induced by electromagnetic quanta. Other transitions which do not appear in the optical spectrum may come to be allowed with electrons. Hence, to find the rate of energy loss and to solve some other problems, the approximation of the energy loss spectrum of electrons should be used. Unfortunately, the latter data for electron energies less than 3 eV are not yet available for the condensed phase.

The thermalization length l is that range of an electron such that the electron energy decreases from a lower excitation potential ω_{in} to the energy of thermal motion ω_{fin}. The thermalization pathlength can be calculated [7] according to the equation

$$l_t = \alpha 4\sqrt{2}\lambda_t(\omega_{in}^{3/2} - \omega_{fin}^{3/2})/(3\sqrt{m_e})   \tag{1.16}$$

where λ_t is the transport pathlength or length of transfer, (that is, the distance at which an electron is scattered by 90° owing to collisions) and a is a constant equal to 4×10^{-20} s^2 g^{-1} cm^{-3}.

The pathlength of slowing down of electrons of energies 7 to 10 eV calculated with Eq. (1.16) is shown in Fig. 1.5 (λ_t has not been determined for lower energies). It is seen to fit fairly well to experimental data obtained from photoelectron spectroscopy [9]. The true pathlength x for electrons of the same energy is also presented in Fig. 1.5. It is evident that $x \gg l_t$. The mode of travel of the electron is determined by the ratio x/λ_t; if it is close to unity, the electron motion through a medium may be considered as quasi-linear. However, if the ratio is much larger than unity, the electron will be heavily scattered and its motion should be represented as diffusive. The motion of electrons of energy below 1 keV may be considered as diffusion [7].

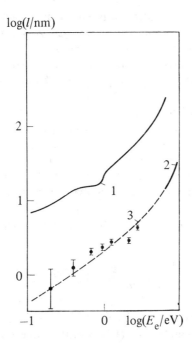

Fig. 1.5. Electron pathlength (curve 1) and thermalization distance (curves 2 and 3) in water as a function of electron energy. 2—calculations using Eq. 1.16 [7]; 3—experimental data [9].

1.3 INTERACTION OF FAST IONS WITH MATTER

Positively charged ions† are conventionally classified as follows:

(1) Accelerated ions which can be produced with any desirable energy from

---

† Though a positron is a positively charged particle, its interaction with matter is the same as that of a normal electron.

any element. However, the larger the mass, the lower the velocity of an ion. Fast ions of moderate size may be multicharged.

(2) α-particles are the doubly charged ions of helium generated by radioactive decay. Their energy depends on the nature of the radionuclide and ranges from 3.98 MeV (^{232}Th) to 8.78 MeV (^{214}At or ^{212}Po). α-particles are monoenergetic, their energy dispersion not exceeding 2%.

(3) Recoil nuclei are charged species emerging from atoms of matter after elastic collisions with accelerated heavy particles or fast neutrons. Their energy depends on the energy of the incident particle. There are also the residual atoms (of energy not exceeding 2 MeV) arising after radioactive emission of an α-particle or proton.

(4) Fission fragments, which are produced on fission of heavy nuclei, their mass ranging from 80 to 160 a.m.u., with energy from 40 to 120 MeV. The charge of fission fragments may be as large as 20.

(5) Cosmic rays, which are composed of 88% neutrons, 9.8% helions, 1.2% nuclei of other elements, and 1% electrons. The energy spectrum of cosmic radiation extends from 0.1 to 10^{10} GeV.

Table 1.2 presented the modes of interaction of fast positive-ion beams of matter. Here and later, all types of positively charge particles, except for those specially mentioned, are meant to be accelerated ions. Fast ions of different energies enter into different types of interaction. The basic type is ionization losses – dE/dx, in eV cm^{-1}. The precondition for ionization is that the velocity of the incident ion must be higher than that of the electrons in the AO (Bohr's principle).

The total ionization losses for ion beams are described with the Bethe–Bloch equation

$$\left(\frac{dE}{dx}\right) = \frac{4\pi z^2 e^4}{m_e v_i^2} n \left[\ln \frac{2 m_e v_i^2}{I(Z)} - \ln(1-\beta^2) - \beta^2 \right] \tag{1.17}$$

where v_i is the velocity of a fast ion.

In calculating the number of particles n, it is necessary to take into account only those electrons of matter which have a velocity less than the velocity of the incident positive ions.

According to the Bethe–Bloch equation, the efficiency of ionization by ion beams is determined only by the charge and the velocity of ions, so this formula can be used for calculation of the linear stopping power $(dE/dx)_{ion}$ for accelerated light ions.

What is the maximum amount of energy $E_{e,max}$ which a fast ion can transfer to an orbital electron? In the classical representation, the atomic electron can gain a doubled momentum, hence we have for a head-on collision

$$E_{e,max} = 4E_i^{m_e}/m_i. \tag{1.18}$$

This is the maximum value (subtracting the binding energy of the electron in the atom or molecule). In grazing collisions, the larger the deflection of the ion from its original course, the less the energy transferred.

As the energy of a fast ion declines to the value corresponding to the speed of electrons in the outer AOs, the capture of an atomic electron by the travelling ion, and the reduction of its positive charge, become possible.

Unlike the path of an electron, that of a fast ion is rectilinear. Deflections caused by scattering at nuclei occur rarely, so the range and the true pathlength coincide in value for a fast ion. The range is usually found experimentally (Tables 1.7 and 1.8).

Table 1.7. Proton ranges [4]

Proton energy/MeV	R/mg cm^{-2}				R/cm (25°C, 0.1 MPa)	
	Be	Al	Cu	Au	air	CO_2
0.05	—	—	—	—	0.071	0.045
0.075	—	—	—	—	0.099	0.065
0.10	0.16	0.26	0.55	1.32	0.127	0.085
0.25	0.42	0.62	1.22	2.61	0.328	0.218
0.50	1.03	1.50	2.55	5.10	0.814	0.544
0.75	1.88	2.84	4.20	8.24	1.48	0.992
1.00	2.92	3.89	6.12	11.9	2.31	1.55
2.50	13.0	16.1	22.7	41.9	10.4	6.83
5.0	42.9	56.0	67.6	118	34.3	22.5
7.5	86.5	103	132	220	70.0	46.1
10.0	140	171	215	345	117	77.2

1.4 INTERACTION OF NEUTRONS WITH MATTER

The principal modes of interaction of neutrons with matter were presented in Table 1.2.

In elastic collisions, part of the kinetic energy of a neutron is transferred to an atomic nucleus of the medium material. As a result, the neutron slows down and changes its direction of motion. The energy E_{nucl} transferred to the nucleus can be calculated from the equation

$$E_{nucl} = \eta E_0 = \frac{4 m_{nucl} m_{neut}}{(m_{nucl} + m_{neut})^2} \cos^2 \theta \, E_0 \tag{1.19}$$

where θ is the angle between the initial direction of motion of the neutron and that of the recoil nucleus.

Table 1.8. Ranges and LET-values for protons in water [10]

Proton energy/ MeV	LET/ eV nm^{-1}	R/μm	Proton energy/ MeV	LET/ eV nm^{-1}	R/μm
0.000 100	5.8	0.0072	0.251	68	3.25
0.000 251	9.2	0.0167	0.500	43.7	8.00
0.000 501	13.0	0.0314	0.794	31.5	16.0
0.000 794	16.2	0.0465	1.00	26.8	22.5
0.001 00	18	0.0558	2.51	14.3	104
0.002 51	29	0.111	5.01	8.3	340
0.005 01	41	0.180	7.94	5.6	790
0.007 94	52	0.242	10.0	4.68	1 180
0.010 00	58	0.279	25.1	2.2	6 250
0.025 1	78	0.480	50.1	1.27	21 800
0.050 1	91	0.730	79.4	0.88	51 000
0.079 4	97	1.05	100.0	0.742	75 700
0.100	91.5	1.25			

Inelastic collisions of neutrons of energy above the minimum excitation energy of the nucleus result in the capture of a neutron followed by decay of the excited nucleus with emission of another neutron of lower energy and of a γ-quantum. In neutron capture (absorption), a compound nucleus is formed which is excited with the kinetic energy of the neutron and is due to evolution of the binding energy. The excited nucleus may return to the ground state in several ways: the emission of one or more γ-photons (radiative neutron capture), the emission of two neutrons or the $(n, 2n)$ reaction, the emission of a proton or helion $((n, p)$ and (n, α) reactions, respectively), and the fission of super-heavy nuclei.

The cross-sections of inelastic scattering and nuclear reactions are individual for each isotope, and they depend on the neutron energy in rather a complicated manner. The linear absorption coefficients and scattering coefficients of thermal neutrons for some chemical elements are given in Table 1.9.

1.5 SPATIAL DISTRIBUTION OF PRIMARY PRODUCTS

The action of radiation or a beam of particles on matter creates reactive primary chemical species (positive ions and electrons, electronically excited and vibrationally excited states, etc.). One of the important features of HEC processes is the inhomogeneous spatial distribution of these particles at their moment of generation [7, 12].

Table 1.9. Linear absorption coefficients and linear coefficients for scattering of thermal neutrons in different substances [11]

Substance	$\rho/g\ m^{-3}$	μ_{ab}/cm^{-1}	μ_{sc}/cm^{-1}
H_2	8.99×10^{-5}	1.78×10^{-5}	2.04×10^{-3}
H_2O	1.00	0.022	3.45
D_2O	1.10	3.05×10^{-5}	0.449
He	1.78×10^{-4}	1.88×10^{-7}	2.14×10^{-5}
Li	0.534	3.29	0.064 9
Be	1.84	0.001 23	0.861
BeO	2.8	0.000 62	0.760
B	2.45	108	0.546
C	1.60	0.000 299	0.385
N_2	0.001 25	0.001 01	0.000 538
O_2	0.001 43	1.04×10^{-8}	0.000 226
F_2	0.001 73	5.39×10^{-7}	0.000 21
Na	0.971	0.013 4	0.102
Al	2.70	0.014 5	0.084 4
Si	2.35	0.008 07	0.085 7
P	1.83	0.007 12	0.178
S	2.1	0.020 5	0.178
Cl_2	3.21×10^{-3}	0.001 84	0.000 87
Cr	6.92	0.248	0.240
Fe	7.86	0.222	0.933
Cu	8.94	0.326	0.610
Zn	7.14	0.072 4	0.237
As	3.70	0.128	0.179
Se	4.5	0.422	0.378
Br_2	3.12	0.158	0.141
Cd	8.65	114	0.155
I_2	4.94	0.164	0.084 4
Sm	7.75	174	0.155
Gd	7.95	1410	–
Pb	11.3	0.005 6	0.361
Bi	9.7	0.000 95	0.252
U	18.7	0.363	0.393

Note: All values refer to natural isotopic contents.

Let us consider first the spatial distribution of the primary reactive species in condensed matter for ionizing radiation, because this problem has been studied in more detail than for other types of energy, beginning with electron beam irradiation. Earlier, we defined the overall parameter for describing the spatial distribution of particles, namely, the LET (see section 1.2.1). With the LET it is possible to determine the mean distance between consecutive ionizations l_i, using the mean energy of ion-pair formation, $l_i = W/LET$. For high-energy electrons l_i exceeds 100 nm, but it is minimal for 50–100 eV electrons, being equal to *ca.* 1.5 nm in water. The LET function shows a change in the mean distance between ionization sites as a fast electron slows down, but here the energy absorbed is referred to the free pathlength, and any deviation from the original direction is not taken into account. In fact, the absorbed energy should be referred not to the free path but to a particular volume (either spherical or cylindrical) related to the electron range, that is, the true distances between ionization sites are larger than those determined from LETs.

The trajectory of an ionizing particle passing through a medium and detected in the form of primary transient reactive species is known as a track. The various types of spatial distribution of the primary transient species are called track patterns. The simplest type of track is a single pair (a parent ion and the electron) situated at such a distance from its neighbouring pairs that their electrostatic interaction may be neglected. This pattern of track is formed by a high-energy electron on head-on collision and in the photoelectric and Compton effects. If an electron (whether secondary or primary) is not highly energetic, ionization events will take place at distances of about a few nm, so that the interaction between the newly-formed pairs cannot be neglected. This pattern of track, involving several (from 2 to 10) ion pairs, is called a spur. It is usually assumed that the formation of one spur of a few nm takes, on average, about 100 eV. Spherical spurs are considered to exhibit a Gaussian distribution of species. When several spurs are produced near one another, then this group, if it can be represented as spherically symmetrical, is called a blob (droplet). When the group is cylindrical it is called a short track. It is generally agreed that an electron with an initial energy of 1 keV forms a short track, but one with an energy of 100 eV forms a spur at the end of its path.

As far as γ-radiation is concerned, most of its transfer of energy to matter occurs via the secondary electrons, hence the tracks created in irradiated materials are the same. The contribution of each pattern of track is not exactly known, but different methods of estimation produce more-or-less similar results.

Below we show the percentage contribution of different track patterns in the interaction of ^{60}Co γ-radiation with water:

Track pattern	[13]	[14]	[15]
Single pairs	64	61.4	26.8
Spurs	12	12.2–20.6	27.4
Blobs	24	18.0–26.4	27.4

A similar picture is also observed for fast electrons.

As long as fast electrons travel rectilinearly and generate electrons of relatively low energy (nearly equal to or below 1 keV), the primary transient products of radiolysis are arranged in the form of a cylinder which comprises the track. It is generally accepted that the radius of the track is 5 to 10 nm and that the concentration of primary intermediates decreases along the radius in accordance with a Gaussian distribution.

The set of track patterns for neutron radiolysis is determined from the secondary radiation (fast ions, γ- and β-rays) which is generated by the interaction of neutrons with the atomic nuclei of a given material.

Transient species formed by the action of near-ultraviolet and visible light are distributed exponentially (in accordance with the Beer–Lambert law) along the depth of the illuminated layer. The same follows for vibrationally excited molecules produced by IR radiation.

Under the action of light, a microheterogeneous distribution of species occurs after photodissociation and photoionization. Radicals are formed on photodissociation in pairs: initially in a solvent cage and then diffusing apart gradually or undergoing so-called geminate recombination (the combination of two particles having the same point of origin). Geminate processes are especially important in magnetic effects in chemistry, when a gradual magnetic-field-dependent evolution of the spin of such a geminate pair takes place, affecting the probability of its recombination and decay. It is possible to increase the lifetime of a geminate pair by preventing substantial separation of the radicals from one another, for example by using micelles as a 'microreservoir'. Photoionization results in the formation of a 'radical cation-solvated electron' pair. In rigid solutions, some kind of distribution of pairs in terms of the distance between these two charged species arises, which complicates the recombination kinetics of these particular geminate pairs.

The track effects are not observed in the gas phase because the distance between the neighbouring transient species induced by ionizing radiation exceeds tens of nm because of the low density of the gas. However, an electric discharge in a gas produces macro-inhomogeneities in the distribution of the primary reactive species (see Chapter 6).

1.6 IONIZATION PROCESS

When considering the mechanism of HEC processes, the determination of the amount of ions formed and their yield in terms of energy absorbed is always a key problem. This problem can be easily solved for a gaseous media because the application of an electric field (for example in an ionization chamber) allows all of the created ion pairs to be collected and thus detected. So, knowing the amount of energy absorbed, their yield per unit absorbed energy can be found.

The entire energy absorbed by a material is consumed via three channels, namely, to the ionization and excitation of molecules, we must add that amount of energy used by delta-electrons (when they become nonionizing) in the energy-loss events. The following equation may then be written:

$$\Delta E = N'_{ion}\overline{E}'_{ion} + N''_{ion}\overline{E}''_{ion} + N_{el}\overline{E}_{el} + N_{vib}\overline{E}_{vib} \qquad (1.20)$$
$$+ N_{rot}\overline{E}_{rot} + N_e\overline{E}_e$$

where ΔE is the absorbed energy, $\overline{E}'_{ion}$ and $\overline{E}''_{ion}$ are the mean ionization potentials for one-electron and two-electron ionization (multiple ionization may be neglected), respectively, $\overline{E}_{el}$, $\overline{E}_{vib}$, and $\overline{E}_{rot}$ are the mean energies of electronic, vibrational, and rotational excitation, respectively, E_e is the mean energy of nonionizing electrons, and N is the corresponding number of species.

Theoretically, the values of N can be calculated if we know the degradation spectrum of ionizing radiation in a material, the cross-section of ionization and excitation for each level, and the ionization potentials and excitation energies for the constituent molecules. Calculations of this type have been reported (see, for example, [16]), though their reliability is not high, because of insufficiently accurate values given for the cross-sections of interaction of low-energy electrons with matter [17].

In principle, the quantities $N_{ion} = N' + N''$, N_{el}, and N_e can also be found experimentally by the scavenger technique, but the procedure requires the use of high solute concentrations, and some assumptions need to be made to account for track effects. This operation is much easier in the gas phase since it is not complicated by these effects.

The number of ions, N_{ion} and, correspondingly, of electrons N_e, can be easily measured in the gas phase, and this makes it possible to calculate the mean energy of ion-pair production W which is the energy consumed in the formation of a pair of ions:

$$W = \Delta E / N_{ion} = \Delta E / N_e \qquad (1.21)$$

W-values have been determined for a large number of compounds and for different types of radiation [18]. Some of these are given in Tables 1.10 and 1.11 from which it is seen that W exceeds by 1.5 to 2.5 times the first ionization potential of the particular gas in every case.

As shown in Table 1.10, there is no noticeable difference in the W-values for different types of ionizing particles. This is related to the fact that a major contribution to ionization by ionizing particles of high energy is made by the δ-electrons. The change in W-value for electrons of energy below 1 keV (Fig. 1.6) is due to the increasing contribution of energy losses consumed in the excitation of molecules. Fast ions exhibit the same effect at essentially larger energies [19].

The W-values in condensed media can be determined experimentally for the very small number of systems where the electron mobility is sufficiently high to enable collection of all the electrons before the breakdown voltage is reached. This turned out to be possible for liquid noble gases (beginning from argon), for

Table 1.10. Mean energy W of ion-pair production in gases for different types of radiation [19]

Gas	W/eV				I_1/eV
	1.8 MeV protons	3.6 MeV protons	5.3 MeV helions	10 keV electrons	
He	—	45.2	42 to 46	41.3	24.58
Ne	—	39.3	36.8	36.3	21.56
Ar	26.6	26.6	26.4	26.4	15.76
Kr	—	23.0	24.1	24.2	13.93
Xe	—	20.5	21.6	22.0	12.13
H_2	—	36.4	—	36.5	15.4
N_2	36.7	36.6	36.4	34.8	15.6
O_2	—	32.2	—	30.8	12.2
Air	35.2	—	35.1	—	—
CO_2	34.4	—	34.2	33.0	13.78

Table 1.11. Mean energy of ion-pair production for different gases and vapours at room temperature [19]

Substance	W_α/eV	W_e/eV	I_1/eV	Substance	W_α/eV	W_e/eV	I_1/eV
CH_4	29.1	27.3	13.12	C_2H_5OH	—	24.8	10.65
C_2H_2	27.4	25.8	11.41	CF_2Cl_2	29.6	—	11.7
C_2H_4	27.9	25.8	10.5	CCl_4	25.8	—	11.1
C_2H_6	26.5	25.0	11.6	NO	28.9	—	9.25
C_3H_8	26.2	24.0	11.2	NH_3	28.6	26.6	10.15
C_3H_6	27.1	—	9.8	BH_3	35.7	—	15.5
cyclo-C_3H_6	25.9	—	9.7	H_2S	23.3	—	10.46
C_4H_{10}	26.0	23.4	10.8	SF_6	35.8	—	19.3
iso-C_4H_{10}	26.3	—	10.4	SO_2	32.4	30.4	12.34
C_5H_{12}	—	23.2	10.4	H_2O	—	29.6	12.61
C_6H_{14}	—	23.0	10.17	CO	34.5	—	14.01
C_6H_6	—	22.1	9.24	N_2O	34.2	32.6	12.72
CH_3OH	—	24.7	10.95	CH_3CHO	—	26.4	10.24
cyclo-C_6H_{12}	25.05	22.7	8.94	CH_3NH_2	—	25.0	9.41
$(C_2H_5)_2O$	—	23.8	9.53	HCl	—	25.3	12.74

some semiconductor solids with a forbidden band gap of less than 3 eV, and for diamond. The mean energy of ion-pair production W in a liquid appears to be less than in the gas phase, which may be explained by a reduction in the first ionization potential on transition to the liquid state.

There is no clearly expressed dependence of W on the value of the first ionization potential for all substances, although a kind of relationship does exist

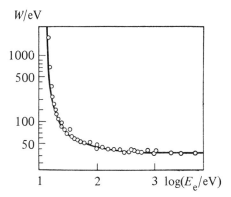

Fig. 1.6. Mean energy of ion production in air as a function of electron energy [19].

for some molecules, as shown in Fig. 1.7. For the noble gases there is a linear dependence, the data for both liquid and gaseous states fitting to the same straight line. More complex molecules usually have larger W-values than simpler ones, and they form a family, which indicates that their W-value increases with increasing ionization potential but also depends on other factors which still

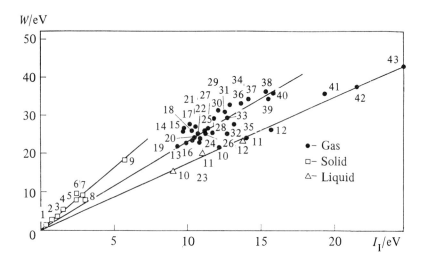

Fig. 1.7. Dependence of the mean energy of ion-pair production on first ionization potential. 1—ZnSb; 2—Ge; 3—Si; 4—GaAs; 5—GaP; 6—CdS; 7—SiC; 8—PbO; 9—diamond; 10—Xe; 11—Kr; 12—Ar; 13—C_6H_6; 14—cyclo-C_3H_6; 15—C_3H_6; 16—C_6H_{14}; 17—NH_3; 18—iso-C_4H_{10}; 19—C_5H_{12}; 20—C_2H_5OH; 21—C_2H_4; 22—C_4H_{10}; 23—H_2S; 24—CH_3OH; 25—CCl_4; 26—C_3H_8; 27—C_2H_2; 28—C_2H_6; 29—CF_2Cl_2; 30—O_2; 31—SO_2; 32—HCl; 33—H_2O; 34—N_2O; 35—CH_4; 36—CO_2; 37—CO; 38—H_2; 39—BH_3; 40—N_2; 41—SF_6; 42—Ne; 43—He.

remain unclear. The larger W-value for complex molecules compared with monoatomic gases points to the fact that internal losses of energy consumed in excitation is significant for these gases. In solids, the relationship between the average energy of ion production and the ionization potential is linear, as shown in Fig. 1.7, but the relative contribution from excitation losses is higher than in gases. In spite of the fact that there is no simple dependence of W on I, the W-values can be determined with 20% accuracy for most organic compounds from their ionization potentials, so the error in determination of the ion-pair yield also does not exceed 20% (which is thought to be acceptable for most purposes).

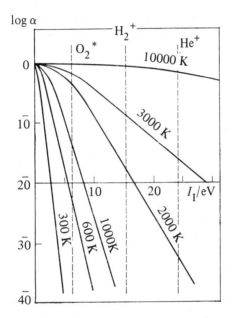

Fig. 1.8. Dependence of degree of thermal ionization of a gas on its ionization potential for different temperatures. Vertical lines show the ionization potentials of hydrogen and helium and the excitation potential of oxygen ($P = 0.1$ Pa).

Values of the first ionization potential for substances in the liquid state $I_{1,l}$ have been reported in recent years (see Table 1.12). Evidently, the ionization potential in a liquid is less by 1 to 4 eV relative to the gas phase ionization potential $I_{1,g}$. This change can be described by the following expression [7]:

$$I_{1,l} = I_{1,g} - \frac{e^2}{2r}\left(1 - \frac{1}{\varepsilon}\right) - (D_2^+ - D_2) + A_g \qquad (1.22)$$

Table 1.12. Ionization potentials in liquid and gas phases [7, 20]

Substance	$I_{1,1}$/eV	$I_{1,g}$/eV	E/eV	Substance	$I_{1,l}$/eV	$I_{1,g}$/eV	E/eV
Argon	13.4	15.76	2.36	Neopentane	8.55	10.35	1.80
Krypton	11.0	13.93	2.93	2,2-Dimethyl	8.49	10.06	1.57
Xenon	9.2	12.13	2.97	butane			
Heptane	8.86	10.35	1.49	Cyclohexane	8.43	9.88	1.45
Cyclopentane	8.82	10.53	1.71	Isooctane	8.38	9.86	1.48
Water	8.76	12.61	3.85	Tetramethyl-	8.05	9.79	1.74
Hexane	8.7	10.18	1.48	silane			
				Toluene	7.75	8.82	1.07

where r is the radius of the average sphere occupied by the molecule, A_g the work function for electron escape from the liquid to the gas phase, and D_2 and D_2^+ are the dissociation energies of the dimer M_2 and the dimer cation M_2^+.

The above expression gives a value for $I_{1,1}$ which coincides with the experimentally-derived quantity within 0.1 eV.

It is interesting to compare the ionization induced by radiation and thermal ionization. The latter occurs owing to the thermal motion of molecules, and the degree of ionization of a gas α can be calculated with the Saha equation [21]

$$\frac{\alpha^2}{1-\alpha^2} = 2\frac{g_i}{g_a}\left(\frac{2\pi m_e}{h^2}\right)^{3/2}\frac{(kT)^{3/2}}{P}\exp\left(-\frac{I_1}{kT}\right) \tag{1.23}$$

where g_i and g_a are the statistical weights (degeneracies) of the ion and atom, respectively.

Some data calculated with this equation, using a total pressure of 0.1 MPa and a temperature ranging from 300 to 10 000 K for a gas, are shown in Fig. 1.8. Since the ionization potential is about 10 eV for most substances, the ionization caused by thermal motion occurs at an appreciable rate only at temperatures above 1000 K. Very few chemical compounds can resist such a temperature without decomposition.

When ionization is caused by the action of high-energy radiation, a non-equilibrium concentration of ions is created. This concentration can be evaluated. For instance, an energy of about 6×10^5 eV is evolved in 1 cm^3 of gas kept under normal conditions and irradiated with 1 MeV-electrons of total power 1 kW. Then, assuming that the ionization of one molecule takes, on average, about 30 eV, we expect 2×10^{14} ions to be produced in 1 cm^3. Ion recombination occurs with a rate constant of 10^{15} M^{-1} s^{-1} in a gas, so a steady-state concentration of ions *ca.* 10^{10} cm^{-3} is created under these conditions. This corresponds to a degree of ionization of 3×10^{-10}. Evidently, the latter value exceeds the degree of thermal ionization by many orders of magnitude.

1.7 RADIATIVE AND RADIATIONLESS ENERGY TRANSFER

The phenomenon of energy transfer consists of the transition of an excited molecule (the donor) to an electronic state of lower energy (most usually, the ground state), the energy being donated to another molecule (the acceptor) having identical or approximately equal energy levels [22]. The transfer of energy can be realized via both radiative and radiationless paths.

Radiative energy transfer consists of emission of a photon by the donor molecule and the subsequent absorption of the photon by the acceptor in the same sample (i.e. reabsorption).

The radiationless transfer of energy [23] occurs owing to the mutual interaction of donor and acceptor molecules located at a distance less than the wavelength of the emitted light. Depending on the nature of this interaction, one differentiates between the induction mechanism and the exchange mechanism of energy transfer; the former is related to the electrostatic dipole or multipole interaction between the oscillators of energy donor and acceptor (Förster mechanism), and the latter to the exchange interaction of the electrons of the donor and acceptor (Dexter mechanism).

The probability of energy transfer, regardless of the mechanism, is proportional to the overlap integral normalized to unit area of the emission spectrum $F(v)$ of the donor and the absorption spectrum $\varepsilon(v)$ of the acceptor

$$\int F_D(v)\varepsilon_A(v)\,dv \tag{1.24}$$

It is related to the requirement to obey the Franck–Condon principle in the processes of emission, absorption, or exchange of energy for each participant molecule. With the exchange mechanism, the selection rules are applied to the pair of donor and acceptor molecules as a whole, and consequently, transitions which may appear to be forbidden if each of the molecules is considered separately may appear to be allowed. This refers especially to triplet–triplet energy transfer, when a triplet donor molecule gives its energy to the singlet ground state of an acceptor molecule and the latter enters the triplet state. To take into account these variations in the selection rules, it is appropriate to use the overlap integrals normalized to unit area of the relevant band in the absorption spectrum

$$\frac{\int F_D(v)\varepsilon_A(v)\,dv}{\int \varepsilon(v)\,dv} \tag{1.25}$$

For radiative energy transfer, the rate constant k for two reacting molecules is inversely proportional to the square of the distance r between the species:

$$k = \frac{0.2303}{4N_A\tau_0 r^2}\int F_D(v)\varepsilon_A\,dv. \tag{1.26}$$

The overall probability for the reabsorption of light depends upon the shape of the sample and on the position of the photon-emitting molecule. For a flat sample, the probability of reabsorption increases with the thickness l:

$$P_A / P_0 = 1 - \exp(-\varepsilon_A [A] l). \tag{1.27}$$

Reabsorption results in distortion of the shape of the emission spectrum, namely, in a decrease in intensity at the short-wavelength edge of the luminescence spectrum. Although reabsorption does not affect in any possible way the rate of photon emission by an energy donor, it can lead to strong perturbation of the kinetics of luminescence decay. The generation of excited molecules continues after terminating the initial exciting pulse owing to luminescence reabsorption, that is, reabsorption acts to increase the decay time. On the excitation of luminescence by light which is absorbed only slightly, the distribution of molecules in the sample is close to uniform and virtually does not change with time in spite of reabsorption. Under these conditions, the luminescence decay is close to exponential even though the decay time may be considerably (even twice) longer than the true lifetime of the excited molecule in the case of strong reabsorption and large quantum yields of luminescence.[†] Here the decay time depends on the emission wavelength and has its lowest value, which is closest to the true value, in the short-wavelength region where strong reabsorption takes place.

During strong absorption of the exciting light, the excited molecules are first formed in a thin, near-surface layer of the sample, and then the excitation energy gradually migrates into the sample owing to reabsorption. The decay turns out to be nonexponential since the probability of reabsorption increases with time and the decay slows down. The apparent decay kinetics depend on the wavelength of observation and, as above, are closest to the true kinetics in the region of strong (short-wavelength) reabsorption.

In inductive radiationless energy transfer, the dipole–dipole interaction usually predominates over the dipole–quadrupole and multipole–multipole interactions since the latter decrease faster with distance. A theory for dipole–dipole energy transfer was developed by Förster [24] and Galanin [25]. The rate constant for the transition is inversely proportional to the distance between the oscillators raised to the sixth power:

$$k = \frac{9000 z^2 \varphi_0 \ln 10}{128 \pi^5 N_A \tau_0 r^6} \int F_D(v) \varepsilon_A(v) \frac{dv}{v^4} \tag{1.28}$$

where z^2 is an orientation factor, its mean value being equal to 2/3 for randomly-arranged and rapidly rotating molecules, φ is the quantum yield of emission, τ_0

[†] The quantum yield of luminescence is equal to the ratio of the number of emitted to absorbed photons.

is the luminescence decay time of the donor, and n is the refractive index of the medium.

The factor $1/v^4$ in the integral accounts for the dependence of the probability of the radiative transition on frequency.

The distance between the donor and acceptor molecules, where the rate constant of energy transfer is equal to the rate constant of spontaneous decay $(1/v_0)$ of excited donor molecules in the absence of the acceptor, is known as the critical distance r_0

$$r_0 = \left\{ \frac{9000 z^2 \psi_0 \ln 10}{128 \pi^5 n^4 N} \int F(v) \frac{\varepsilon_A(v)}{v^4} \, dv \right\}^{1/6} . \tag{1.29}$$

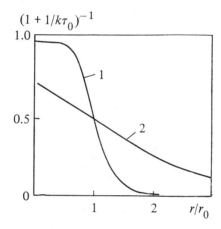

Fig. 1.9. Probability of radiationless energy transfer as a function of distance between energy-donor and energy-acceptor for induction (1) and exchange (2) mechanisms.

Because the rate constant of energy transfer depends on the sixth power of the distance, the probability of energy transfer from an excited donor molecule at a distance r_0 is close to unity (Fig. 1.9). Thus the concept of a 'quenching sphere' has been introduced; the probability of energy transfer in the sphere is considerably higher than that of the spontaneous decay of the excited donor molecule.

The value of r_0 is equal to about a few nm for allowed transitions in the donor and acceptor, that is, the energy transfer can be realized inside a sphere incorporating 10^3 to 10^4 molecules. In solutions, there exists a kind of random distribution by distance for the donor and acceptor molecules. For a uniform distribution

$$N(r) = 4\pi r^2 [A] N_A . \tag{1.30}$$

The effects observed for luminescence quenching and luminescence kinetics of a donor, and for luminescence sensitation of the reactions of an acceptor caused by energy transfer, are averaged over these distances. With the dipole–dipole mechanism of energy transfer, the decay turns out to be nonexponential and the dependence of the quantum yield of donor luminescence on the concentration of acceptor in viscous solutions (that is, on discounting the translational diffusion of molecules) is nonlinear

$$I(t) = \exp\left(-\frac{t}{\tau_0} - 2b\sqrt{t/\tau_0}\right),$$

$$\frac{\varphi_0}{\varphi} = 1 + \frac{\pi}{2}\frac{[A]}{(A_0]} + \frac{\pi}{4}(\pi - 2)\left(\frac{[A]}{[A_0]}\right)^2 + \dots, \tag{1.31}$$

where

$$b = \frac{\sqrt{\pi}}{2}\frac{[A]}{(A_0]} \quad \text{and} \quad [A_0] = \frac{3}{4000\,\pi r_0^3 N_A}.$$

For low-viscosity solvents, where the diffusion of donor and acceptor molecules is rather higher with respect to the lifetime of the excited state, even more complicated equations are obtained.

The exchange mechanism of energy transfer is realized at comparatively short distances between the molecules when overlapping of the electronic wave functions of these molecules takes place. The rate constant for exchange transfer of energy decreases exponentially with the distance between the molecules:

$$k = \frac{2\pi}{h}z^2 \exp\left(-\frac{2r}{\alpha}\right)\int F_D(v)\varepsilon_A'(v)\,dv, \tag{1.32}$$

where $\varepsilon_A'(v)$ is the absorption spectrum of the acceptor normalized to unit area, and a is the decay parameter for the wave functions of donor and acceptor.

The factor z^2 takes into account the selection rules, in particular, selection by spin. To describe the effects observed in the kinetics of quenching and luminescence, it is also necessary to carry out averaging of the distance. The simplest representation of luminescence decay of the energy donor by the exchange mechanism is based on the Perrin model [26]. According to this, only those molecules luminesce which have no neighbouring acceptor molecules within a volume V ('the interaction sphere'):

$$\frac{\varphi}{\varphi_0} = \lim_{N\to\infty}\left[1 - \frac{V[A]}{N}\right]^N = \exp(-V/[A]). \tag{1.33}$$

As regards the decay kinetics of donor luminescence, there is also a nonexponential law in the case of energy transfer by the exchange mechanism:

$$I(t) = \exp\left\{-\frac{t}{\tau_o} + \gamma^3 \frac{[A]}{[A]_o} \frac{t}{\tau_o} e^\gamma \int \exp\left(-\frac{t}{\tau_o} e^\gamma\right) [\ln y]^3 \, dy\right\} \qquad (1.34)$$

where

$$\gamma = \frac{2r_o}{\alpha} \quad \text{and} \quad [A]_o = \frac{3}{4000 \pi r^3 N_A}.$$

2

Elementary events in high-energy chemistry

The final stable products of photolysis, of radiolysis, of plasma reactions, and of other HEC processes are generally formed through a complex sequence of fast reactions of transient species of different types carrying an excess potential energy and thus possessing a high reactivity. These species also take part in conventional chemical processes, although they can barely be observed because of their low concentration. The conditions of HEC provide a sufficiently high concentration of transient species making it possible to obtain reliable information on the physical chemistry of all the intermediates and on their physico-chemical properties by means of the pulse technique of generation and detection (basically, spectroscopic and electrical methods), the technique of scavenging of transients, and other methods. In radiation chemistry and photochemistry, the reactions of individual intermediate species are conventionally called elementary processes; in our opinion, this may be extended to other branches of HEC.

Let us try to outline our views on the basic types and reactions of intermediate species insofar as many of them have been discovered only in HEC processes. Naturally, the properties of transient species do not depend on their method of generation. Thus, in spite of the fact that we will rely mainly on data obtained from photochemistry, radiation chemistry and plasma chemistry, the information detailed here concerns every branch of HEC and chemistry as a whole. There is a huge body of data available about transient species, and we shall concentrate therefore only on some of their most representative types.

2.1 TRANSIENT SPECIES

On acquiring high energy, a substance generates primary reactive species or transient products of the first generation TP_1. Then, in a series of consecutive transformations, they are converted to transient products of subsequent generations TP_2, TP_3, etc., at last, to the final products FP. This type of classification, first made for the products of methanol radiolysis [27], has proved useful.

The intermediate products of HEC processes are quite different in terms of their properties. Their origin is related to the gain or loss of an electron (ionization), with the gain of additional energy by orbital electrons (excitation), with the excitation of vibrational and rotational levels, and with bond rupture and the formation of other chemical bonds (decomposition, solvation, etc.). Some types of transient species are thermodynamically stable and can exist indefinitely if isolated; however, all of them are highly reactive with respect to intermediates of the same or other types and with many stable molecules. Below, we try to systematize the basic types of transient product.

2.1.1 Gas-phase species

Every molecule or atom in the gas phase under ambient conditions may be considered as a single particle. The following types of intermediate can be distinguished for single molecules and atoms.

Ions can be categorised as (i) molecular (*primary, parent*) ions, that is, a singly-charged positive ion produced in the primary act of interaction of ionizing radiation with a molecule (doubly-charged ions can be also formed in the primary act, but their relative yield is usually very small), (ii) *secondary* ions produced by reaction of the primary ion with the atoms and molecules of the medium, (iii) *excited* ions, those undergoing electronic or vibrational excitation, and (iv) *fragment* ions produced on decomposition of an excited ion (fragment ions can also be considered as secondary ions).

Free electrons can possess both thermal (*thermal electrons*) and higher energies. *Hyperthermal electrons* carry an energy exceeding $(3/2) kT$ but less than the energy of the first vibrational level E_{vib}, *subexcitation electrons* are those with an energy exceeding E_{vib} but below the first triplet level, and *nonionizing electrons* are those whose energy is less than the first ionization potential.

Excited molecules and atoms are species which have undergone electronic, vibrational, and rotational excitation. Electronically excited states (see section 1.5.1) may be of different multiplicities (*singlet, triplet*, etc.). Excited states corresponding to a large quantum number n are termed Rydberg states regardless of their multiplicity.

Superexcited states (SES) are those excited states whose potential energy is higher than the first ionization potential of a given atom or molecule.

Free radicals are chemical species having an unpaired electron in their outermost MO and classified as *radical ions* (*radical cations* and *radical anions*) having both an unpaired electron and an electronic charge, *excited free radicals* which feature electronic or vibrational excitation in addition to the unpaired electron, *biradicals* featuring *two* unpaired electrons, and *stable free radicals* where the unpaired electron is strongly delocalized so that their reactivity is dramatically reduced.

Clusters are multiparticle aggregates formed through the forces of intermolecular interaction; they can be either *neutral* or *charged*.

2.1.2 Species in the condensed phase

Atoms and molecules in condensed media cannot be considered as isolated particles because each molecule interacts with its neighbouring species. In general, the forces of intermolecular interaction (van der Waals' forces) are of an electrostatic nature and comprise both attractive forces (orientation, induction, and dispersion forces) and repulsive forces. These forces are distance-dependent and are usually described by an interaction energy potential $U(r)$. If a particle carries an electric charge, it is also necessary to include the Coulombic interaction.

Owing to the intermolecular interactions, so-called quasi-particles can be formed in a condensed phase in addition to the transients typical of the gas phase. These particles are elementary excitations of a group of interacting species. The quasi-particles may be described as having momentum, energy, and mass ('apparent mass'). The clearest picture of a quasi-particle as an aggregate, behaving in some respect like a quantum particle, is known for solids. In recent years, however, this concept also appears to be have been used successfully in a description of the interactions in liquids. There are two groups of quasi-particles; bosons and fermions. Bosons appear and decay singly, while fermions operate in pairs. We shall consider in some detail the most important of the quasi-particles found in HEC.

An **exciton** is a hydrogen-like bound state of a conduction electron and a hole. It corresponds to electronic excitation in a dielectric or in a semiconductor. It can migrate in a crystal but without charge or mass transfer. An exciton in a molecular crystal (the Frenkel exciton) is the excitation of the electronic system of one molecule which propagates through the crystal in the form of a wave. The radius of an exciton in an ionic crystal is larger than the corresponding value for the lattice constant (an exciton of intermediate radius); an exciton in a semiconductor has a very large radius (the Hoine–Mott exciton). The lifetime of excitons is short, thus an electron and a hole recombine, emitting a photon within 10^{-7} to 10^{-5} s. Excitons can also undergo radiationless decay, for example on being captured by a lattice defect. Excitons display a characteristic optical absorption.

A **plasmon** is a quantum of plasma oscillation or a quasi-particle representing the oscillation of electrons around heavy ions in a plasma. It has an energy of a few dozen eV and a lifetime of 10^{-16} to 10^{-15} s. On decay, it generates a molecular superexcited state with an excitation energy ranging from 15 to 25 eV.

A **polaron** is an excited state of an electron moving in a crystal inside a potential well created as a result of polarization and deformation of the crystal lattice by the electron field. The polaron may also be considered as a compound particle (electron + phonon) which can travel in a crystal as a complete entity. In crystals it can be a charge carrier. The apparent mass of a polaron is larger than that of the electron.

A **phonon** is a quantum of crystal lattice vibration. It propagates as a wave in a crystal. The vibrational energy of a crystal is approximately equal to the sum of the phonon energies (the energy of the zero-point vibrations of the lattice is not included). The higher the temperature, the larger the number of phonons. Phonons interact with each other, with different quasi-particles, and with crystal lattice defects.

A **hole** is a particle lacking an electron, the particle being an atom, molecule, or an ion constituting a crystal. In crystals it behaves like a positively-charged species. The hole propagates by a charge-transfer mechanism. The apparent mass of a hole is generally higher than that of a conduction electron. The hole can be localized in the vicinity of a lattice imperfection.

A **conduction electron** (quasi-free electron, mobile electron) is a particle formed simultaneously with a hole and travelling independently of the motion of the hole. The conduction electrons in metals are permanent, while in other types of crystal they appear on excitation.

Conduction electrons and holes are fermions, while excitons, plasmons, polarons, and phonons are bosons.

In addition to quasiparticles, there are some other species existing in the condensed phase, which include neutral molecules of the medium, which are usually considered as normal chemical particles.

Solvates (clusters) are aggregated species in a liquid where a chemical species is bound by intermolecular forces to several solvent molecules. The solvate is a trivial state of existence of ions (the most stable form of solvate), atoms, molecules, and excited species in solutions. Solvation causes a significant change in the dynamic and energetic properties of a particle.

A **solvated electron** is an electron delocalized over several solvent molecules. When in a solid phase, it is called a *stabilized* or *trapped* electron.

Ion pairs are particles with opposite charges separated by solvent molecules (anion+cation, cation+solvated electron, hole+quasi-free electron). Under the action of Coulombic forces an ion pair can move in a concerted way without recombination for a certain time.

A **radical pair** consists of two radicals stabilized in a solid matrix at a small distance (*ca.* few tenths of a nm) from each other.

Donor–acceptor complexes (molecular complexes, charge-transfer complexes) are formed from a molecule of an electron donor D and that of an electron acceptor A because of overlap of the orbitals of A and D, so the donor–acceptor bond distance is less than the sum of the van der Waas' radii of the bound atoms.

Donor–acceptor complexes can be composed of molecules in their ground state, of excited molecules, radicals, or ions. Binary donor–acceptor complexes consisting of different particles are called exciplexes if one of the species is in its electronically excited state (either singlet or triplet). The charge in these complexes is almost completely localized on the acceptor (in charge-transfer complexes). A donor–acceptor complex can originate from an electronically excited molecule and the same molecule in the ground state, when it is termed an *excimer*. The excimer is a product of excitonic interaction.

2.1.3 Energy characteristics of transient species

To obtain a quantitative description of the processes involving transient species, it is necessary to know their energetic and thermodynamic properties. Exchange of different types of energy between the transients and the bulk material occurs at rather different rates. For instance, the lifetime of electronically excited states of molecules ranges from 10^{-9} to 10 s, while their thermalization takes 10^{-12} s in a condensed phase and 10^{-11} to 10^{-8} s in the gas phase, depending on pressure. Therefore, many reactions of unstable species can be treated from the standpoint of conventional chemical kinetics and thermodynamics, assuming these species to be in partial thermal equilibrium with the medium. Because of the short lifetime of transient species, normal experimental methods cannot be used for determination of their thermodynamic properties although they can be evaluated sufficiently accurately by using thermodynamic cycles [28].

For electronically excited singlet molecules, the entropy is believed to change slightly upon excitation (for triplet states one must take into account the degeneracy, equal to 3). This assumption is rather firmly supported by comparison of the equilibrium constants for some reactions of electronically excited molecules determined by different methods. The enthalpy H and the Gibbs free energy G increase by the value of the excitation energy, the latter being easily found from relevant spectral data:

$$H^{o*} = H^o + E^*$$ (2.1)

$$G^{o*} = G^o + E^*$$ (2.2)

Thus the enthalpies of formation of excited molecules and the enthalpies of their reactions can be easily determined. The primary product of reaction can be formed either in its excited state or in the ground state. (Reactions of this sort are called, respectively, adiabatic and diabatic reactions[†]). In the former case, the

---

† Strictly speaking, the terms 'adiabatic' and 'diabatic' in quantum mechanics refer to processes occurring on the same potential surface and to those accompanied by a transition from one surface to another, respectively.

expression for the change of energy in a reaction in the excited state includes the difference between the excitation energy of the reactant E^* and that of the primary product $E^{*\prime}$:

$$\Delta H^{\circ *} = \Delta H^{\circ} - (E^* - E^{*\prime}). \tag{2.3}$$

In the latter case it includes only E^*. For example, many protolytic reactions of excited molecules [29, 30] occur adiabatically and their acidity constants K_a^*'s are calculated from the acidity constants for the ground state molecules, K_a, and the excitation energies for the acid (E^*) and the conjugate base $(E^{*\prime})$:

$$pK_a^* = -\log K_a^* = \Delta G^*/2.3\,RT = pK - (E^* - E^{*\prime}) \tag{2.4}$$

Depending on the structure of the compound, the excitation energy of an acid may be higher or lower than for its conjugate base, and thus the acidity can correspondingly increase or decrease upon excitation.

The detachment (or attachment) or an electron from (or to) excited and unexcited molecules produces the same electronic state of the radical ion; hence the ionization potential I_e always decreases

$$I_e^* = I_e - E^* \tag{2.5}$$

and the electron affinity A_e increases

$$A_e^* = A_e + E^* \tag{2.6}$$

by the value of the excitation energy of the excited species.

In the condensed phase it is better to use electrochemical potentials instead of ionization potentials since the former take into account the quite significant effect of solvation of charged particles:

$$E^{\circ}(M^+/M^*) = E^{\circ}(M^+/M) - E^*, \tag{2.7}$$

$$E^{\circ}(M^*/M^-) = E^{\circ}(M/M^-) + E^*. \tag{2.8}$$

Similarly, the thermodynamic properties can be calculated for all the other types of electronically-excited molecules.

The enthalpies of formation of inorganic and organic radicals are usually found from data on the bond energies for the corresponding molecules. Thus the enthalpy of formation of a radical R can be determined from the enthalpies of formation of compound RX and of atom X, knowing the bond energy D_{R-X}

$$\Delta H_f^{\circ}(R) = \Delta H_f^{\circ}(RX) - \Delta H_f^{\circ}(X) - D^{\circ}(RX) \tag{2.9}$$

The standard enthalpies of formation of atoms have been determined sufficiently precisely for most elements of the periodic table.

The bond energies can be found from calorimetric, mass spectrometric, and spectroscopic measurements. When it is possible to use approximate values, semi-empirical methods of calculation are used. Bond energy and enthalpy of formation data for atoms and radicals are available in tabulated form [31, 32]. It may be assumed that the enthalpy of formation of a radical cation is equal to the sum of the enthalpy of formation for the corresponding molecule and its ionization potential, while the enthalpy of formation of a radical anion is equal to the difference between the enthalpy of formation and the electron affinity of the molecule (taking the state of the free electron as standard):

$$\Delta H_f^0(M^{+\cdot}) = \Delta H_f^0(M) + I_e(M) \tag{2.10}$$

$$\Delta H_f^0(M^{-\cdot}) + \Delta H_f^0(M) - A_e(M). \tag{2.11}$$

Many radicals are sufficiently stable to be studied experimentally, for example by means of mass spectrometry. Thus, the ionization potentials of free radicals are determined from electron impact or photoelectron spectroscopy studies. Since radicals carry unpaired electrons in their nonbonding orbitals, their ionization potentials are appreciably lower than those of the parent molecules (see Table 2.1).

Table 2.1. Enthalpies of formation, ionization potentials, and electron affinities for some atoms, molecules, and radicals in their different electronic states [31, 32]

Species	ΔH°_{298} /kJ mol^{-1}	I_e/eV	A_e/eV	Species	ΔH°_{298} /kJ mol^{-1}	I_e/eV	A_e/eV
H(^2S)	218	13.60	0.75	$O_2(X^3\Sigma_g^-)$	0	12.08	0.44
O(^3P)	249	13.62	1.47	$O_2(a^1\Delta_g)$	94	11.10	1.42
O(^1D)	439	11.65	3.44	$\cdot OH(X^2\Pi)$	39	13.2	1.83
I($^2P_{3/2}$)	107	10.45	3.1	$HO_2(X^2A'')$	20	11.5	3.0
I($^2P_{1/2}$)	198	9.51	4.0	$H_2O(X^1A_1)$	−239	12.61	−5.0

The bond energies for radical ions are calculated from the energies of similar bonds in molecules, using the ionization potentials for the molecule and a relevant radical:

$$D(RX^{+\cdot}) = D(RX) - \left\{ I_e(RX) - I_e(R^\cdot) \right\}. \tag{2.12}$$

2.2 ELECTRONICALLY EXCITED STATES OF MOLECULES

The primary chemical reactions of excited molecules are subdivided into electron transfer, dissociation, etc., which will be considered later in separate sections. Together with this chemically-based classification, the systematization of their reactions in terms of their physical mechanism is also very important. Unlike the majority of thermal reactions occurring adiabatically on the potential surface of the ground electronic state, at least three types of mechanism are possible for excited molecules. These are (i) adiabatic transformation on the potential surface of an excited state, (ii) diabatic transition to the potential surface of the ground or some other state, and (iii), reactions in the so-called 'hot' (non-thermalized or highly vibrationally excited) ground electronic state resulting from internal conversion of electronically excited states.

Considering the reactivity in adiabatic reactions, conventional approaches may be adopted which are based mainly on transition state theory, the quantum yield for the primary reaction being determined by the ratio of the rate constant (activation energy) of the reaction to the sum of the rate constants for radiative and radiationless decay.

The key feature of diabatic reactions is the transition between the potential surfaces of different electronic states. Thus the quantum yield of reaction depends on the ratio of the probabilities of crossing to the ground state potential surfaces related to the products or the reactants. These probabilities, in turn, are determined by the form of the potential surfaces for the excited and ground states and by the velocity of the system travelling on the potential surface.

The decisive factor for reactions in 'hot' ground electronic state is the ratio of the rates of thermalization to that of surmounting the potential barrier. The closer the arrangement of the atomic nuclei in a molecule where conversion from the ground state takes place to the transition state of the reaction, the greater the probability of formation of products.

Thus, the nature of the chemical reactions of excited molecules is determined by the form of the potential surface for the corresponding excited state, and by the existence of energy barriers or 'funnels' on the surface leading to a transition to lower electronic states. The calculation of potential surfaces for polyatomic molecules in their excited states appears rather complicated. Meanwhile, there are some rather simple ideas enabling the probability of these or other processes to be estimated qualitatively. These approaches are principally based on symmetry considerations, especially the spatial symmetry of the electronic wave functions (of electron spin) for different states, depending on the position of the nuclei in a molecule.

2.2.1 Basic types of reaction of excited molecules

Electron-transfer reactions of excited molecules with electron acceptors or donors are of prime importance in high-energy chemistry. If a molecule is excited, its ionization potential is reduced but its electron affinity increases by the magnitude of the excitation energy. Therefore, excited molecules appear to be both strong electron donors and strong electron acceptors simultaneously, and the direction of electron transfer is determined mainly by the properties of the co-reactant molecules. For electron transfer to take place, it is necessary that the excitation energy exceeds the difference in redox potentials of the electron donor D and acceptor A

$$E^o > E^o(D/D^*) - E^o(A^-/A). \qquad (2.13)$$

The nature of the primary products of electron transfer reactions depends on the particular properties of the reactant molecules and on the medium polarity. In a polar medium ($\varepsilon \geq 10$), radical-ions of the reactant are most frequently formed, although their yield is usually much less than unity because of the degradation of the electronic excitation of the reactant (quenching) competing with the electron transfer and because of the cage recombination of the radical ions. In the case of triplet state reactions, the spin-forbiddenness retards crossing to the ground singlet state and the recombination of radical ions, so the yield of radical ions is close to unity. In nonpolar and weakly polar media some compounds form excited donor–acceptor complexes, or exciplexes. Singlet exciplexes are characterized by the existence of a typical emission band. A photo-induced electron-transfer reaction in fluid solutions occurs via two stages: diffusion followed by the electron transfer itself

$$A^* + D \rightleftharpoons A^*D \rightleftharpoons A^-D^+ \rightarrow A^{-\cdot} + D^{+\cdot} \, (\text{or } A + D).\qquad(2.14)$$

Resulting from the isoenergetic electron transfer inside the collision complex $A + D$, the radical-ion pair $A^{-\cdot}D^{+\cdot}$, with a nonequilibrium configuration of nuclei and nonequilibrium solvation sphere surrounding it, is formed. Relaxation leads to the formation of an exciplex, radical ions, or other products, or to a transition to the ground state (the degradation of the excitation energy), depending on the conditions. The electron transfer in the collision complex proceeds by tunnelling: its rate is determined both by the activation energy (the energy of the change in the nuclear configuration and in the solvent environment needed to equalize the energy levels in the initial and final states) and by the probability factor which depends on the overlap of the electronic wave functions of the donor and acceptor.

For many electron-transfer reactions there exists a correlation between the activation energy and the Gibbs free energy of the electron transfer. Thus a characteristic dependence of the rate constant for a process related to electron transfer on the ΔG value of the transfer is observed in liquid solutions (Fig. 2.1):

$$k = \frac{k_D}{1 + (k_{-D}/k_0)\exp(\Delta G^{\neq}/RT)}\qquad(2.15)$$

where k_D and k_{-D} are the rate coefficients for the diffusive approach and diffusive separation of the reactants, respectively, k_0 is the pre-exponential factor for electron transfer in the collision complex, and $\Delta G^{\neq}$ is the free energy of activation which is related to the Gibbs free energy ΔG of the electron transfer

$$\Delta G = E^{\circ}(A^-/A) - E^{\circ}(D/D^+).$$

Generally, correlation equations of the Marcus type

$$\Delta G^{\neq} \approx \Delta G_0^{\neq}(1 + \Delta G_0/\alpha \Delta G_0^{\neq})^2,\qquad(2.16)$$

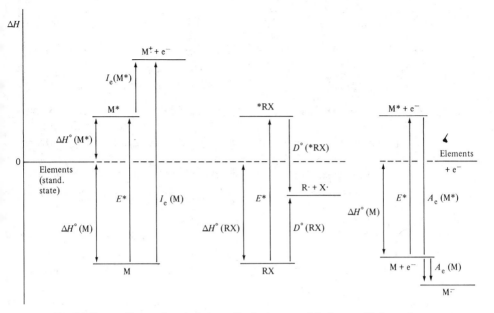

Fig. 2.1. Energy diagram for calculations of ionization potential, electron affinity, and dissociation energy of excited molecules.

or of the Rehm–Weller type

$$\Delta G^{\neq} \approx \frac{\Delta G}{2} \pm \sqrt{[(\Delta G_0^{\neq})^2 + (\Delta G/2)^2]},$$ (2.17)

or of the linear type (Polanyi)

$$\Delta G^{\neq} \approx \Delta G_0^{\neq} - \alpha \Delta G$$ (2.18)

are used which produce quite good agreement with experimental values in the region $0 < \Delta G < 50$ kJ mol^{-1} essential for reactions of excited molecules. If $\Delta G < 0$, the reaction rate constant for electron transfer is close to the diffusion limit k_D.

Relationships typical of electron tunnelling can be observed in frozen solutions where diffusion does not interfere with the kinetics of the elementary process. The rate constant of tunnelling decreases exponentially as the distance between the donor and acceptor increases:

$$k = v \exp(-2r/a)$$ (2.19)

where v is the frequency factor and a a parameter describing the decay of the electronic wave function outside the molecule.

In frozen glassy solutions, a homogeneous distribution of molecules is observed which results in the following distance distribution function of the reactants:

$$\rho(r) = 4\pi r^2 [Q] N_A / 1000, \tag{2.20}$$

where $[Q]$ is the concentration of quencher and in the rather complicated decay kinetics of excited molecules following pulsed excitation:

$$N(t) = N_o \exp\{-k_o t - N_A \pi a^3 [Q] [\ln vt]^3 / 6000\} \tag{2.21}$$

Experimental data on the kinetics of fluorescence quenching with different quenchers in frozen glassy solutions are described fairly well by Eq. (2.21), which enables the parameters a and v to be found. Typical values for these parameters are 0.1 nm and 10^{14} s^{-1}, respectively.

Dissociation of electronically excited molecules (bond cleavage) can proceed by different mechanisms: direct photodissociation, predissociation, induced predissociation, and dissociation through a hot ground state.

Direct photodissociation takes place when the excited electronic state has a repulsive potential energy curve (surface). The photon energy necessary for direct photodissociation from the singlet excited state must normally exceed 1.5–2 times the bond energy. A continuous (structureless) absorption spectrum is the distinctive feature of direct photodissociation. Dissociative triplet states can be produced both by intersystem crossing from excited singlet states and by energy transfer from a triplet sensitizer, which needs less photon energy:

$$^3M + X_2 \rightarrow {}^1M + {}^3X_2 \rightarrow {}^1M + 2X \cdot$$

In most cases homolytic bond cleavage occurs to yield atoms or radicals:

$$HCl + hv \rightarrow H + Cl$$

$$C_2H_5I + hv \rightarrow C_2H_5 \cdot + I \cdot.$$

Heterolytic bond cleavage is known for some metal halides and certain organic molecules:

$$TlBr + hv \rightarrow Tl^+ + Br^-.$$

Direct photodissociation can yield products either in the ground or excited states:

$$CS_2 + hv \rightarrow CS^*(A^1\Pi) + S.$$

Predissociation is the diabatic conversion (by intramolecular energy transfer) of the initially formed 'attractive' excited state of the molecule into the repulsive excited state, which dissociates, for example

$$S_2\left({}^3\Sigma_g^-\right) + hv \rightarrow S_2\left({}^3\Sigma_u^-\right) \rightarrow S_2\left({}^3\Pi_u\right) \rightarrow 2S({}^3P).$$

Predissociation causes broadening of some lines in the absorption spectra of simple molecules. This broadening is barely noticeable ($10^{-3} - 1$ cm^{-1}) if the rate constant for conversion is relatively small ($10^7 - 10^{10}$ s^{-1}).

Table 2.2. Photodissociation of some molecules in the gas phase [32]

Molecule	λ/nm	Products	Quantum yield
H_2	<111	2H (1s)	
	< 85	H(1s) + H(2p, 2s)	
HI	<280	H + I($^2P_{3/2}$)	ca. 0.9
		H + I($^2P_{1/2}$)	ca. 0.1
I_2	500–930	I($^2P_{3/2}$) + I($^2P_{1/2}$)	ca. 0.05–1
	500–840	2I($^2P_{1/2}$)	
O_2	<242	2O(^3P)	ca. 1
	<175	O(^3P) + O(^1D)	
	<133	O(^3P) + O(^1S)	
O_3	<900	O_2($^3\Sigma$) + O(^3P)	ca. 1
TlBr	<280	Tl + Br	
	<201	Tl$^+$ + Br$^-$	
	<192	Tl* + Br	
H_2O	<200	H + OH($^2\Pi$, $^2\Sigma$)	ca. 1
	147	H_2 + O(^1D)	ca. 0.1
H_2S	200–250	H + SH	ca. 1
	<165	2H + S	
NH_3	<217	H + NH$_2$	0.14
	<155	2H + NH($^3\Sigma^-$)	
	<129.5	H_2 + NH($^1\Pi$)	
	<123	NH$_3$ + e$^-$	
CH_3I	<360	CH$_3$ + I($^2P_{1/2}$)	ca. 0.8
		CH$_3$ + I($^2P_{3/2}$)	
	<266	CH$_2$ + HI	

Predissociation is very common in organic photochemistry, where the primary excitation is usually localized in an aromatic ring or other group (for example, carbonyl), but the cleavage occurs in a side chain:

$$C_6H_5Br + h\nu \rightarrow {}^*C_6H_5Br \rightarrow C_6H_5 \cdot + Br$$
$$R_2C{=}O + h\nu \rightarrow R_2C{=}O^* \rightarrow R\dot{C}{=}O + R \cdot$$
$$Ar_3CCN + h\nu \rightarrow {}^*Ar_2CCN \rightarrow Ar_3C^+ + CN^-$$

The formation of excited organic ions was observed in adiabatic heterolytic photodissociation of some rigid aromatic molecules:

Induced predissociation is caused by collisions (in the gas phase) of an excited molecule with some other molecules, which increase the probability of conversion into the repulsive state.

A clear example is the photodissociation of the iodine molecule, which has a bound excited state ($^3\Pi$) and fluoresces in the gas phase at very low pressures. The addition of foreign gases (such as Ar at $ca.$ 4 kPa) was found to induce a normally forbidden radiationless transition to another (repulsive) excited state $^3\Sigma_u^+$ (which crosses the $^3\Pi$ state), which quenches the fluorescence and results in formation of two ground-state I atoms:

$$I_2 + h\nu \rightarrow I_2\left(^3\Pi\right) \rightarrow I_2\left(^3\Sigma_u^+\right) \rightarrow 2I(^2P_{1/2}).$$

A magnetic field can also induce radiationless transitions and quench iodine fluorescence and promote photodissociation.

Hot ground-state dissociation is assumed for some photoreactions in the gas phase. Two mechanisms of such dissociation are proposed [280]. In both cases the excitation energy must greatly exceed the energy of the dissociating bond, and the probability of internal conversion (radiationless transition from the excited electronic state to the ground state) must be high. Vibrational energy released in this process is usually localized on certain chemical bonds (acceptor modes) which have essentially different equilibrium bond lengths in the excited and ground states. Scission occurs if the energy received by the particular bond exceeds its dissociation energy. The cleavages of other, weaker bonds can be caused by redistribution of the excess vibrational energy in the ground electronic state (quasi-equilibrium dissociation).

Excited electronic states of molecules can be formed not only as a result of photon absorption but also by, for example, electron impact. The characteristic time of excitation of molecules by electron impact is $ca.$ 10^{-16} s, while the characteristic lifetime of molecule is $ca.$ 10^{-14}–10^{-13} s in an unstable state and even more in a stable state. Hence, this type of dissociation is a two-stage process including the excitation of the state and its decay.

The process of exciting unstable states, or states inclined to predissociate, means that the probability of degradation of the molecule does not depend on pressure, temperature, or composition of the plasma, and is close to unity.

However, if the probability of spontaneous predissociation is low, for example because of the violation by such transitions of the selection rule, then the probability of dissociation will depend on the composition and temperature of the gas as well as on its pressure.

To illustrate the above we shall consider the cross-sections for electron-impact induced dissociation of an unexcited molecule of H_2 via electronically excited states. On analysing the energy-level diagram of the hydrogen molecule (Fig. 2.2) it is evident that the $b^3\Sigma_u^-$ state is unstable. Besides, an avalanche of transitions to this level from the $a^3\Sigma_g^+$ and $C^3\Pi_u$ states takes place, consequently excitation of all of these transitions leads to dissociation. The cross-sections of dissociation via these electronically excited states are presented in Fig. 2.3. The dissociation can occur on transition to the repulsive branches of the potential curves of the $B^3\Sigma_u^+$, $C^1\Pi_u$, and $B^1\Sigma_u^+$ states, resulting in formation of excited $2p$ and $2s$ hydrogen atoms. The cross section of formation of hydrogen atoms makes a major contribution to the dissociation of H_2 at energies over 50 eV. As the pressure changes, the probability of dissociation from the $a^3\Sigma_g^+$ and $C^3\Pi_u$ states may change. Thus, the dissociation from the $a^3\Sigma_g^+$ state is significant only at a pressure of 133 Pa since this state is deactivated effectively with increasing pressure owing to collisions with heavy particles.

Thus, our example of the dissociation of molecular hydrogen shows that this is a multichannel process. As the number of atoms in a molecule increases, dissociative ionization becomes more significant. Besides, most of the excited vibronic states result in effective dissociation.

Fragmentation reactions are those when an excited molecule is decomposed, giving two (or more) species in one step. In contrast to dissociation, this type of reaction is supposed to involve rearrangement of some bonds (usually by a concerted mechanism) rather than cleavage of one bond. Typical examples are the photodecompositions of ethyl iodide and diazomethane:

$$CH_3CH_2I + h\nu \rightarrow CH_2{=}CH_2 + HI$$

$$CH_2N_2 + h\nu \rightarrow {:}CH_2 + N_2.$$

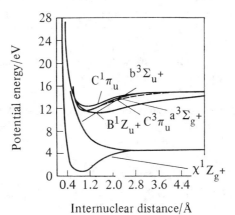

Fig. 2.2. Electronic terms of molecules H_2 and H_2^+.

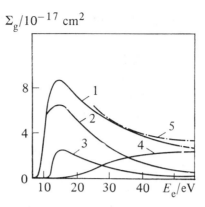

Fig. 2.3. Total and partial cross-section for electron-impact induced dissociation of electronically excited H_2 molecules ($X^1\Sigma_g^+, v = 0$): 1. total cross-section; 2. via $b^2\Sigma_g^+$; 3. $a^3\Sigma_g^+$; 4. with formation of H(2p) and H(2s) atoms; 5. sum of partial cross-sections 2–4.

Addition reactions of excited molecules to ground state species also usually proceed by a concerted mechanism, and they conform to the rules of conservation of orbital symmetry (Woodward–Hoffmann rules) as, for example, in the cycloaddition of an excited alkene to a ground-state alkene molecule:

$$\begin{array}{ccc} CR_2^* & CR_2 \\ \| & + & \| \\ CR_2 & CR_2 \end{array} \longrightarrow \begin{array}{cc} R_2C-CR_2 \\ | \quad\quad | \\ R_2C-CR_2 \end{array}$$

These reactions occur by the diabatic mechanism, that is, the product is formed in the ground state. The transition from the excited state potential surface to the ground state potential surface takes place along the reaction pathway. The excited-state potential surface has no significant potential barrier before the intersection of the potential surfaces, and so the addition occurs with high efficiency.

However, many photoaddition reactions are complex, multistage processes which can include different types of primary process. For instance, the addition of alcohols to alkenes yields different (Markownikoff and anti-Markownikoff) products in the presence of electron-donating and electron-withdrawing sensitizers, since the primary step is an electron transfer yielding radical-anions or -cations as transients:

$$Ph_2C{=}CH_2 + ROH + h\nu \begin{array}{c} D \nearrow Ph_2C{<}^{OR}_{CH_3} \\ \\ A \searrow Ph_2CHCH_2OR. \end{array}$$

Abstraction reactions of atoms by excited molecules are typical for n, π^* excited states (for example, of carbonyl or heteroaromatic compounds) which have the electronic structure and properties typical of σ-radicals:

$$^3R_2C=O(n, \pi^*) + HR' \rightarrow R_2\overset{\cdot}{C}OH + \cdot R'.$$

Sometimes, substituents or a solvent can alter the nature of the lowest excited state and hence crucially influence its reactivity. Reaction rate constants for abstraction by n, π^* states depend on the exothermicity of the reaction in accordance with the Polanyi equation.

Isomerization reactions include *cis-trans*-isomerization, stereoisomerization, valence isomerization, etc. *Cis-trans*-photoisomerization is typical for all compounds containing double bonds: alkenes, azo-compounds, etc. It can proceed both from the excited singlet and triplet states. Many compounds have a near-orthogonal equilibrium configuration in their excited states, in contrast to a planar ground state. This accounts for the high efficiency both of *cis-trans* and *trans-cis*-photoisomerizations.

Valence isomerizations and rearrangements are common for aromatic compounds and polyenes. Many of them are concerted reactions. Most aromatic and heteroaromatic compounds undergo photoisomerization yielding unstable isomers such as benzvalene, Dewar benzene, or prismane, and, finally, positional isomers:

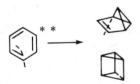

Electrocyclic photoreactions of polyenes lead to cyclic isomers:

$$\text{(diene)} \xrightarrow{h\nu} \text{(cyclobutene)}$$

Sigmatropic shifts are typical for polyenes. They proceed by a concerted mechanism and lead to photoisomers:

$$\text{(pentadiene)} \xrightarrow{h\nu} \text{(pentadiene isomer)}$$

Irradiation of 1,4-dienes or 3-arylalkenes yields vinylcyclopropanes or arylcyclopropanes (di-π-methane rearrangement), probably by a 1,2-shift of the vinyl or aryl group accompanied by ring closure:

$$\text{(1,4-diene)} \longrightarrow \text{(vinylcyclopropane)}$$

2.2.2 Exciplexes and excimers

Many excited molecules can interact with other ground-state molecules forming excited complexes (exciplexes). If the excited complex is formed by identical molecules, then it is called an excimer (excited dimer). The bonding of two molecules in exciplexes is usually caused by electron donor–acceptor interaction (Fig. 2.4). The exciplex wave function can be approximated by the sum of the wave functions of locally excited and charge-transfer states:

$$\Psi(AD)^* = a\Psi(A^*D) + (1-a)\Psi(A^-D^+). \tag{2.22}$$

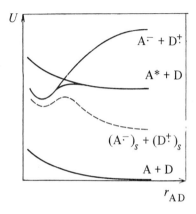

Fig. 2.4. Potential-energy diagram for formation of an exciplex in nonpolar media via photoinduced electron-transfer.

Usually $a \ll 1$ for exciplexes. Exciplexes have large dipole moments (μ $ca.$ 5×10^{29} C m). The enthalpy of exciplex formation depends on the difference between the oxidation potential of the donor and reduction potential of the acceptor (and the excitation energy of the initial donor or acceptor molecules). The entropy decrease is $ca.$ 80 J K^{-1} mol^{-1}, which is typical for rigid complex formation.

Bonding in excimers is caused by delocalization of the excitation (exciton interaction):

$$\Psi(A_2^*) = (1/\sqrt{2})\Psi(A^*A) + (1/\sqrt{2})\Psi(AA^*). \tag{2.23}$$

Excimers are nonpolar species ($\mu = 0$). The enthalpy of excimer formation strongly depends on the transition moments. Exciplexes and excimers of aromatic molecules are thought to have a sandwich structure with a distance between the planes of $ca.$ 0.33 nm.

The formation of singlet exciplexes (and excimers) is evinced by the appearance of a new emission band in the presence of an electron donor or electron acceptor (for excimers, at a high concentration of the fluorescent compound).

Triplet exciplexes can be observed from their absorption spectra. The emission and absorption spectra of exciplexes are strongly sensitive to the solvent polarity (which is used for the experimental determination of their dipole moments). Exciplexes are usually observed in nonpolar solvents ($\varepsilon < 20$), while they are unstable in polar solvents. In many cases exciplexes and excimers act as the intermediates of photochemical reactions, and their formation can affect both the reactivity of excited molecules and the mechanism of their photoreactions.

2.3 SUPEREXCITED STATES

As noted in section 1.1, electron impact of a molecules causes its ionization and excitation. It was generally believed that electronically excited states thus formed are of an energy less than the first ionization potential. However, there is some possibility of excitation of inner electrons having an energy higher than I_1. (It does not matter for an atom which of its electrons is excited.) So, if the excitation energy exceeds I_1 then autoionization necessarily takes place. With a molecule, however, competing processes of degradation of the excitation energy are possible, therefore the following processes can occur for a superexcited state (SES) of this type, that is, a state of the molecule of energy exceeding I_1 [33, 34].

$$
\text{M}^{**} \begin{cases} \text{M}^+ + e^- & \text{Autoionization} \\ \text{R}_1 + \text{R}_2\,(\text{R}_2^*) & \text{Predissociation} \\ \text{M} + \Delta E'_{\text{vib}} + h\nu' & \text{Radiative decay} \\ \text{M} + \Delta E_{\text{vib}} & \text{Radiationless decay} \end{cases}
$$

so that the yield of autoionization is less than the total yield of SES (denoted hereafter as M^{**}).

There have been no direct observations of SES insofar as their lifetimes are very short, although there are many indirect indications of their formation on irradiation of different substances. The strongest demonstration is the production of lines due to the Balmer series of H atoms after irradiation of CH_4 with mono-energetic electrons at a threshold of 21.8 eV, whereas $I_1 = 12.7$ eV [35].

The following types of SES may exist: (i) the Rydberg-type one-electron excitation of a large quantum number, (ii) electron excitation into orbitals of higher energy than the ionization potential I_1 and (iii) the simultaneous excitation of two or more electrons in a molecule.

Energy transfer between the electronic levels of all types of SES takes a certain time (autoionization is assumed to be instantaneous). The time may transpire to be longer than a half-period of vibration so that decay processes become probable.

The formation of SES can be caused both by the direct action of ionizing radiation and by the degradation of plasmons in a condensed phase. The G-value of SES is nearly unity [33].

2.4 IONS

When formed, positive ions are highly reactive and undergo various reactions [36], of which the following are most typical.

(i) **Fragmentation of primary ions**—the primary excited ions which have received an excess of internal energy in the act of ionization dissociate, producing a fragment ion of smaller mass and a neutral particle (molecules or radical). The rate constant k for this unimolecular process can be found from the equation [36]:

$$k = k' \left(\frac{E - E_{min}}{E} \right)^{n-1} \tag{2.24}$$

where k' is the normalization factor, E is the internal energy of the ion, E_{min} is the minimum energy required for the fragmentation and, n the number of effective vibrations of the ion.

From Eq. (2.24) it is clear that the higher the internal energy, the greater the rate constant for fragmentation. On the other hand, the greater the number of effective vibrations, the lower the rate constant, that is, the probability of decomposition of more complex ions is smaller. The quantitative description of the fragmentation of excited ions is still hindered by the lack of reliable data on the values of E, E_{min}, and n.

Excited ions have different lifetimes. Experimentally it was found that the formation of fragment ions usually takes less than 10^{-8} s (that is, the resolution time of a mass spectrometer). Since molecules have a set of excitation levels, a diversity of fragment ions is generally found, for example methanol gives rise to the following fragment ions: CH_3O^+, $\cdot CH_2O^+$, CH_3^+, $\cdot CH_2^+$, $\cdot CH^+$, CHO^+, and H^+. A similar picture is observed for other substances.

General consideration of mass spectrometric data [37] leads to the following conclusions:

(a) The precursors of fragment ions are excited molecular ions. Molecular ions are observed in mass spectra but for every compound, and, when detected, are not usually among the most abundant ions in the spectrum. This is an indication of the fact that the electron impact and similar interactions mainly produce excited molecular ions.

(b) Extensive degradation of a molecular ion, which requires much energy, is normally less probable. This corresponds to the observation (see section 1.2) that the higher the energy to be transferred in the interaction of ionizing radiation with matter, the lower the probability of energy transfer.

(c) Increasing the concentration of substance in the gas phase causes both the probability of production of fragment ions to decrease and the extent of fragmentation. These decreases are related to collisional deactivation and to the emergence of competing processes, namely various ion–molecule reactions.

(d) Some of the excited parent ions are characterized by a lifetime so short that they do not undergo deactivation but decompose even in the condensed phase. For example, the fragment ion $C_2H_5^+$ is produced with a yield of 0.25 in liquid propane, whereas G is $ca.$ 0.4 in the gas phase; the ions HCO^+ and CO^+ are formed with a yield of 0.8 in gaseous methanol at low pressures, but this value is 0.1 in solid or liquid alcohol. However, there is an alternative explanation of these facts, namely the fragment ions are the products of degradation of superexcited states.

(e) Since the precursors of fragment ions possess a different excess energy, their decomposition rate constants are different, as follows from Eq. (2.24). Hence, for each excited molecular ion there exists a distribution of reaction rates and, correspondingly, a distribution in the efficiencies of inhibition of fragmentation through collisional deactivation.

(ii) **Isomerization of ions** is the second unimolecular process competing with dissociation of excited ions. The fragment ions can also undergo isomerization. Here are some examples of isomerization processes:

for a primary ion:

$$\text{cyclo} - C_4H_8^+ \rightarrow (1 - C_4H_8^+)^* - \begin{array}{l} \rightarrow 2 - C_4H_8^+ \\ \quad (9.02 \text{ eV}) \\ \rightarrow \text{iso} - C_4H_8^+ \\ \quad (9.06 \text{ eV}) \end{array}$$

$$(I_1 = 10.84 \text{ eV}) (9.58 \text{ eV})$$

and for a secondary ion:

$$(CH_3)_2 C\dot{H}C_2H_5^+ \rightarrow (CH_3CHCH_2CH_3^+)^* + \cdot CH_3$$

$$(CH_3CHCH_2CH_3^+)^* \rightarrow (CH_3)_3C^+.$$

The reaction rate constant for isomerization is defined by the same Eq. (2.14) as for fragmentation, the only difference being that of E_{min}, which is the minimum energy needed for a species to isomerize. Exact values of E, E_{min}, and n are also unknown in this case; that is, isomerization rate constants have to be found experimentally. The dependence of the reaction rate of isomerization on the concentration of substance in a system is similar to that for ion-fragmentation; the fraction of ions isomerized decreases as the pressure in the system increases.

Although mass spectrometry provides direct information on the process of ion fragmentation, it cannot produce direct data on the isomerization of ions. Direct information on the processes of isomerization of radical-cations and radical-anions can be obtained by the ESR technique.

(iii) **Ion–molecule reactions** refer to a large family of reactions of highly reactive ions [36], which embraces charge-transfer processes, proton-transfer reactions, and ion–molecule reactions proper, that is, processes in which an atom or a group of atoms (simple or complex ion) is transferred. A large body of data on the cross-sections for gas-phase ion–molecule reactions and on the rate constants for some of the reactions has been collected by Virin *et al.* [38], although the mass spectrometric determination of reaction rate constants is not easy. Nevertheless, it should be noted that most of the above cross-sections and rate constants have been obtained for ions accelerated in an electric field, whereas ions involved in the processes of high-energy chemistry do not usually gain a significant excess of kinetic energy (even under the conditions of a low-temperature plasma). Therefore it is better to use the values of cross-sections and rate constants which have been obtained mass spectrometrically, but with as low a kinetic energy of the ions as possible. In general, there is no simple dependence of the cross-section and the reaction rate constant on the kinetic energy of the ion.

Charge transfer. The reaction of charge transfer signifies two types of process: transfer between identical species

$$A^+ + A \rightarrow A + A^+$$

and transfer to an acceptor of positive charge (which is, in fact, an electron donor)

$$A^+ + B \rightarrow A + B^+.$$

Charge transfer in gas phase occurs at the high rate typical of diffusion-controlled reactions. In the condensed phase, a primary ion is represented as a hole (before its localization) and migration of the hole is essentially the transfer of charge from one group of molecules to another.

The transfer of positive charge to an acceptor is possible only in the case when the ionization potential of the solute is less than that of the positive-charge donor. Some rate constants for charge-transfer reactions are given in Table 2.3.

One very important instance of charge transfer worth mentioning is when a molecule obtains some additional excitation energy in the act of ionization, causing its dissociation, the latter being able to occur through different pathways. This type of charge transfer is called *indirect ionization* [39]. The process of charge transfer with the simultaneous fragmentation of the ions formed is quite typical for the gas phase. Some processes of indirect ionization are shown in Table 2.4. The presence of fragment ions in a mass spectrum seems to be inherent since the excitation of an ion generated in the chamber of a mass spectrometer takes place in the very act of ionization. However, the charge transfer also results in fragment ions, their formation in some cases being the major pathway of the transfer. Quantum-chemical calculations on the reaction $H_2O^{+\cdot}$ ion with methanol show [7] that the electron is transferred to the water

Table 2.3. Rate constants for charge-transfer reactions in the gas phase [38]

Reaction	Kinetic energy of ions/eV	$k/10^{-10}$ $cm^3\ s^{-1}$	ΔI_1 /eV
$\cdot H_2^+ + CO_2 \rightarrow \cdot CO_2^+ + H_2$	≤ 0.1	20	1.62
$\cdot H_2^+ + CO_4 \rightarrow \cdot CH_4^+ + H_2$	≤ 0.1	14	2.28
$\cdot H_2^+ + C_2H_2 \rightarrow \cdot C_2H_2^+ + H_2$	≤ 0.1	48	3.99
$\cdot H_2^+ + C_2H_4 \rightarrow \cdot C_2H_4^+ + H_2$	≤ 0.1	22	4.9
$\cdot H_2^+ + C_2H_6 \rightarrow \cdot C_2H_6^+ + H_2$	≤ 0.1	29	3.8
$OH^+ + H_2O \rightarrow H_2\dot{O}^+ + \cdot OH$	≤ 0.1	15.6	0.57
$H_2\dot{O}^+ + H_2S \rightarrow H_2\dot{S}^+ + H_2O$	≤ 0.1	8.9	2.15
$H_2\dot{O}^+ + NH_3 \rightarrow \dot{N}H_3^+ + H_2O$	≤ 0.1	5.6	0.31
$\cdot N_2^+ + O_2 \rightarrow \cdot O_2^+ + N_2$	≤ 0.06	0.7	3.4
$\cdot N_2^+ + H_2O \rightarrow H_2\dot{O}^+ + N_2$	$ca.\ 0.03$	22	2.99
$\cdot N_2 + NO \rightarrow \cdot NO^+ + N_2$	$ca.\ 0.03$	3.3	6.35
$\cdot N_2^+ + NH_3 \rightarrow \cdot NH_3^+ + N_2$	$ca.\ 0.03$	18.3	5.45
$\cdot CH_4^+ + H_2S \rightarrow H_2\dot{S}^+ + CH_4$	≤ 0.1	0.59	2.66
$\cdot CH_4^+ + NH_3 \rightarrow \cdot NH_3^+ + CH_4$	≤ 0.1	0.69	2.97

Table 2.4. Ion abundances arising from methanol with different ionization techniques [38]

Ion produced from methanol	Chemical ionization by				Electron impact	Appearance potential/eV
	Kr^+	Xe^+	CH_4^+	H_2O^+		
$\cdot CH_3OH^+$	1	2	74	21	19	10.84
$CH_2OH^+ + H$	81	96	–	–	35	11.66
$CH_3O^+ + H$	–	–	2	3	3	11.66
$CH_2\dot{O}^+ + H$	8	2	24	72	0.7	12.35
$CHO^+ + H + H_2$	6	–	–	3	20	14.6
$CH_3 + H + H_2$	4	–	–	–	10	13.5
ΔI_1/eV	3.30	1.47	2.34	1.96		

ion, not from the first electron orbital of the alcohol ($I_2 = 10.84$ eV and 10.65 eV, respectively) but from the second orbital ($I_2 = 11.8$ and 11.6 eV for methanol and ethanol, respectively). The ion thus produced receives an excess energy in the form of electronic excitation, which causes its subsequent dissociation.

Owing to the development of pulse radiolysis in the last 20 years, there have appeared some data on charge-transfer reactions in liquids (Fig. 2.5). It is seen that the diffusion limit for the CCl_4^+ ion is reached rapidly (with ΔI_1 of 1 eV). In the case of hole-scavenging in hydrocarbons, the limiting value is reached at a higher difference in the ionization potentials (ca. 3 eV), and the rate constant for the hole is by one and half orders of magnitude larger than for the ion. This testifies to the fact that the mass of a hole is considerably less than that of the ion.

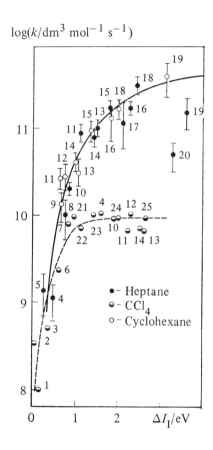

Fig. 2.5. Dependence of reaction rate constant for charge transfer to solute in heptane, cyclohexane, and tetrachloromethane on difference between the first ionization potentials of solvent and solute [40]. Solutes: 1—dichloromethane, 2—trichloro-methane, 3—dichloroethane, 4—cyclohexane, 5—3-methylheptane, 6—tert-butyl chloride, 7—1-chlorobutane, 8—decalin, 9—benzene-D_6, 10—heptene-1, 11—benzene, 12—hexene-2, 13—toluene, 14—cyclohexane, 15—2,3-dimethylbutene-2, 16—tetramethylbenzene, 17—biphenyl, 18—triethylamine, 19—N,N,N',N'-tetra-methyl-p-phenylenediamine, 20—pyrene, 21—cyclopentene, 22—heptane, 23—methylhexane, 24—hexene-1, and 25—methylcyclohexane.

Proton transfer. This reaction may be written as follows:

$$M_{(1)}H^+ + M_{(2)} \rightarrow M_{(1)} + M_{(2)}H^+.$$

A necessary, but insufficient condition for the occurrence of a reaction of this kind is that the proton affinity of molecule $M_{(2)}$ must be higher than that of $M_{(1)}$ [41]. Unfortunately, values for proton affinities in the gas phase cannot be transferred to the liquid state, as shown in Table 2.5, where the differences in the proton affinities (*PA*) of ammonia and some amines in the gaseous and liquid states are presented. It is obvious that even the sign of the difference is not the same throughout the substances considered. Proton affinities are usually obtained from mass spectrometry by ion cyclotron resonance measurements.

Table 2.5. Change in Gibbs free energy $\Delta G°$(in eV) for reaction $NH_4^+ + M \rightarrow NH_3 + MH^+$ in the gas phase and in aqueous solution [36]

M	Gas phase at 600 K	Aqueous solution
NH_3	0	0
$C_6H_5NH_2$	−0.39	+0.27
CH_3NH_2	−0.47	−0.08
$C_6H_5NHCH_3$	−0.66	+0.26
$(CH_3)_2NH$	−0.80	−0.08
C_5H_5N	−0.81	+0.23
$(CH_3)_3N$	−1.01	−0.03

Reactions of proton transfer are normally very fast because they are all exothermic. Thus the reactions

$$CH_3X^+ + CH_3X \rightarrow CH_3XH^+ + CH_2X$$

(where X = H, F, Cl, OH, SH, and NH_2) have a rate constant of 10×10^{-10} to 25×10^{-10} cm^3 s^{-1}, which means that almost every collision results in reaction. A large number of cross-sections and rate constants for proton transfer reactions have been determined for ions in the gas phase at different energies and pressures (Table 2.6) and for ions in some liquid systems (Table 2.7) [31, 36].

Negative ions can also take part in reactions of proton transfer, for example:

$$OH^- + C_3H_6 \rightarrow H_2O + C_3H_5^- \qquad k = 2.6 \times 10^{-10} \text{ cm}^3 \text{ s}^{-1}$$

$$O^- + C_2H_5OH - \begin{cases} \rightarrow \, ^\cdot OH + C_2H_5O^- & k = 3.6 \times 10^{-10} \text{ cm}^3 \text{ s}^{-1} \\ \rightarrow OH^- + C_2H_5O^\cdot & k = 4.4 \times 10^{-10} \text{ cm}^3 \text{ s}^{-1} \end{cases}$$

$$NO_2^- + HBr \rightarrow HNO_2 + Br^- \qquad k = 19 \times 10^{-10} \text{ cm}^3 \text{ s}^{-1}$$

Table 2.6. Rate constants for gas-phase proton-transfer reactions [38]

Reaction	Ion energy/eV	$k/10^{-10}\,\mathrm{cm^3\,s^{-1}}$
$HD^+ + HD \longrightarrow H_2D^+ + D$	0.03	4 to 9
$ \longrightarrow HD_2^+ + H$	0.03	3 to 11
$D_2^+ + D_2 \rightarrow D_3^+ + D$	0.03	6 to 17
$D_2^+ + H_2O \rightarrow H_2DO^+ + D$	0.1	17.3
$H_2^+ + CO \rightarrow COH^+ + H$	0.1	11 to 22
$H_2^+ + CO_2 \rightarrow CO_2H^+ + H$	0.1	12 to 49
$H_2^+ + CH_4 \rightarrow CH_5^+ + H$	0.1	0.11
$H_2\dot{O}^+ + H_2 \rightarrow H_3O^+ + H$	0.03	14
$H_2\dot{O}^+ + H_2O \rightarrow H_3O^+ + \dot{O}H$	0.03	5 to 18
$H_2\dot{O}^+ + H_2S \rightarrow H_3O^+ + H\dot{S}$	0.1	5.9 to 7.0
$D_2\dot{O}^+ + D_2O \rightarrow D_3O^+ + \dot{O}D$	0.03	12 to 37
$N_2^+ + HD \longrightarrow N_2H^+ + D$	0.1	5.6
$ \longrightarrow N_2D^+ + H$	0.1	5.4
$NH_3^+ + NH_3 \rightarrow NH_4^+ + \dot{N}H_2$	0.03 to 0.1	10 to 21
$CH_4^+ + H_2O^+ \rightarrow H_3O^+ +\ \dot{C}H_3$	0.01	24
$CH_4^+ + CH_4 \rightarrow CH_5^+ +\ \dot{C}H_3$	0.1	9.5 to 13
$\dot{C}_2H_6^+ + H_2O \rightarrow H_3O^+ +\ \dot{C}_2H_5$	0.03	12

Table 2.7. Rate constant, $k/10^4\,\mathrm{s^{-1}}$ for liquid-phase proton transfer from alcohols to arene anions [42]

Alcohol	Biphenylide ion	Anthracenide ion	p-Terphenylide ion
CH_3OH	6.9	8.1	0.04
C_2H_5OH	2.6	2.3	0.03
C_3H_7OH	3.2	2.4	–
$CH_3CH(OH)CH_3$	0.55	0.36	–

Transfer of an ion or group of atoms. The composition of the group transferred can be different; that is, for the same pair of reacting species, the reaction may proceed via several pathways simultaneously. For example:

$$C_2H_6^+ + C_2H_6 \rightarrow C_4H_{10}^+ + H_2 \qquad\qquad k = 0.04 \times 10^{-10}$$

$$\rightarrow C_3H_9^+ + H_2 + H \qquad\qquad k = 0.1 \times 10^{-10}$$

$$\rightarrow C_3H_9^+ + CH_3 \qquad\qquad k = 0.17 \times 10^{-10}$$

$$\rightarrow C_3H_8^+ + CH_4 \qquad\qquad k = 0.23 \times 10^{-10}$$

$$\rightarrow C_3H_7^+ + CH_3 + H_2 \qquad\qquad k = 0.14 \times 10^{-10}$$

$$\rightarrow C_2H_5^+ + C_2H_5 + H_2 \qquad\qquad k = 3.2 \times 10^{-10}$$

$$\rightarrow C_2H_4^+ + C_2H_6 + H_2 \qquad\qquad k = 1.85 \times 10^{-10}$$

(here the ion energy is *ca.* 0.1 to 0.2 eV and the rate constant is given in $cm^3 s^{-1}$ [38]).

As shown, there are condensation reactions with the elimination of either an H atom or a hydrogen molecule, condensation reactions with C—C bond cleavage and elimination of CH_3 or CH_4, and hydrogen atom abstraction reactions from one or both of the reactants. Some examples of ion–molecule reactions involving the transfer of other groups and species are given in Table 2.8.

An interesting subset of ion–molecule reactions is that of condensation reactions. We have seen examples of these for benzene and ethane. The degree of condensation of different unsaturated compounds increases as the pressure in the system increases (owing to competition with neutralization) (Fig. 2.6). Condensation is also observed for alkanes although here the yield of condensed ions is only a few per cent (however, the degree of condensation is rather high). Thus, $C_2H_7^+$, $C_3H_9^+$, $C_4H_{11}^+$, $C_5H_{13}^+$, and $C_6H_{16}^+$ ions appear in this sequence (that is, with a CH_2 increment) from methane as the pressure is raised; the curves pass through a maximum at the corresponding pressures.

Ion–molecule reactions are generally considered to occur via a transition state (activated complex) which has an excess of electronic and vibrational energy owing to the difference in ionization potentials:

$$A^+ + B \rightarrow (A^+.B)^{\neq} \rightarrow \Sigma C^+ + \Sigma D$$

(the existence of condensation reactions confirms formation of the activated complex). Depending on the excess energy and the degrees of freedom, one or other degradation reaction of the transition state takes place. If the complex lives long enough, it can be deactivated by collisions with other molecules or even decompose back to the reactants, and therefore a change in pressure will vary the contributions of different pathways to the overall reaction.

The rate constant largely depends on the energy of the ions (though the kinetic energy of the ions is perhaps not wholly included in the excess energy of the activated complex). Figure 2.7 shows that the rate constants for the reactions of H_2O^+ and NH_3^+ ions with their parent molecules decrease as the energy of the ions increases. However, this relationship is not the same for all

Table 2.8. Rate constants for gas-phase ion–molecule reactions [38]

Reaction	Ion energy /eV	$k/10^{-10}\ cm^3\ s^{-1}$
$SO_4^+ + H_2O \rightarrow H_2\dot{O}_3^+ + SO_2$	ca. 0.03	1
$S^+ + H_2S \rightarrow HS_2^+ + H$	< 0.1	0.64
$\rightarrow \dot{S}_2^+ + H_2$	< 0.1	0.24
$H\dot{I}^+ + CH_3I \rightarrow CH_3I_2^+ + H$	< 3	3.6
$\dot{S}F_4^+ + O_2 \rightarrow SF_3^+ + (\dot{F}O_2)$	< 0.3	0.21
$N^+ + N_2O \rightarrow \ \dot{N}O^+ + N_2$	1.1	5.5
$N^+ + NH_3 \rightarrow N_2H^+ + H$	< 3	0.6
$\dot{N}_2^+ + H_2O \rightarrow N_2H^+ + \ \dot{O}H$	ca. 0.03	21
$\dot{N}_2O^+ + CH_4 \rightarrow N_2OH^+ + \ \dot{C}H_3$	< 0.1	10
$NO_2^+ + \ \dot{N}O_2 \rightarrow NO^+ + \ \dot{N}O_3$	< 0.1	4.6
$C^+ + O_2 \rightarrow \ \dot{C}O^+ + O$	1.25	9 to 11
$\dot{C}O_2^+ + CH_4 \rightarrow HCO^+ + CH_3\dot{O}$	ca. 0.1	0.14
$\rightarrow HCO_2^+ + \ \dot{C}H_3$	ca. 0.1	23
$\dot{C}O_2^+ + C_3F_8 \rightarrow CF_3^+ + (C_2F_5CO_2^{\cdot})$	< 0.1	6.7
$\rightarrow \ \dot{C}_2F_4 + (CF_4 + CO_2)$	< 0.1	2.3
$\rightarrow C_2F_5^+ + (CF_3\dot{C}O_2)$	< 0.1	0.4
$CH_3^+ + CH_4 \rightarrow C_2H_5^+ + H_2$	< 0.1	9 to 13
$CH_3^+ + C_2H_6 \rightarrow C_2H_5^+ + CH_4$	1.25	17
$\dot{C}H_4 + C_2H_2 \rightarrow C_3H_3^+ + H_2 + H$	ca. 1.8	5.4
$\dot{C}_2H_2 + CH_4 \rightarrow C_3H_5^+ + H$	ca. 0.03	1.6
$\rightarrow C_2H_3^+ + \ \dot{C}H_3$	< 0.1	6.5
$\dot{C}_6H_5^+ + C_6H_6 \rightarrow \ \dot{C}_{12}H_{11}^+$	ca. 0.03	3 to 4.3
$\rightarrow C_{12}H_{10}^+ + H$	ca. 2	0.7
$\rightarrow \ \dot{C}_{12}H_9^+ + H_2$	ca. 0.03	0.5 to 0.8
$\rightarrow \ \dot{C}_{10}H_9^+ + C_2H_2$	ca. 0.03	1 to 24
$CCl_3^+ + CH_3CHO \rightarrow CH_3CHCl^+ + (CCl_2O)$	< 0.1	0.3
$CH_4\dot{S}iH^+ + (CH_3)_2 SiH_2 \rightarrow$		
$\dot{S}i_2(CH_3)_2^+ + CH_4 + H_2$	ca. 0.03	1.6

Note: Tentative products are indicated in parentheses.

Ion abundance

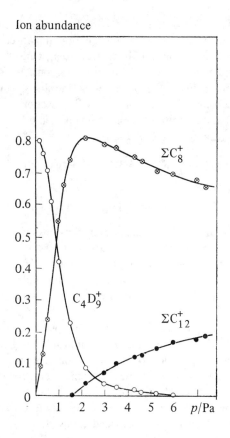

Fig. 2.6. Ion composition in perdeuterated isobutene as a function of pressure [36].

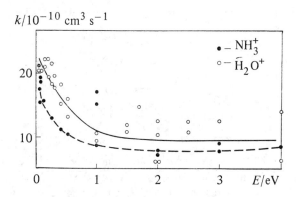

Fig. 2.7. Dependence of rate constant on the ion energy for ion–molecule reactions of NH_3^+ with NH_3 and of H_2O^+ with H_2O [38].

ion–molecule reactions. Thus, the rate constant for the reaction of O^+ ion with N_2 increases over identical ranges of ion energy. If two competing reactions occur, the rate constant for one of them decreases, while the other increases, with increasing energy of the ions. Thus the rate constant for charge transfer from deuterium ions to hydrogen increases as the energy increases, but conversely, the rate constant for the ion–molecule reaction producing HD_2^+ decreases. The contributions from different channels also depend on the velocities of the ions. With two competing processes

$$\text{cyclo–}\dot{C}_6H_{12}^+ + C_nH_{2n} \rightarrow C_6H_{11} + \dot{C}_nH_{2n+1}$$

and

$$\text{cyclo–}\dot{C}_6H_{12}^+ + C_nH_{2n} \rightarrow C_6H_{10}^+ + \dot{C}_nH_{2n+2},$$

the contribution from the first reaction, which consumes more energy, increases with increasing energy of the cyclohexane ion.

Current theoretical models [36] can provide more-or-less accurate determination of the rate constants for diffusion-controlled ion–molecule reactions, but it is impossible to predict the direction of a particular ion–molecule reaction with these models.

Negative ions can be produced in the following processes: (i) electron attachment to a molecule, (ii) an ion–molecule reaction involving negative ions, (iii) the collision of neutral molecules, (iv) the collision of a fast positive ion with a molecule, and (v) interaction of a molecule with an electronically-excited molecule, metastable atom, or Rydberg state.

Electron attachment may be represented, according to Christophorou [43], by the following scheme, including the activated complex:

$$e^-(E) + AB...CD(E_0, v \approx 0) \xrightarrow{\sigma_0(E)} [AB...CD]^- (E_0 \text{ or } E) \quad (1)$$

$$[AB...CD]^{-*} \longrightarrow AB...CD^{(-)} + e^-(E') \quad (2)$$

$$\longrightarrow AB^* + CD^- \quad (3)$$

$$\longrightarrow [AB...CD]^* + D^- \quad (4)$$

$$AB^* + CD^{-*} \begin{cases} \longrightarrow CD^* + e^-(E') & (5) \\ \longrightarrow C^* + D^- & (6) \end{cases}$$

$$[AB...CD]^{-*} \longrightarrow [AB...CD]^- + Q \quad (7)$$

where $e^-(E)$ is an electron with a given kinetic energy, E_0 and E are the ground and excited states of the particle, respectively, $v \approx 0$ is the lowest vibrational excitation, $\sigma_0(E)$ is the cross-section of electron capture depending on the electron energy and E' is the kinetic energy of the electron after dissociation; the asterisk in parentheses indicates possible excess internal energy.

(An electron can also become attached to vibrationally and electronically excited molecules, but these processes have not been investigated in detail.)

Reactions (2) to (7) represent different pathways of degradation of the activated complex. Reaction (2) is the indirect elastic or inelastic scattering of electron, E' being equal to or less than E. Reactions (3) to (6) are different types of electron capture, that is, the formation of ions of a smaller mass and of molecular fragments which can also be in their excited states. Under these conditions, multiple dissociation of polyatomic molecules can occur. Some examples of dissociative electron capture for electrons with a predetermined kinetic energy are shown in Fig. 2.8, from which multiplicity of degradation pathways is clearly seen. The cross-section of formation of a fragment ion goes through a maximum depending on the electron energy, which indicates the resonance nature of the capture. The large half-width of the resonance band in the above examples is evidence that the lifetime of the activated complex is very short ($\tau \leq 10^{-12}$ s).

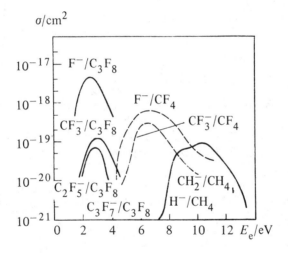

Fig. 2.8. Dependence of dissociative capture (dissociative attachment) of an electron by perfluoropentane, tetrachloromethane, and methane on the electron energy in the gas phase at room temperature [43] (formulae to the curves denote the ions appearing with respect to their parent molecules).

Reaction (7) corresponds to the production of a metastable negative ion, which is possible for molecules with a positive electron affinity. The excess energy evolved is consumed mainly in collisions with other molecules. Ions of this sort may have rather different lifetimes before dissociation yielding simpler ions:

$\dot{C}_6F_6^-$ $(12-13) \times 10^{-6}$ s

cyclo-$\dot{C}_6F_{10}^-$ $(106-113) \times 10^{-6}$ s

cyclo-$\dot{C}_6F_{12}^-$ $(236-450) \times 10^{-6}$ s

$C_6H_5\dot{N}O_2^-$ 18×10^{-6} s

$o-C_6H_4(CH_3)\dot{N}O_2^-$ 13×10^{-6} s

$m-NO_2C_6H_4\dot{C}H_3^-$ 19×10^{-6} s

$p-NO_2C_6H_4\dot{C}H_3^-$ 14×10^{-6} s

$\dot{C}_6H_6^{-*}$ $(0.2-1) \times 10^{-12}$ s

$\dot{O}_2^-$ 2×10^{-12} s

$\dot{S}O_2^{-*}$ 2×10^{-10} s

In addition, they can take part in various ion–molecule reactions such as electron transfer, protonation, etc.

Negative ions can also be formed on the collision of neutral species [44] as, for instance, in the reaction of alkali metal atoms with halogen molecules

$$K + Br_2 \begin{cases} \rightarrow K^+ + \dot{B}r_2^- \\ \rightarrow K^+ + Br^- + Br \end{cases}$$

Occurrence of this type of reaction is determined by its threshold energy. If the energy of the positive ion is high enough, the abstraction of two electrons in one act becomes possible. Thus, a proton with an energy of 1 keV to 1 MeV can abstract two electrons from a helium or argon atom

$$H^+ + He \rightarrow H^- + He^{2+}$$

The formation of ions from electronically excited states is considered in Chapter 4, which deals with photochemistry.

2.5 SOLVATED AND QUASI-FREE ELECTRONS

In polar liquids an excess electron created by radiation undergoes a transition to a localized state resulting in the solvated electron [42]. The solvate electron is considered as an independent chemical entity which is the electron settled in a spherical cavity surrounded by solvent molecules. The cavity radius is *ca.* 0.25 to 0.3 nm depending on the solvent. The electron is delocalized over all the neighbouring molecules of solvent (for water, different theoretical models give this number of neighbouring molecules as 3, 4, or 6). Appropriate models for the solvated electron are discussed by Feng and Kevan [45].

The formation of solvated electron is believed to occur in accordance with the following scheme. Thermal fluctuations cause two solvent dipoles to turn towards one another by their positive poles. A potential well is thus formed between the dipoles, its depth being determined by the size of the dipoles and by their angle of direction (water molecules arranged in a line with a head-to-head orientation of the positive charges provide a potential-well depth of >20 eV). When approaching sufficiently near, the excess electron is trapped at this well, creating in it a field of negative charge; the solvent molecules gradually reorientate, adjusting their dipoles to assume the optimum position corresponding to the minimum potential energy. The formation of the solvated state of the electron takes place if the following condition is fulfilled [7]:

$$\Delta A = - (U - E_{min}) + w_r < 0 \qquad (2.25)$$

where $(U - E_{min})$ is the binding energy of the electron in the potential well of depth U, E_{min} is the energy of the ground-state solvated electron with respect to the bottom of the potential well and w_r is the work performed in rearrangement of the medium (reorientation of the dipoles and setting the optimum distance between them).

Since the concentration of traps for the excess electrons (groups of solvent dipoles with randomly-disturbed orientations capable of trapping the excess electron) in liquids is sufficiently large (it is equal to 0.01 M in frozen alkali glasses [46], and the mobility of quasi-free electron[†] is very high, solvation proceeds quite rapidly, that is it takes picoseconds. As shown in Table 2.9, the solvation time τ_s is considerably less than the time of dielectric relaxation τ_1 of the corresponding fluid; that is, the process of solvation appears to occur during a period when the rearrangement of the medium has not yet been completed, and so the medium needs to be characterized by its high-frequency permittivity ε_∞. Hence we have to consider not τ_1 but the relaxation time of the excess electron $\tau'_1 = \tau_1 \varepsilon_\infty / \varepsilon_0$ [49] to compare with the solvation time. From Table 2.9 it follows that τ_s is close to τ'_1.

The quasi-free state of an electron in a liquid is determined by the properties of the latter, mainly, by the energy V_0 needed for the transition of an electron from a vacuum to the liquid (the conduction bandwidth). For most liquids V_0 is less than zero (Table 2.10).

Let us define a quantity $\tilde{V}_0$ as suggested by Byakov & Nichiporov [7]:

$$\tilde{V}_0 = -\left[\frac{1.3}{\pi}\left(1 - \frac{1}{\varepsilon_\infty}\right) + \frac{1}{\pi}\left(\frac{1}{\varepsilon_\infty} - \frac{1}{\varepsilon_0}\right)\right]^2 \text{Ry.} \qquad (2.26)$$

---

† The excess electron cannot be considered as a free electron before its solvation since its electrostatic field in the condensed medium interacts with the electrostatic field created by permanent and transient dipoles. Thus it is called a quasi-free electron.

Table 2.9. Characteristic solvation times for excess electrons in liquids [47, 48]

Solvent	T/K	τ_s/ps	τ_1/ps	τ'_1/ps
Water	296	0.3	9.2	0.2
Methanol	293	10.7 ± 1.0	52	9
Ethanol	293	23 ± 2	191	30
Propanol	293	34 ± 3	430	81
	240	107 ± 10	2500	360
	145	5.9×10^4	2.2×10^7	2.6×10^6
Pentanol	293	34	927	200
Decanol	293	51	2019	623
Ammonia	203	0.2	–	–
Methylamine	193	0.2	–	–

Then the value of $(V_0 - \tilde{V}_0)$ will distinguish two types of liquid: ones where electrons can be solvated $(V_0 > \tilde{V}_0)$ and others where only the quasi-free state of the electron $(V_0 < \tilde{V}_0)$ is possible. As shown in Fig. 2.9, the solvated electron can be formed in most liquids, but it cannot exist in liquid xenon, krypton, tetramethylsilane, or methane.

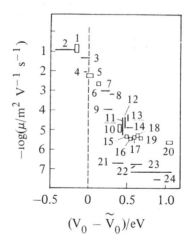

Fig. 2.9. Mobility of excess electrons as a function of $V_0 - \tilde{V}_0$ [7] (to the left of the dashed line is the region where only quasi-free electrons can exist; to the right is that of the localized electrons): 1–xenon; 2–krypton; 3–argon; 4–tetramethylsilane; 5–neopentane; 6–2,2,4,4-tetramethylpentane; 7–2,2-dimethylbutane; 8–isooctane; 9–cyclpentane; 10–pentane; 11–toluene; 12–benzene; 13–cyclohexane; 14–hexane; 15–heptane; 16–dodecane; 17–octane; 18–nonane; 19–helium; 20–decane; 21–water; 22–neon; 23–methanol; and 24–ethanol.

Table 2.10. Mobility of excess electron μ, yield of free ions G_{fi}, and conduction bandwidth V_0 for different liquids [7, 50, 51]

Liquid	T/K	$\mu/10^{-4}\,m^2\,V^{-1}\,s^{-1}$	G_{fi}, pair per 100 eV	V_0/eV
Helium	4.2	0.020	–	+1.05
Neon		0.002	–	+0.5 to +0.7
Argon	85	475	2.5	–0.18
Krypton		1800	4.0	–0.35 to –0.16
Xenon	163	2200	4.5	–0.65
Methane	111	400	–	0.00 (109 K)
Ethane	200	1	–	–0.02
Propane	238	0.55	–	–0.07 (222 K)
Butane	296	0.4	0.22	+0.12
Pentane	296	0.08 to 0.16	0.14	–0.02 to +0.04
Neopentane	296	70	0.86	–0.43 to –0.35
Cyclopentane	296	1.1	0.16	–0.21 to –0.17
Hexane	296	0.7 to 0.09	–	0.00 to +0.16
Cyclohexane	296	0.22 to 0.45	0.15	+0.01
2,2,4-Trimethyl-pentane	296	5.3 to 7.0	0.33	–0.18
2,2,4,4-Tetra-methylpentane	296	24	0.83	–0.33
Benzene	295	0.11 to 0.60	0.053	–0.14 to –0.05
Toluene	295	0.06 to 0.54	–	–0.22
Tetramethyl-silane	295	90 to 100	–	–0.62 to –0.51
Water	295	$(1.6\text{–}2.0) \times 10^{-3}$	4.7	–1.3 to –1.0
Methanol	294	0.0006	3.2	–1.0 to –1.4
Ethanol	298	0.0003	–	–0.66
Propanol-1		0.0005	–	–0.3
Butanol-1		0.0007	–	0.0

The properties of the solvated electron are also determined by the properties of the liquid [52]. The solvated electron is bound to different matrices to different extents. This is manifested, for example, by the position of the maximum in its absorption spectrum, which is at *ca.* 600–800 nm for polar compounds and at *ca.* 2000 nm for nonpolar liquids. Another characteristic indicating the degree of binding of the electron by medium is the mobility of the excess electron in an electric field. Values for the mobility are given in Table 2.10 from which it is clear that the mobilities differ by six to seven orders of magnitude for different substances, the lowest values being observed for highly polar compounds (water and alcohols). This difference may be easily account for in terms of an equilibrium between the quasi-free and localized states of the electron. If this assumption is true, then the mobility of the excess electron μ will be equal to the sum of the mobilities μ_{qf} and μ_{loc} for its quasi-free and localized states, respectively [7].

$$\mu = \frac{\mu_{qf}\, \exp(\Delta A/kT) + \mu_{loc}}{1 + \exp(\Delta A/kT)} \qquad (2.27)$$

where ΔA is the change in Helmholtz free energy.

The solvated and quasi-free electrons are chemical species which play a very important part not only in HEC processes but in many other processes, see [53]. Their physicochemical properties have been studied extensively. Compared with the quasi-free electron, much more information has been obtained about the solvated electron (e_s) since the latter can be observed directly by means of spectrophotometry and ESR spectroscopy. Data on quasi-free electrons (which are usually present in a system together with the localized form of the electron) are obtained indirectly, in particular, by conductivity and luminescence measurements (obtained via the formation of excited states on recombination). The properties of the solvated electron in water, that is, the hydrated electron e_{aq}^- have been studied in most detail [42, 54].

The solvated electron is a highly reactive species, with values for its reaction rate constants in many system approaching the upper limit for diffusion-controlled reactions (*ca.* 10^{-10} M^{-1} s^{-1}). The solvated electron can take part in addition reactions (reduction):

$$C_6H_5NO_2 + e_s^- \rightarrow C_6H_5NO_2^- \cdot,$$

$$C_6H_5 - C_6H_5 + e_s^- \rightarrow C_6H_5 - C_6H_5^- \cdot,$$

$$NO_3^- + e_s^- \rightarrow NO_3^{2-},$$

$$Cd^{2+} + e_s^- \rightarrow Cd^+.$$

The ions thus formed may turn out to be metastable, decaying after some time via a process of dissociative ionization (see section 2.4):

$$RCl + e_s^- \rightarrow R + Cl^-.$$

Reactions of this type are often called the radical transformation reactions, that is, one radical (e_s^-) is transformed into another (R).

Rate constants for some reactions of solvated electrons in polar liquids are listed in Table 2.11. Quasi-free electrons are assumed to present in negligible amounts in polar media because of the low mobility of excess electrons (see Table 2.10). It should be noted that the solvated electron behaves almost as a 'normal' chemical species, that is, its reaction rate constant decreases as the viscosity increases, although the decrease is not as steep as might be expected.

Table 2.11. Rate constant for reactions of solvated electron with solutes in water and alcohols at room temperature, in $M^{-1} s^{-1}$ [49]

H^+	2.2×10^{10}	$(3.9–6.8) \times 10^{10}$	$(2–4.5) \times 10^{10}$	2×10^9
Cd^{2+}	5.2×10^{10}	—	—	5.6×10^9
O_2	2×10^{10}	1.9×10^{10}	1.9×10^{10}	—
CH_3COCH_3	$(5.6–6.5) \times 10^9$	$(2–3) \times 10^9$	6×10^9	1.3×10^8
$C_6H_5NO_2$	4.2×10^{10}	—	1.5×10^{10}	—
$C_{10}H_8$	5.4×10^9	$(2–2.7) \times 10^9$	4.3×10^9	—
$C_6H_5CH_2OH$	2.3×10^8	—	—	5.5×10^7
$C_6H_5CH_2Cl$	5.5×10^9	5.0×10^9	5.1×10^9	—
$ClCH_2COO^-$	1.2×10^9	1.7×10^8	—	1.9×10^8
NO_3^-	8.5×10^9	$(0.4–2.1) \times 10^8$	3×10^7	—
NO_2^-	3.7×10^9	9.1×10^7	—	—
$\eta_{20}/MPa\ s$	0.010	0.0060	0.0120	14.99

According to contemporary kinetic theories [7], a bimolecular reaction proceeds in two stages: by diffusion of the species A and B until they meet in a cage (with rate constant k_D) and then by interaction of A and B in the cage (with rate constant k_b). The apparent rate constant k is related to the rate constants for the separate stages as follows:

$$k = k_D k_b / (k_D + k_b). \tag{2.28}$$

If the species are charged, then k_b depends on the ionic strength of the solution μ_i

$$\log (k_b / k_o) = -1.02 z_A z_B \sqrt{\mu_i} / (1 + a \sqrt{\mu_i}) \tag{2.29}$$

where z_A and z_B are the charges of the reactant ions, k_o is the rate constant when the ionic strength equals zero, and a is a proportionality factor close to unity.

The rate constant k_D is determined with the Debye equation

$$k_D = 4\pi DR \left\{ \frac{z_A z_B r_o}{R} \middle/ \left[\exp\left(\frac{z_A z_B r_o}{R} \right) - 1 \right] \right\} \tag{2.30}$$

where r_o is the Onsager radius (see section 2.6); $R = R_A + R_B$, i.e. the sum of the radii of the reactants, and $D = D_A + D_B$, i.e. the sum of the diffusion coefficients of the reactants.

For electrically neutral reactants, Eq. (2.9) is transformed to the Smoluchowski equation

$$k_D = 4\pi RD \tag{2.31}$$

A considerable change in the reaction rate constant of the solvated electron in aqueous solutions of high ionic strength was observed when the salt concentration was increased (Fig. 2.10). To account for this phenomenon it was assumed that the rate constant k_b is time-dependent. Later, however, it was shown [7] that the experimental dependence of the rate constant for a reaction of the solvated electron with a solute can be described quantitatively by taking into consideration the effects of ionic strength on both k_b and k_D

$$k_D = 4\pi DR \left[\frac{L_D}{L_D + r_0} \frac{z_A z_B (r_0/R) \exp\{z_A z_B r_0/(L_D + r_0)\}}{\exp(z_A z_B\, r_0/R) - 1} + \frac{r_0}{L_D + r_0} \right] \quad (2.32)$$

where L_D is the Debye screening distance, equal to $\sqrt{8\pi_0\mu}$.

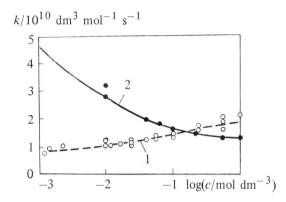

Fig. 2.10. Concentration-dependence of the rate constant for reactions of the hydrated electron with nitrate ion (1) in KNO_3 solution with copper(II) ion (2) in $Cu(ClO_4)_2$ + $Ca(ClO_4)_2$ solution [7].

The curves shown in Fig. 2.10 have been calculated with Eq. (2.32), and they satisfactorily describe the experimental data over a wide range of salt concentrations for negatively- and positively-charged electron scavengers.

Rate constants for the reaction of the hydrated electron with various solutes in aqueous solutions can be calculated from Eqs. (2.30) and (2.31). The experimental values of these constants for most anions and some uncharged species turned out to be considerably higher than the values calculated based solely on the collisional mechanism (see Fig. 2.11 where rate constants for the reactions with charged solutes are given at zero ionic strength). This difference leads to the conclusion that the hydrated electron can react with a solute not by direct contact but by means of the sub-barrier transfer of the electron to the solute molecule [7]. Tunnelling is found to be even more typical of the reactions of the trapped electron [55–57].

In nonpolar solvents, the observed rate constants for reactions of the solvated electron with solutes significantly exceeds the diffusion limit (*ca.* $10^{10}\ M^{-1}\ s^{-1}$)

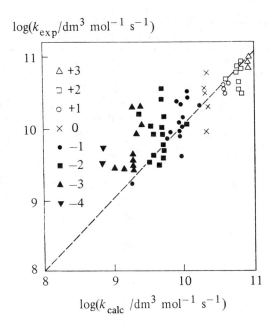

$\log(k_{exp}/\mathrm{dm}^3 \mathrm{mol}^{-1} \mathrm{s}^{-1})$

$\log(k_{calc}/\mathrm{dm}^3 \mathrm{mol}^{-1} \mathrm{s}^{-1})$

Fig. 2.11. Comparison of experimental rate constants of the hydrated electron with values calculated from the Debye equation (2.30) [7].

at a viscosity of 1 mPa s, as it is shown for cyclohexane at room temperature (Table 2.12). As seen, there are some values above 10^{12} M^{-1} s^{-1} for the rate constant. Evidently, this is due mainly to the quasi-free, rather than the localized electron, taking part in the reaction with the solutes, since the mobility and, consequently, the diffusion coefficient of the quasi-free electron are considerably higher than those of the localized electron. It might be expected that the rate constant for the reaction of an excess electron with solute would increase with increasing mobility of the electron (the diffusion coefficient of the solute may be neglected). This relation, however, appears to be more complex in reality (see Fig. 2.12 where the dependence of the rate constant for some solutes on the mobility of the excess electrons in different nonpolar liquids is shown). Thus the rate constant for oxygen is almost independent of the mobility, but it is higher than in polar solvents (see Table 2.11). On the other hand, the rate constant for SF_4 increases with increasing mobility and approaches a value of 2×10^{14} M^{-1} s^{-1} in neopentane and tetramethylsilane where the mobility is highest. The rate constant for CCl_4 also increases, but for N_2O and benzoquinone it passes through a maximum in isooctane. Thus, the reaction rate constant for the excess electron is determined by other factors in addition to mobility. At the same time, the value of the rate constant indicates that only the quasi-free electron is taking part in the reaction.

Table 2.12. Rate constants for reactions of excess electron with solutes in cyclohexane at room temperature [58]

Solute	$k/10^{12}\,M^{-1}\,s^{-1}$	Solute	$k/10^{12}\,M^{-1}\,s^{-1}$
O_2	0.02 to 0.17	C_6H_5COOH	0.2
CO_2	4.3 to 4.8	$C_6H_5CH_2OCOCH_3$	4.8
N_2O	2.4	$C_6H_5NH_2$	0.2
SF_6	4.0; 4.6	$C_6H_5NNC_6H_5$	1.0
NH_3	0.26	C_6H_5CN	2.6
$(C_6H_5)_3CH$	3.7	Galvinoxyl	9.0
C_6H_6	0.008; 0.12	Diphenylanthracene	6.6
$C_6H_5CH_3$	0.004	CH_3Cl	0.5; 1.4
Biphenyl	2.3 to 3.3	CH_3Br	4.1
p-Terphenyl	2.4	CH_3I	2.0
cis-Stilbene	1.2	CH_2Cl_2	2.4
$trans$-Stilbene	0.86	$CHCl_3$	2.9
Naphthalene	2.0	CCl_4	2.4 to 4.3
Anthracene	1.6	C_2H_5Br	2.0; 2.6
Phenanthrene	2.0	n-C_3H_7Cl	0.15
Pyrene	2.1	C_6H_5Cl	2.6
C_6H_5OH	0.1	C_6H_5Br	3.1
$C_6H_5OCH_3$	0.005	C_6H_5I	3.4
C_6H_5CHO	2.6	C_6H_5F	0.077
$C_6H_5CH_2OH$	0.4	$C_6H_5CH_2Cl$	3.8
$C_6H_5COCH_3$	2.4	$C_6F_{11}CF_3$	0.35

The activation energies of the rate constants are also close to the activation
energies of the mobilities. For example, the activation energy for the reaction of
CCl_4 with the excess electron in hexane, methylcyclohexane, and isooctane is
equal to 20.9, 19, and 2.0 kJ mol^{-1}, respectively, and the activation energy of the
mobility in the same sequence is 18–23, 19.7, and 4.9 kJ mol^{-1}. A similar picture is also observed for oxygen [40].

Since the time for solvation in polar liquids is small (see Table 2.9) and,
hence, the lifetime of quasi-free electron is also small, then effects attributed to
quasi-free electrons in water and alcohols can be observed only in very concentrated solutions. Such data were obtained for the technique of picosecond pulse
radiolysis [59]. Fig. 2.13 shows the change in the radiation yield of the solvated
electron Ge_s^-/G^0e^- on addition of acetone to a solvent (the detection time for
Ge_s^- was 30 ps). With the acetone concentrations used, its reaction with e_s^- may
be neglected compared with the detection time on considering the value for the
reaction rate constant in the solvents concerned (see Table 2.11). Consequently,

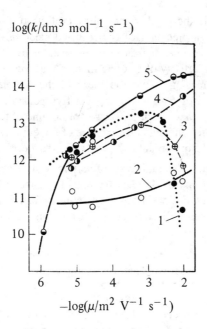

$\log(k/\mathrm{dm}^3 \ \mathrm{mol}^{-1} \ \mathrm{s}^{-1})$

$-\log(\mu/\mathrm{m}^2 \ \mathrm{V}^{-1} \ \mathrm{s}^{-1})$

Fig. 2.12. Rate constants for the quasi-free electron in hydrocarbons and tetramethyl-silane [58]: 1–oxygen; 2–dinitrogen oxide; 3–sulphur hexafluoride; 4–tetrachloro-methane; 5–p-benzoquinone.

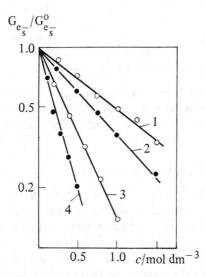

$G_{e_s^-}/G_{e_s^-}^{o}$

$c/\mathrm{mol} \ \mathrm{dm}^{-3}$

Fig. 2.13. Relative yields of the solvated electron in water (1), ethylene glycol (2), methanol (3), and ethanol (4) as a function of acetone concentration [59].

the observed decrease in the yield of the solvated electron results from interaction of the solute with a precursor of the solvated electron, namely, with the quasi-free electron in polar liquids is not yet possible, the rate constant cannot be determined. Thus, the reactivity of a solute in such reactions is characterized as that concentration decreasing e-fold the yield of the solvated electron. This concentration is denoted c_{37}. Values of c_{37} for five polar liquids are given in Table 2.13. It is obvious that the larger the solvation time of the excess electron in these liquids, the more efficient is the capture of the quasi-free electron in the series (see Table 2.9). Similar data on the effective decrease in $G(e_s^-)$ were obtained by means of ns pulse radiolysis of solutions of electron scavengers in water, propanol, glycerol, N,N-dimethylformamide, N-methylpyrrolidone, and tetrahydrofuran [60] where the solutes used (N-methylacetamide, pyrrolidone, benzene, toluene, and phenol) have reaction rate constants $< 4 \times 10^7$ M^{-1} s^{-1}.

Table 2.13. The $1/c_{37}$-value describing the reactivity of quasi-free electrons in different solvents at room temperature [59]

Solute	Water	Ethylene glycol	Methanol	Ethanol	Propanol
H^+Ion	0.1	0.8	1.4	1.7	2.0
Acetone	0.7	1.0	2.0	3.33	4.0
Acetophenone	—	2.5	4.0	5.9	6.8
Trichloromethane	—	2.5	4.0	6.3	6.8
Tetrachloromethane	—	2.8	6.3	7.1	9.0
Methyl ethyl ketone	0.9	1.25	2.9	3.3	4.8
Biphenyl	—	—	3.3	5.3	5.6

The linear relationship obseved when plotting $\log [G(e_s^-)/G(e_s^-)]$ against solute concentration may be an indication of the fact that the species reacting with the scavenger is not just the quasi-free electron but one possessing a hyperthermal kinetic energy [7].

2.6 CHARGE RECOMBINATION

The action of radiation or the application of an electric field on a gaseous substance produces charged pairs: an electron and a positive ion. Further, they can become transformed to a negative ion and a secondary positive ion, respectively, following interaction with molecules of the substance. Under the same conditions in a condensed phase, a hole and a conduction electron (the quasi-free electron) are created which can be later localized in the form of a positive ion and a solvated (trapped) electron. In all of these cases, the formation of pairs of opposite charges is observed, their concentration considerably exceeding that expected from thermodynamic equilibrium. If so, the excess charges must disappear through neutralization (recombination):

$$A^+ + B^- \begin{cases} \to AB \\ \to A + B. \end{cases}$$

The theory of charge recombination in the gas phase has been developed fairly well [41]. For a homogeneous distribution of ions by volume, the rate of recombination may be written as a second-order kinetic equation

$$-\frac{dn_+}{dt} = -\frac{dn_-}{dt} = an_+n_- \tag{2.33}$$

where n_+ and n_- are the concentrations of the positive ion and the electron respectively, and a is the reaction rate constant (recombination coefficient) for charge neutralization.

The value for a at room temperature and normal pressure is *ca.* 10^{-6} cm^3 s^{-1} (or *ca.* 10^{15} M^{-1} s^{-1}) [21] for the recombination of positive ions and electrons and *ca.* 10^{-10} cm^3 s^{-1} (or 10^{11} M^{-1} s^{-1}) for the recombination of positive and negative ions. The a-value is pressure-dependent. For low pressures, a increases with increasing pressure, approaching a limiting value at atmospheric pressure and then decreasing with further increase in pressure [21, 44].

The recombination of positive ions and electrons (negative ions) in the gas phase can be completely suppressed by applying an electric field with an intensity of 1000 V cm^{-1} or by addition of efficient electron scavengers, for example 1% SF_6 or N_2O in water vapour and in gaseous hydrocarbons.

Let us consider the process of charge recombination in liquids. The G_{fi}-value, called the 'yield of free ions', or the number of ion pairs produced per 100 eV of absorbed energy, was determined for many liquids in pulse radiolysis measurements within 10^{-8}–10^{-7} s [51]. Some values for G_{fi} are given in table 2.9. It is seen that G_{fi} ranges from *ca.* 10^{-2} to several ion pairs for different substances. Inasmuch as the mean energy of ion pair production is nearly the same for most substances (see section 1.7), the initial yield of ion pairs has to be about 5 for all of the compounds listed in Table 2.10, hence charge recombination in liquids, unlike gases, proceeds at essentially different rates. If liquid substances react slowly with the electron (water, alcohols) or do not react at all (hydrocarbons, noble gases, tetramethylsilane), then transient negative species are not formed within the period of neutralization, so the excess electron is the only negative species which takes part in the recombination. From this it might be expected that there would be a certain relation between G_{fi} and the mobility μ of the excess electron. The corresponding data are presented in Fig. 2.14. As shown, there is a distinct relationship between G_{fi} and μ for nonpolar liquids where the solvated electron is weakly bound to the matrix (see section 2.5). The G_{fi}-value increases with increasing μ in nonpolar matrices, while polar liquids do not exhibit such an effect.

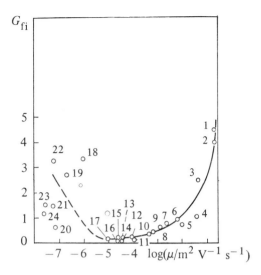

Fig. 2.14. Dependence of radiation yield of free ions G_{fi} in liquids on the mobility of the excess electrons [7, 50, 51]:1–xenon; 2–krypton; 3–argon; 4–methane; 5–tetramethylsilane; 6–neopentane; 7–2,2,4,4,-tetramethylpentane; 8–2,2,5,5-tetramethylhexane; 9–2,2,3,3-tetramethylpentane; 10–isooctane; 11–cyclopentane; 12–ethane; 13–cyclobutane; 14–benzene; 15–cyclohexane; 16–toluene; 17–pentane; 18–ammonia; 19–water; 20–methanol; 21–butanol-1; 22–propanol-1; 23–ethanol; and 24–ethylene glycol.

The problem of the recombination of an electron with its parent positive ion in a nonpolar medium was first considered by Onsager. His equation for the probability of ion recombination P looks as follows

$$P = 1 - \exp{(\rho/r_0)}, \qquad (2.34)$$

where r_0 is the initial distance between the ions and ρ is the Onsager radius

$$\rho = \frac{e^2}{4\pi\varepsilon_0\varepsilon kT} \qquad (2.35)$$

where ε is the relative dielectric permittivity of the medium.

The Onsager radius is the distance at which the action of the electric field of a point charge becomes comparable with the thermal action of the neighbouring molecules. Being at a distance $r > \rho$, an electron does not experince the attractive force of the positive ion, while at $r < \rho$ it does.

Fig. 2.15 shows the relative yield of ions escaping recombination as calculated from Eq. (2.34). It is seen that an increase in ε inhibits recombination. In nonpolar liquids $\varepsilon \approx 2$, so it might be expected from Eq. (2.34) that a significant fraction of the electrons would not recombine with their parent ions. However, as seen from Fig. 2.14, the fraction of ions escaping recombination in normal

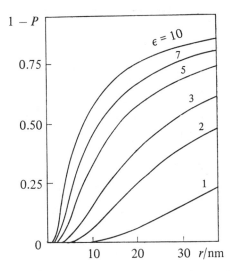

Fig. 2.15. Dependence of probability $1 - P$ of escaping charge recombination on distance between the electron and its positive ion for different permittivities (ε is indicated by numbers near the curves).

hydrocarbons is equal to *ca.* 0.15:3 = 0.03. Such a low value as this is caused by the situation that the excess electrons produced on irradiation are nonuniformly distributed around their positive ions. When expelled from molecules, electrons possess different kinetic energies and are thermalized at different distances from the parent ions. The higher the energy of the electron, the lower its probability of production. The relation between the number of electrons and the distance they travel from their point of origin is the distribution function $F(r)$. The function itself is unknown, although attempts have been made to find the form of the function by using the effects of temperature, of solute concentration, and of applied electric field (G_{fi} increases with increasing field intensity or with increasing temperature). However, the above approaches do not provide an unequivocal solution of the problem of $F(r)$. The most productive approach is that based on calculation of the degradation spectra of electrons while taking into account thermalization rates [16]. The form of the distribution function found [15] for two different rates of thermalization is shown in Fig. 2.16.

It follows that quite a small fraction of the electrons escapes beyond the limits of the Onsager radius (the latter is equal to 27.8 nm for $\varepsilon = 2$ and room temperature). However, even this approach needs some corrections to be made for the slowing down of electrons in the field of the parent ion to produce the best results, but the problem has yet to be solved. Nevertheless, the appearance of the distribution curves for excess electrons in Fig. 2.16 seems to be close to reality.

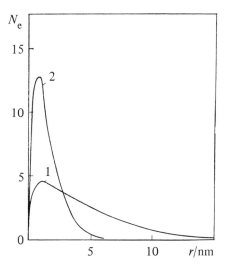

Fig. 2.16. Distance distribution for electrons for thermalization rates of 10^{13} eV s^{-1}
(1) and 10^{14} eV s^{-1} (2) [15].

Increasing mobility of the excess electrons will tend to spread $F(r)$; the higher
the mobility, the larger the fraction of electrons escaping the limits of the
Onsager radius. An increase in G_{fi} with increasing μ is observed for nonpolar
compounds, exactly in accordance with the above. However, it is impossible for
all electrons to escape recombination, whatever the case; even in krypton and
xenon, which exhibit the largest ratio, G_{fi}/G_0 does not exceed 0.7 (G_0 is the
total ion yield).

According to Fig. 2.14 there is no direct relation between μ and G_{fi} for polar
compounds. This may be accounted for by the fact that the mobility in polar
liquids corresponds mostly to solvated electrons, the contribution of the quasi-
free state being negligibly small; this contrasts with nonpolar media where the
excess electrons exists chiefly in the quasi-free state, and it is the latter which
determines the high mobility of the electrons. The solvated electrons do not
participate in charge recombination since neutralization is complete within a
period less than the solvation time (see Table 2.8), that is, only the quasi-free
electrons take part in the recombination. Unfortunately, nothing is known about
the mobility of these in polar liquids.

The process of fast recombination of radiation-induced charged pairs is conventionally called geminate (spur) recombination. It is not easy to evaluate the
rate of geminate recombination because of the uncertainty in such parameters
such as the distribution function $F(r)$, the rate of electron thermalization, and the
effect of charge on the rate of thermalization. However, the recombination rate
constant can be found from data on the kinetics of neutralization of free ions
which have escaped geminate recombination (Fig. 2.17).

$\log(k/\mathrm{dm}^3\,\mathrm{mol}^{-1}\,\mathrm{s}^{-1})$

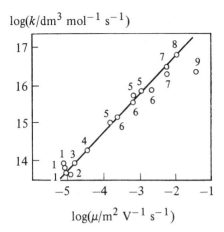

$\log(\mu/\mathrm{m}^2\,\mathrm{V}^{-1}\,\mathrm{s}^{-1})$

Fig. 2.17. Rate constant for charge recombination as a function of mobility of excess electrons in polar liquids [40, 61, 62]: 1–hexane;   2–heptane; 3–pentane; 4–cyclohexane; 5–isooctane; (200, 296, and 300 K); 6–neopentane + hexane mixtures; 7–neopentane; 8–tetramethylsilane; 9–methane.

As seen from Fig. 2.17, the rate constant is proportional to the mobility of the excess electrons. In fact, the rate constant for charge recombination has to be proportional to the sum of the mobilities of the excess electrons and the holes but the mobility of the excess electrons is always higher than that both of holes and ions (Table 2.14), so it is only the mobility of the electron which determines the observed value for the rate constant of charge neutralization and geminate recombination. The maximum value of the recombination rate constant is $5 \times 10^{16}\,\mathrm{M}^{-1}\,\mathrm{s}^{-1}$ which is the highest figure among all the rate constants for bimolecular reactions. In liquids with a high mobility of the excess electrons, the value of the recombination rate constant is comparable with that of charge recombination in the gas phase.

Table 2.14. Mobility of charged species μ in alkanes in $10^{-4}\,\mathrm{m}^2\,\mathrm{V}^{-1}\,\mathrm{s}^{-1}$ [40]

Alkane	$h_+{}^a$	e^-	Ion
Heptane	0.02	0.046	0.00026
Decane	0.02	0.038	0.00026
Cyclohexane	0.02	0.35	0.00026
Decalin	0.01	0.013	—

^{a} There are only a few liquids where the hole mobility has been determined.

2.7 SOLVATES AND CLUSTERS

When writing down chemical equations, we normally include only the reacting species themselves:

$$A + B \rightarrow C + E. \tag{8}$$

Strictly speaking, however, this notation is exact only for the gas phase under low pressure. In the condensed phase, the molecules are bound to each other by intermolecular forces, that is, the reactions are those of species aggregated with the solvent rather than of separate particles. These aggregates used to be called solvates, meaning molecules or atoms surrounded by nearest-neighbour particles (a solvate is denoted with the subscript 's')

$$A_s + B_s \rightarrow C_s + E_s. \tag{9}$$

Some tasks, including those covered in this book (determination of ionization potentials in the condensed phase, investigation of the behaviour of excess electrons) require solvation to be taken into consideration. But in many problems, in particular on kinetics, the solvation effect is incorporated into the corresponding value for a reaction rate constant, so the simplified notation of type (8) can be used in these cases.

The association of molecules around a separate molecule due to intermolecular interaction, or around an ion because of Coulombic interaction, is called a cluster, as currently accepted in quantum chemistry [63]. By definition, a cluster is a multiparticle aggregate consisting of identical (homonuclear) or different (heteronuclear) species (atoms, molecules, or ions) so that the interactions between the species are either equivalent (neutral clusters with the edge effect neglected) or nonequivalent (charged clusters). Clusters and solvates are to some extent analogous concepts, but the term 'solvate' is normally used for condensed media, while the term 'cluster' refers to the gaseous state.

One distinguishes between physical and chemical clusters. Physical clusters exist only in the medium where they have been formed, and a change of medium results in their spontaneous dissociation (for example a neutral cluster on transfer from the liquid to the gas phase dissociates into separate molecules at the same temperature). A chemical cluster can be isolated from its medium, that is, it exists as a separate entity for some time (an ionic cluster transferred from the liquid to the gaseous phase is conserved if the temperature is kept unchanged, the simplest example of this is an aerosol).

All properties of clusters depend on the number of particles involved. Thus, the stability of a cluster expressed as its dissociation energy can either increase (Na_n^+)decrease (($O_2)_n^+$) with increasing n. The greater the number of particles bound in a cluster, the lower is the ionization potential of the cluster. Some examples of the change in ionization potential for metal and nonmetal homonuclear clusters are given in Table 2.15. It is probably solvation which causes

Table 2.15. Ionization potentials of neutral homonuclear clusters M_n, (in eV) [63]

M	I_{M_1}	I_{M_2}	I_{M_3}	M	I_{M_1}	I_{M_2}	I_{M_3}
Li	5.69	5.13	—	NO	9.25	8.74	8.49
Na	5.60	5.10	4.11	CO_2	13.79	13.05	12.9
K	4.63	4.23	3.57	N_2	15.5	14.69	14.64
Cs	4.11	4.02	3.32	NH_3	10.3	9.54	—

the decrease in ionization potential on transition from the gas to the liquid phase (see section 1.6). Positively charged clusters show the same tendency. In negatively-charged clusters, the electron affinity increase. The change in ionization potential on clustering may be ca. 1 to 2 eV. The thermodynamic functions of some clusters, including entropy, have been determined from mass spectrometric date.

Table 2.16 presents some examples of positively- and negatively-charged clusters. Ionic clusters can probably be obtained for all compounds because even such weakly polarized atoms as He and Ar form clusters with a large number of constituent particles. One of the most interesting examples is the homonuclear negatively charged clusters with an excess electron, namely $(H_2O)_n^-$ and $(NH_3)_n^-$, which in the liquid state are normally termed solvated electrons. These species are found in the vapour phase even at supercritical temperatures. Theoretical calculations show that a stable negative cluster can exist even in such nonpolar compounds as ethane, involving, however, a sufficiently large number of molecules: $n \geq 18$ [64]. Correspondingly, the excess electron in liquid ethane can be localized because of its low mobility (see Table 2.10). Metal atoms can also form negatively charged clusters, for example K_2^- and K_3^-.

Ionic clusters, as well as separate ions in the gas phase, participate rather effectively in ionic reactions of all possible types. A reaction specific for clusters is the 'sticking' of neutral molecules or atoms to the aggregate to form a larger particle. For the cluster $H_3O^+(H_2O)_n$ and some other clusters the equilibrium represented by the sequence of reactions

$$H_3O^+ \underset{-H_2O,-k_1}{\overset{+H_2O,k_1}{\rightleftharpoons}} H_3O.H_2O \underset{-H_2O,-k_2}{\overset{+H_2O,k_2}{\rightleftharpoons}}$$

$$H_3O^+(H_2O)_2 \underset{-H_2O,-k_3}{\overset{+H_2O,k_3}{\rightleftharpoons}} H_3O^+(H_2O)_3 \quad (10)$$

has been studied in detail at different temperatures, water contents, total pressures, etc. [65, 66]. It is interesting to note that the rate constant for direct reactions is nearly constant for any value of n, while the rate constant for cluster dissociation tends to decrease as n increases (that is, the cluster is stabilized).

Table 2.16. Positive and negative ionic clusters

Cluster	n	Cluster	n	Cluster	n
$He^+(He)_n$	10^5	$NO^+(H_2O)_n$	1–4	$NO_2^-(H_2O)_n$	1–15
$Ar^+(Ar)_n$	10^4	$NO^+(N_2O)_n$	1,2	$NO_2^-(CO_2)_n$	1–4
$N^+(N_2)_n$	5×10^3	$H_3O^+(H_2O)_n$	1–51	$NO_3^-(H_2O)_n$	1–3
$H^+(H_2)_n$	10^5	$H_2S^+(H_2O)_n$	1	$CO_3^-(CO_2)_n$	1–4
$H^+(N_2)_n$	1–7	$H_2S^+(H_2S)_n$	1–4	$Cl^-(H_2O)_n$	1–4
$H^+(CO)_n$	1–7	$CH_5^+(H_2O)_n$	5, 10	$Cl^-(CH_3Cl)_n$	1–3
$H^+(O_2)_n$	1–7	$CH_5^+(CH_4)_n$	1–5	$Cl^-(C_2H_5Cl)_n$	1–3
$S^+(S)_n$	1–7	$C_6H_5OH_2^+(H_2O)_n$	1–3	$Cl^-(CO_2)_n$	1–4
$Se^+(Se)_n$	1–7	$(H_2O)_n^-$	2–44	$I^-(CO_2)_n$	1–4
$Te^+(Te)_n$	1–7	$(H_2O)_2^-(Ar)_n$	1–3	$CH_5^-(H_2O)_n$	5, 10
$O_2^+(O_2)_n$	1–9	$(NH_3)_n^-$	26–90	$OH^-(H_2O)_n$	1–6

2.8 FREE RADICALS

Free radicals (or simply radicals) are chemical species possessing an unpaired electron in their outermost electronic orbital.

There are two types of radical: those with an electron deficiency with respect to the parent species (e.g. $\dot{C}lO_4$ from ClO_4^- or $CH_4^{+\cdot}$ from CH_4) and those with an additional electron ($C_6H_6^{-\cdot}$ from C_6H_6 or $NO_3^{2-\cdot}$ from NO_3^-). A radical can have the same number of atoms ($\dot{N}O_3$ from NO_3^-, $C_{10}H_8^-$ from $C_{10}H_8$) or fewer ($\dot{C}H_3$ from CH_4, $\dot{O}H$ from H_2O, or $\dot{H}$ from H_2) compared with the parent molecule.

Radicals play a very important role in HEC, a significant part of the final products being formed through a radical intermediate. Radical cations are produced in the primary process of irradiation. Neutral radicals are secondary species, appearing via the transformation of transient species of other types. The only exceptions are for ionic crystals, melts, and solutions of salts which are characterized by the formation of neutral radical following the ionization of ions

$$AB^- \overset{\text{\textasciitilde}}{\longrightarrow} AB + e^-.$$

Radicals are usually highly reactive entities which enter various first- and second-order processes. Even those radicals referred to as stable (triphenyl-methyl, chlorine dioxide, or nitroxyls) participate actively in chemical reactions (radical scavenging, electron transfer, etc.) although they are thermodynamically stable. We shall now consider the main radical reactions.

2.8.1 First-order reactions

First-order reactions are typical of thermodynamically unstable radicals and of radicals in electronically or vibrationally excited states. They include isomerization and dissociation (fragmentation) reactions.

Isomerization includes *cis–trans* isomerization, cyclization, and rearrangements. Isomerization in all cases results in a more stable radical.

There are many examples of *cyclization* of radicals, with the double or the triple bond, two, four, or five carbon atoms distant from the radical centre [67]

An equilibrium between the cyclic and open forms has been established for some radicals

$$CH_2 = CH_2 - CH_2 - \dot{C}H_2 \rightleftharpoons \dot{C}H_2 - \triangleleft.$$

For a number of radicals, the open form is energetically favoured, and so the cyclic radical undergoes bond cleavage (*ring-opening reaction*)

Rearrangements resulting in the reorganization of the atomic framework of a radical, or in a change in position of the functional group, are also common [67], some of them proceeding via the formation of transient cyclic species. A well-known process is the migration of an aryl group from a tertiary to a secondary carbon atom or to an oxygen atom

$$Ph_3\dot{C}O \rightarrow Ph_2\dot{C}OPh.$$

It should be noted that migration can occur over rather large distances, for example

$$CH_3-\underset{\underset{CH_3}{|}}{\overset{\overset{Ph}{|}}{C}}-CH_2CH_2\dot{C}H_2 \rightarrow CH_3-\underset{\underset{CH_3}{|}}{\overset{|}{\dot{C}}}-CH_2CH_2CH_2Ph.$$

Vinyl, acyl, and acetoxyl groups and halogen atoms can migrate via formation of an intermediate bridged radical (1,2-shift)

$$CH_2=CHCH\dot{C}H_2 \rightarrow \underset{\underset{\displaystyle CH_2}{|}}{\overset{\displaystyle CH_3}{}}\ \overset{HC-\dot{C}H_2}{\underset{CH_2-CHCH_3}{\diagup\diagdown}} \rightarrow CH_3\dot{C}HCH_2CH=CH_2,$$

$$CH_3CO-\underset{\underset{\displaystyle CH_3}{|}}{\overset{\overset{\displaystyle Ph}{|}}{C}}-\dot{C}H_2 \rightarrow \overset{H_3C \quad O\cdot}{\underset{\underset{\displaystyle CH_3}{|}}{Ph-\overset{\diagdown\diagup}{\underset{\diagup\diagdown}{C}}-CH_2}} \rightarrow \underset{\underset{\displaystyle Ph}{|}}{\overset{\overset{\displaystyle CH_3}{|}}{\cdot C}}-CH_2-COCH_3,$$

or

$$CCl_3\dot{C}HCH_2Br \rightarrow \overset{\cdot Cl}{\underset{\underset{\displaystyle Cl}{|}}{Cl-\overset{\diagup\diagdown}{C}-CHCH_2Br}} \rightarrow \dot{C}Cl_2CHClCH_2Br.$$

A fluorine atom which has no d-orbitals and, hence, no chance to form the bridging bond, cannot undergo the 1,2-shift.

The migrations of hydrogen atom are, strictly speaking, not rearrangement reactions according to the definition of the latter, but they are usually considered together. Examples include 1,3,-, 1,4,-, and 1,5-shifts to carbon and oxygen atoms:

$$\underset{\underset{\displaystyle PhCHCH_2CH_2CH_2\dot{C}H_2}{}}{\overset{\overset{\displaystyle H}{|}}{}} \rightarrow Ph\dot{C}CH_2CH_2CH_2CH_3$$

or

$$\underset{\underset{\displaystyle CH_3CHCH_2CH_2C(CH_3)_2\dot{O}}{}}{\overset{\overset{\displaystyle H}{|}}{}} \rightarrow CH_3\dot{C}CH_2CH_2C(CH_3)_2OH.$$

It should be noted that the migration of an H atom can be masked behind reactions of radicals with molecules. When a radical abstracts a hydrogen atom from the weakest C—H bond, it is very difficult to distinguish kinetically the corresponding migration.

Fragmentation of radicals. The degradation of radicals is often encountered in HEC especially if a process is carried out at high temperature. Fragmentation reactions have been studied quite extensively [67], thus the decarbonylation of acylic radicals and decarboxylation of acyloxyl radicals

$$R\dot{C}O \rightarrow \dot{R} + CO \qquad \text{and} \qquad R\dot{C}O_2 \rightarrow \dot{R} + CO_2$$

occur quite easily. Rate constants for the homolytic dissociation of these radicals, in s^{-1}, are given below (benzene solvent, 25°C) [68]:

$CH_3\dot{C}O$	$n\text{-}C_3H_7\dot{C}O$	$i\text{-}C_3H_7\dot{C}O$	$(CH_3)_3C\dot{C}O$
ca. 1	ca. 1	3.9×10^3	5.2×10^4

$PHCH_2\dot{C}O$	$PhC\dot{O}O$
7×10^6	5×10^5

The fragmentations of alkoxyl and α-alkoxyl radicals occur in another way:

$$(CH_3)_3 C\dot{O} \rightarrow (CH_3)_2 CO + \dot{C}H_3$$

and

$$C_6H_5\dot{C}HOR \rightarrow C_6H_5CHO + R\cdot$$

An interesting case is the reversible addition of halogen atoms and thioalkyl radicals to alkenes followed by dissociation of the resulting radical

For radicals containing two functional groups X and Y, the fragmentation resulting in the release (or elimination) of HY

occurs via formation of the intermediate cyclic radical due to intramolecular hydrogen bonding [69]. The simultaneous scission of two bonds proceeds via a self-consistent mechanism. All of the conditions which prevent cyclization (lack of a hydrogen atom in the group –XH, complexation with transition metal ions, etc.) will also prevent the elimination of HY.

The fragmentation of peroxide radicals proceeds similarly via the formation of a five- or six-membered ring due to intramolecular hydrogen bonding and

$$R - \underset{\underset{OH}{|}}{\overset{\overset{\displaystyle \cdot O_2}{|}}{C}} - CH(OH)R' \longrightarrow R - \underset{\underset{OH}{|}}{C} \overset{\overset{\displaystyle \overset{\cdot}{O} \text{---} O}{\diagdown \diagup}}{+} CH - R' \longrightarrow RCOOH + R'CHO + \overset{\cdot}{O}H$$

2.8.2 Second-order reactions

Reactions of radicals with molecules. There are three basic types of reaction of radicals with molecules: addition, abstraction, and electron transfer.

Radical addition. Molecules having a double or triple bond can attach radicals to themselves. A classical example is the formation of peroxide radicals when a radical is captured by a molecule of oxygen

$$R + O_2 \rightarrow R\overset{\cdot}{O}_2.$$

This is a necessary step in the oxidation of organic compounds. Another example is the addition of radicals to a monomer M, for example ethylene

$$R \cdot + CH_2 = CH_2 \rightarrow R - CH_2 - \overset{\cdot}{C}H_2$$

to produce so-called propagating radicals RM·, which lead to radical polymerization. Addition reactions often proceed at a high rate, that is, close to that of diffusion (Table 2.17). Thus, the rate constants for scavenging of linear alkyl radicals by oxygen in liquid solutions at room temperature are *ca.* $10^9\,M^{-1}\,s^{-1}$. However, values as high as these are not typical for all addition reactions.

The addition of free radicals usually occurs at both the carbon atoms of a double bond , although the probabilities are not the same. Contrary to the Markownikoff rule for the ionic mechanism, a radical is attached with higher probability to a less-substituted carbon atom. For instance, $\overset{\cdot}{C}F_3$ attacks

Table 2.17. Arrhenius parameters for radical addition reaction to ethylene in the gas phase [67]

Radical	$\log (A$ $/M^{-1}\,s^{-1})$	E_a $/kJ\,mol^{-1}$	Radical	$\log (A$ $/M^{-1}\,s^{-1})$	E_a $/kJ\,mol^{-1}$
$\overset{\cdot}{C}F_3$	8.4	10.1	$n-\overset{\cdot}{C}_3H_7$	7.4	21.4
$\overset{\cdot}{C}Cl_3$	8.3	26.4	$iso-\overset{\cdot}{C}_3H_7$	8.2	29.0
$\overset{\cdot}{C}H_2F$	7.6	18.1	$n-\overset{\cdot}{C}_4H_9$	7.3	28.1
$\overset{\cdot}{C}H_3$	8.1	29.4	3-Methylbutyl	7.5	26.9
$\overset{\cdot}{C}_2H_5$	8.2	31.9	$\overset{\cdot}{N}F_2$	7.6	65

trifluoroethylene, producing $CF_3CHF\dot{C}F_2$ and $CF_3CF_2\dot{C}HF_2$ in a ratio of 2.1:1. Data from Table 2.18 illustrate this tendency for a series of ethylene derivatives.

Table 2.18. Rate constant ratio for radical addition to the 1 and 2 position of the sp^2 atoms of alkenes in the gas phase [67]

Alkene \ Radical	$\cdot CF_3$ (200°C)	$\cdot CCl_3$ (150°C)	$\cdot C_3F_7$ (150°C)
$CH_2=CHF$	1:0.12	1:0.077	1:0.050
$CH_2=CF_2$	$1:0^a$	1:0.012	1:0.009
$CH_2=CHCl$	1:0	1:0	–
$CH_2=CHCH_3$	1:0.12	1:0.071	–
$CHF=CF_2$	1:0.48	1:0.29	1:0.25
$CHCl=CF_2$	1:11.5	1:25	–
$CF_2=CFCl$	1:0.25	1:0	–
$CF_2=CFCF_3$	1:0.25	1:0	–

a Zero means the absence of the corresponding radical from the addition products.

The solvated electron, functioning as a radical, can also add to a multiple bond and, in addition, it can be captured in a lower unoccupied molecular orbital. The correspondingly rate constants follow a Hammett-type relation for aromatic compounds [70],

$$\log k = \log k_o + \rho\sigma \tag{2.36}$$

where k and k_o are the rate constants for addition to a given molecule and the unsubstituted compound, respectively; ρ is the reaction constant and σ is the substituent constant.

Fig. 2.18 shows an example of such a relationship for aqueous solutions of monosubstituted benzenes [54]. A similar dependence is also observed for some halo-substituted aliphatic compounds, using the Taft equation [54].

Abstraction reactions. When interacting with organic molecules, a radical can abstract atoms of hydrogen, chlorine, bromine, and iodine from these molecules, but never fluorine atoms because of the high C—F bond strength [67]. The abstraction of a hydrogen atom from inorganic hydrogen-containing molecules is rarely observed. The abstraction of a hydrogen atom is the abstraction reaction which has been studied in most detail (Table 2.19). For a radical attacking different alkanes the Polanyi–Semenov rule is fulfilled

$$E_a = \alpha\Delta H^o + C \tag{2.37}$$

where E_a is the activation energy of the reaction, ΔH^o is the reaction enthalpy, and a and C are constants. This is illustrated in Fig. 2.19.

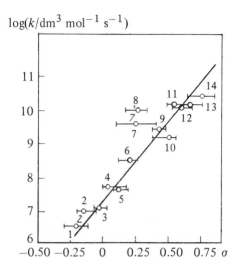

Fig. 2.18. Dependence of rate constant for reaction of the hydrated electron with monosubstituted benzenes in aqueous solution at room temperature on the Hammett σ-constant [54]: 1–OH; 2–CH_3; 3–H; 4–F; 5–SH; 6–Cl; 7–CF_3; 8–COOH; 9–Br; 10–I; 11–SO_2NH_2; 12–$CONH_2$; 13–CN; 14–NO_2.

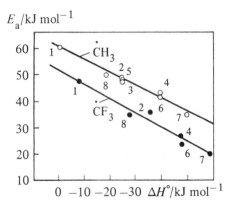

Fig. 2.19. Activation energy for the abstraction of hydrogen atoms from alkanes by $\overset{\bullet}{C}H_3$ and $\overset{\bullet}{C}F_3$ radicals as a function of the enthalpy of reaction [67]. 1–methane; 2–ethane; 3–propane (primary carbon atom); 4–propane (secondary carbon atom); 5–n-butane (primary carbon atom); 6–n-butane (secondary carbon atom); 7–isobutane (tertiary carbon atom); 8–neopentane.

As regards hydrogen halides, bromine and iodine atoms are abstracted more easily, but chlorine atoms less readily, than hydrogen. Hydrogen abstractions show a kinetic isotope effect, with a value close to that predicted from the difference in the zero-point energies of the C–H and C–D bonds. Values for the

Table 2.19. Arrhenius parameters and enthalpies of reaction of hydrogen
abstraction from ethane by alkyl radicals in the gas phase [67]

Radical	$E_a/\text{kJ mol}^{-1}$	$\log (A/\text{M}^{-1}\text{ s}^{-1})$	$\Delta H^{\circ}_{298}/\text{kJ mol}^{-1}$
$\dot{C}H_3$	50.0	8.9	-25
$\dot{C}HF_2$	52.5	9.6	-8
$\dot{C}F_3$	35.3	7.7	-34
$\dot{C}_2F_5$	38.2	8.3	-21
$\dot{C}_3F_7$	38.6	7.4	-21

rate constant of abstraction differ from each other quite considerably (Table
2.20). It is interesting to note, however, that the reactivity toward a given radical
is nearly the same within a homologous series as shown in Table 2.20 for linear
alcohols. This enables an unknown reaction rate constant for a radical to be
estimated provided that the reactivity of some other radical toward the same
solute is known.

Electron-transfer reactions (oxidation–reduction or redox reactions). Redox
transformations of different radicals in aqueous solution of a wide variety of
inorganic and organic radical scavengers have been widely studied, for example

$$H + Ce^{4+} \rightarrow H^+ + Ce^{3+},$$

$$\dot{O}H + Ce^{3+} \rightarrow OH^- + Ce^{4+},$$

$$H\dot{O}_2 + Fe^{2+} \rightarrow HO_2^- + Fe^{3+},$$

$$Ar\dot{O} + Ce^{4+} \rightarrow ArO^+ + Ce^{3+},$$

$$\dot{C}H_2OH + Fe^{3+} \rightarrow Fe^{2+} + H^+ + CH_2O,$$

and

$$(CH_3)_2\dot{C}OH + NO_3^- \rightarrow \dot{N}O_2 + OH^- + CH_3COCH_3.$$

Electron-transfer reactions have been also observed in other systems, as exemplified (with the corresponding rate constants) in Table 2.21. The standard
redox potentials for a large number of radicals in aqueous (Table 2.22) [73] and
nonaqueous [18] media have been determined.

Radical–radical reactions. There are two possible types of reaction between
radicals: combination and disproportionation. These are the only processes
leading to the decay of radicals via coupling of their unpaired electrons.

In combination (dimerization) reactions, radicals combine to give a molecule:

Table 2.20. Rate constants for reactions of radicals with alcohols in aqueous solutions at room temperature (in $M^{-1} s^{-1}$) [71, 72]

Radical / Alcohol	CH_3OH	C_2H_5OH	$n\text{-}C_3H_7OH$	$i\text{-}C_3H_7OH$	$(CH_2OH)_2$	Relative† reactivity
H	2.8×10^6	2.8×10^7	3.1×10^7	8.2×10^7	1.9×10^7	0.003–0.02
$\dot{O}H$	8.4×10^8	1.6×10^9	2.8×10^9	2.1×10^9	1.6×10^9	1
$\dot{C}O_3^-$	5×10^3	1.6×10^4	1.9×10^4	3.9×10^4	–	$(0.6$–$1.8) \times 10^{-5}$
$\dot{S}O_4^-$	2×10^7	3×10^7	4.8×10^7	8.8×10^7	–	$(1.7$–$4.2) \times 10^{-2}$
$\dot{N}O_3$	1.1×10^6	3×10^6	–	3×10^6	1.6×10^6	$(1.0$–$1.6) \times 10^{-2}$
$\dot{P}O_4^{2-}$	1×10^7	1.9×10^7	–	1.8×10^7	–	$(0.9$–$1.2) \times 10^{-2}$
$H\dot{P}O_4^-$	1×10^7	4×10^7	–	4×10^7	–	$(1.2$–$2.1) \times 10^{-2}$
$H_2\dot{P}O_4$	4.1×10^7	7.7×10^7	–	2.5×10^7	–	$(1.2$–$4.9) \times 10^{-6}$
Cl_2^-	3.5×10^3	4.5×10^3	–	1.9×10^3	–	$(0.9$–$2.4) \times 10^{-6}$
$CH_3\dot{O}$	2.6×10^5	5×10^5	–	–	6×10^5	$(3$–$4) \times 10^{-4}$

† Rate constant for linear alcohols with respect to reactivity towards OH radical.

Table 2.21. Reaction rate constants for electron transfer from radical anions to aromatic molecules in propanol-2 at 25°C [42]

Donor	Acceptor	$k/10^9 \ M^{-1} \ s^{-1}$
Biphenylide ion	Naphthalene	0.26 ± 0.08
	Phenanthrene	$0.6 \ \pm 0.3$
	o-Terphenyl	$3.2 \ \pm 0.3$
	Pyrene	$5.0 \ \pm 1.8$
	Anthrecene	$6.4 \ \pm 1.1$
p-Terphenylide ion	Pyrene	$3.6 \ \pm 1.1$
	Anthracene	$5.5 \ \pm 0.9$
m-Terphenylide ion	Ryrene	$3.5 \ \pm 1.2$
o-Terphenylide ion	Pyrene	$4.0 \ \pm 1.8$

$$\dot{C}_2H_5 + \dot{C}_2H_5 \rightarrow C_4H_{10}$$

$$\dot{C}H_2OH + \dot{C}H_2OH \rightarrow HOCH_2CH_2OH$$

$$\dot{H} + \dot{O}H \rightarrow H_2O$$

and

$$Ph\cdot \ + \dot{C}H_3 \rightarrow PhCH_3$$

In disproportionation reactions, the transfer of a hydrogen or halogen atom, or oxygen molecule (or any other group of atoms) from one radical to another in the transition state formed upon collision of the radicals, occurs in such a way that *two* or more products are formed, in contrast to combination which gives only one:

$$\dot{C}_2H_5 + \dot{C}_2H_5 \rightarrow C_2H_4 + C_2H_6,$$

$$\dot{C}l_2^- + \dot{C}l_2^- \rightarrow Cl_3^- + Cl^-,$$

or

$$H\dot{O}_2 + H\dot{O}_2 \rightarrow H_2O_2 + O_2.$$

The total rate constant for both reactions, disproportionation and combination, is usually determined and is called the (re)combination constant. Values for the recombination constants in gas and liquid phases are given in Table 2.23. This information is more useful if augmented by data on the disproportionation-to-combination rate constant ratio, which is often determined (Table 2.24). The ratio k_d/k_c is governed by the entropy factor (the difference ΔS between the entropies of the products of disproportionation and combination as is shown in Fig. 2.20.

Table 2.22. Standard electrode potentials of free radicals in aqueous solutions [73]

Radical	$E°/V$
$\dot{O}H$	+2.0
$\dot{C}H_3$, $R\dot{C}H_2$, $R\dot{C}HR$, $R\dot{O}$; $A\dot{r}$, $\dot{P}h$. $Ph\dot{C}H_2$, and $RCH=CH-\dot{C}H_2$	0
$CH_2\dot{C}(OH)COOH$	0.16
$\dot{C}H_2COH$, $\dot{C}H_2CO_2H$, $\dot{C}H_2COCH_3$, and $CH_3\dot{C}HCOCH_3$	–0.1
$\dot{C}H_2(CH_3)_2COH$	–0.1
$\dot{C}l_2^-$, $\dot{B}r_2^-$, $\dot{I}_2^-$, and $(\dot{S}CN)_2^-$	–0.3 to –0.2
$CH_3\dot{C}(OH)COO^-$	–0.2
$\dot{C}H_2COO^-$	–0.3
$^-O\dot{C}HCOO^-$	–0.48
$HOCH_2\dot{C}HOH$	–0.7 to –0.3
$\dot{C}H_2OH$	–0.73
$CH_3\dot{C}HOH$, $CH_3\dot{C}HOC_2H_5$, $NH_2\dot{C}HCOO^-$, and $\dot{C}H_2N(CH_3)_2$	–0.9 to –0.77
$(CH_3)_2\dot{C}OH$	–0.9 to –0.83
$\dot{P}O_3^{2-}$	–1.0
$\dot{C}O_2^-$	–1.03
$\dot{C}H_2O^-$, $CH_3\dot{C}HO^-$, and $CH_3\dot{C}OCH_3^-$	–1.3
$\dot{H}$	–2.0
e_{aq}^-	≤ –2.7

Table 2.23. Gas-phase (*ca.* 100°C) and liquid-phase (25°C) rate constants for radical combination [67, 71, 72, 74, 75]

Radical	Solvent	$2k/10^9 M^{-1} s^{-1}$
$\dot{C}H_3$	Gas phase	20
$\dot{C}F_3$	–"–	23
$\dot{C}F_2Cl$	–"–	12
$\dot{C}Cl_3$	–"–	4–8
$\dot{C}_2H_5$	–"–	0.4–20
$\dot{C}_2Cl_5$	–"–	0.5
$(CH_3)_2\dot{C}H$	–"–	60
$(CH_3)_3\dot{C}$	–"–	3

Table 2.23. (continued)

Radical	Solvent	$2k/10^9 \text{ M}^{-1} \text{ s}^{-1}$
$\dot{O}H$	Gas phase	0.5
$\dot{N}H_2$	$-"-$	8
$CH_3\dot{S}$	$-"-$	50
$\dot{C}H_3$	$cyclo\text{-}C_6H_{12}$	4.4
$\dot{C}Cl_3$	$-"-$	0.1–0.6
$\dot{C}_2H_5$	C_2H_6	3.2
$n\text{-}\dot{C}_3H_7$	$cyclo\text{-}C_6H_{12}$	1.7
$(CH_3)_2\dot{C}H$	C_6H_6	9.6
$(CH_3)_3\dot{C}$	$cyclo\text{-}C_6H_{12}$	1.1
$(CH_3)_2\dot{C}CN$	C_6H_6	0.9
$n\text{-}C_5H_{11}$	$cyclo\text{-}C_6H_{12}$	1.2
$n\text{-}C_6H_{13}$	$-"-$	1.1
$cyclo\text{-}\dot{C}_6H_{11}$	$-"-$	1.3
$Ph\dot{C}H_2$	C_6H_6	4.1
$Ph_2\dot{C}H$	$-"-$	2.4
$Ph\dot{C}(CH_3)_2$	$-"-$	8
H	H_2O	20 ± 6
$\dot{O}H$	$-"-$	10.6 ± 1.4
$\dot{O}^-$	$-"-$	2–25
$H\dot{O}_2$	$-"-$	$(16.2 \pm 1.0) \times 10^{-4}$
$\dot{O}_2^-$	$-"-$	$\leq 0.7 \times 10^{-9}$
$\dot{C}H_2OH$	CH_3OH	2.85
$CH_3\dot{C}HOH$	C_2H_5OH	1.20
$C_2H_5\dot{C}HOH$	$n\text{-}C_3H_7OH$	0.63
$(CH_3)_2\dot{C}OH$	$i\text{-}C_3H_7OH$	0.62
$n\text{-}C_3H_7\dot{C}HOH$	$n\text{-}C_4H_9OH$	0.51
$n\text{-}C_7H_{15}\dot{C}HOH$	$n\text{-}C_8H_{17}OH$	0.17
$CO_3^{\dot{-}}$	H_2O	0.012
$SO_4^{\dot{-}}$	$-"-$	1
$SO_3^{\dot{-}}$	$-"-$	0.4
$SO_2^{\dot{-}}$	$-"-$	10
$\dot{N}O_3$	$-"-$	0.84
$\dot{N}O_2$	$-"-$	0.027–0.076

Table 2.23. (continued)

Radical	Solvent	$2k/10^9$ M^{-1} s^{-1}
$H_2\dot{P}O_4$	H_2O	0.25–1.0
$H\dot{P}O_4^-$	–"–	0.14–0.47
$\dot{P}O_4^-$	–"–	0.1–1.0
$Cl_2^- \cdot$	–"–	14 ± 3
$Br_2^- \cdot$	–"–	1.5–5.3
$I_2^- \cdot$	–"–	6–14
$(SCN)_2^- \cdot$	–"–	2.2–3.9
e_{aq}^-	–"–	10.2 ± 1.4

Table 2.24. Disproportionation to recombination rate constant ratio for reactions of aliphatic radicals in gas and liquid phases at room tempeature [67, 76]

Radical	Gas phase	Liquid phase
$CH_3\dot{C}H_2$	0.13	0.12–0.26
$CH_3CH_2\dot{C}H_2$	0.15	0.13–0.15
$C_2H_5CH_2\dot{C}H_2$	0.14	0.13–0.14
$(CH_3)_2CH\dot{C}H_2$	0.076	—
$(CH_3)_2\dot{C}H$	0.66	1.2
$CH_3CH_2\dot{C}HCH_3$	0.63–0.67	1.0–1.1
$(CH_3)_3\dot{C}$	2.3–3.1	4.5
$cyclo\text{-}\dot{C}_5H_9$	1.0	1.0
$cyclo\text{-}\dot{C}_6H_{11}$	0.5	1.1
$\dot{C}H_3 + \dot{C}_2H_5$	0.04	—
$\dot{C}H_3 + n\text{-}\dot{C}_3H_7$	0.06	—
$\dot{C}H_3 + i\text{-}\dot{C}_3H_7$	0.16	—
$\dot{C}_2H_5 + t\text{-}\dot{C}_4H_9$	0.50	—
$\dot{C}_2H_5 + i\text{-}C_4H_9$	0.04	—

The stability of free radicals is affected by a number of factors: (i) by the degree of delocalization of the unpaired electron over the group of atoms bound to the radical centre, (ii) by steric hindrance, restraining bimolecular reactions, and (iii) by the ability of the radical centre to adopt a planar configuration. No absolute criterion of the stability of free radicals has yet been found. For HEC purposes it may be accepted that a short-lived radical is considered to be stable

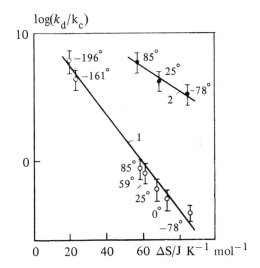

Fig. 2.20. The ratio k_d/k_c as a function of the difference in entropies of disproportionation and recombination products for primary (1) and secondary (2) heptyl radicals [77]. (The temperatures by the points refer to the conditions under which the constants have been measured.)

if it does not undergo any change before entering into some of its bimolecular reactions (abstraction, electron transfer, recombination, etc.), that is, if its lifetime as regards dissociation and isomerization is much greater than the sum of the lifetimes in bimolecular reactions. Otherwise we must take into account either the complete or partial consumption of radicals via unimolecular channels.

The reactivity of short-lived radicals of different types has been the subject of numerous studies, and copious information on the topic is available. Data on the reaction rate constants of e_{aq}^-, H atom, $\dot{O}H$, and $H\dot{O}_2$ radicals [71], on the reactivity of organic radicals [72, 79, 80], and of metal ions in unusual valency states [81] have been published. Some aspects of the kinetics of liquid-phase radical reactions have been reviewed [82], as have the reactions of electronically-excited molecules [83] and spin effects in the chemistry of free radicals [84–86].

2.9 MOST TYPICAL NONEQUILILBRIUM REACTIONS

Given below is a classification of the types of elementary reaction of excited species as concluded from numerous experimental and theoretical studies.

(i) Spontaneous dissociation

$$AB^* \rightarrow A + B$$

where A and B are atoms, radicals, or molecules. The predissociation of, for example, electronically-excited alkenes

$$C_2H_4^* \rightarrow C_2H_2 + H_2$$

or

$$C_3H_6^* \rightarrow C_3H_5 + H$$

is a raction of this type.

(ii) Spontaneous isomerization

$$AB^* \rightarrow BA.$$

Some reactions of this type are caused by photochemical or thermal action through primary electronic excitation.

The above two processes include all possible types of unimolecular reaction. Other elementary reactions considered are bimolecular.

(iii) Electronic energy transfer

$$AB^* + B \rightarrow A + B^*.$$

This type of process is represented by sensitized fluorescence, for example

$$Hg(^3P) + Tl(^2P_{1/2}) \rightarrow Hg(^1S_0) + Tl^*.$$

(iv) Collisional quenching without reaction

$$A^* + B \rightarrow A + B,$$

for example the quenching of electronically-excited sodium by hydrogen

$$Na(^3P) + H_2\left(^1\Sigma_g^+\right) \rightarrow Na(^2S) + H_2\left(^1\Sigma_g^+\right) + h\nu.$$

(v) Quenching accompanied by dissociation of the quencher molecule with the formation of fragmented products

$$A^* + BC \rightarrow A + B + C$$

for example

$$Hg\,(^3P_1) + H_2 \rightarrow Hg(^1S_0) + H + H.$$

(vi) Quenching with dissociation of the quencher and formation of a new molecule

$$A^* + BC \rightarrow AB + C$$

for example

$$Cd(^3P_1) + H_2 \rightarrow CdH\left(^2\Sigma^+\right) + H.$$

(vii) Quenching with dissociation of an excited substrate

$$AB^* + C \rightarrow A + B + C$$

for example the induced predissociation

$$\dot{O}H\left(^2\Sigma^+\right) + H_2 \rightarrow O(^3P) + H(^2S) + H_2.$$

(viii) Quenching with isomerization of an excited substrate

$$AB^* + C \rightarrow BA + C,$$

for example

$$1\text{-}C_4H_8^* + 1\text{-}C_4H_8 \rightarrow 2\text{-}C_4H_8 + 1\text{-}C_4H_8.$$

(ix) Association of colliding excited and unexcited molecules in the presence or in absence of a third body

$$A^* + B(+\,M) \rightarrow AB\ (+\,M),$$

for example

$$O(^1D) + CO\left(^1\Sigma_g^+\right) + M \rightarrow CO_2\left(^1\Sigma_g^+\right) + M.$$

(x) Formation of excited molecules as a result of exchange reactions involving atoms and polyatomic molecules

$$A + BC \rightarrow AB^* + C.$$

Rupture of the bond B—C and formation of the bond A—B takes place in ա reaction of this type. These reactions have the following features: (i) vibrational excitation of the products, especially of the molecule where the bond A—B is formed (the excitation of Product C is less); (ii) possible electronic excitation, in addition to vibrational, if the enthalpy of the exothermic reaction is sufficient. A large number of exchange reactions with the participation of H, O, N, S, halogen atoms, and alkali metal atoms are reviewed in [87].

(xi) Electron impact-induced stepwise dissociation via excitation of the vibrational levels of the ground electronic state of a molecule

$$N_2\left(X^1\Sigma_g^+, v\right) + e^- \longleftrightarrow N_2\left(X^1\Sigma_g^+, v+1\right) + e^-$$

$$N_2\left(X^1\Sigma_g^+, v\right) + N_2\left(X^1\Sigma_g^+\right) \longleftrightarrow N_2\left(X^1\Sigma_g^+, v\pm1\right) + N_2\left(X^1\Sigma_g^+, v\pm1\right)$$

$$N_2\left(X^1\Sigma_g^+, v_2\right) + N_2\left(X^1\Sigma_g^+, v_3\right) \longleftrightarrow N_2\left(X^1\Sigma_g^+, v\pm1\right) + N_2\left(X^1\Sigma_g^+, v\pm1\right)$$

$$N_2\left(X^1\Sigma_g^+, v\right) + N_2\left(X^1\Sigma_g^+, v\right) \longrightarrow 2N(^4S) + N_2\left(X^1\Sigma_g^+, v\right)$$

(xii) Dissociation after electron impact-induced stepwise excitation of electronic levels

$$N_2\left(X^1\Sigma_g^+, v\right) + e^- \rightleftharpoons N_2\left(A^3\Sigma_u^+, v\right) + e^-,$$

$$N_2\left(A^3\Sigma_u^+, v\right) + e^- \rightleftharpoons N_2(B^3\pi_g, v') + e^-,$$

$$N_2(B^3\Pi_g, v \geq 13) \rightleftharpoons 2N(^4S).$$

(xiii) Dissociation of molecules via electronically-excited levels in the collisions of heavy particles

$$AB^* + CD(X, v) \longrightarrow AB + B + CD(X, v - \Delta v),$$

$$N_2\left(A^3\Sigma_u^+, v\right) + N_2\left(X^1\Sigma_g^+, v\right) \longrightarrow N_2(B^3\pi_g, v_1) + N_2\left(X^1\Sigma_g^+ + \Delta v, v\right),$$

$$N_2(B^3\Pi_g, v \geq 12) \longrightarrow 2N(^4S).$$

(xiv) Dissociative ion-electron recombination and associative ionization

$$Ar(\sigma^1S_0) + Ar(4d; 6s; 5d) \rightarrow Ar_2^+ \cdot + e^-,$$

$$Ar_2^+ \cdot + e^-, \rightarrow Ar(\sigma^1S_0) + Ar(3p; 3p'; 3s; 3s'),$$

$$N(^2D) + O(^3P) \rightleftharpoons NO^+ + e^-.$$

(xv) Dissociative electron attachment to molecules

$$HI + e^- \rightleftharpoons HI^- \cdot$$

$$HI^- \cdot \rightarrow I^- + H.$$

(xvi) Combination reactions

$$A + B \rightarrow AB(v)$$

$$AB(v) + M \rightarrow AB + M.$$

Reactions of type (xvi) are exothermic and yield excited products, for example

$$H + NO \rightarrow HNO(v)$$

$$HNO(v) + M \rightarrow HNO + M.$$

If M is a molecule of the reaction product then the vibrational exchange of energy must accelerate the process of energy redistribution. Radical combination, for instance, involves the following processes

$$\dot{C}H_3 + \dot{C}H_3 \leftrightarrow C_2H_6(v)$$

$$C_2H_6(v) + M \rightarrow C_2H_6 + M$$

with evolution of the C—C bond energy and a change in configuration of the methyl groups on collisional stabilization.

To give some more examples. Excited alkyl fluoride molecules can be formed by

$$\text{exchange;} \quad \dot{C}_2H_5 + F_2 \to C_2H_5F(v) + F,$$

$$\text{combination;} \quad \dot{C}H_3 + \dot{C}H_2F(v) \to C_2H_5F(v),$$

and

$$\text{insertion of methylene;} \quad :CH_2 + CH_3F \to C_2H_5F(v)$$

with excess energies of 294, 478, and 483 kJ mol^{-1}, respectively.

(xvii) Radiative collisional combination

$$2N(^4S) \to N_2\left(^5\Sigma_g^+\right) - \begin{array}{l} \longrightarrow N_2(B^3\pi_g, v_k = 12, \ j_{min} = 34) \\ \longrightarrow N_2(a^3\pi_g, v_k = 6, \ j_{min} = 14) \end{array}$$

$$2S(^3P) \to S_2^* \to S_2(3B^3\Pi_g^-, v_k \geq 10);$$

$$O(^3P) + N(^4S) \to NO(a^4\Pi) \to NO(C^2\Pi, v_k = 0);$$

$$O(^3P) + H(^2S) \to \dot{O}H\left(^2\Sigma^-\right) \longrightarrow \dot{O}H\left(A^2\Pi^+, v_k = 0; 1\right);$$

$$O(^3P) + CO\left(X^2\Sigma_g^+\right) \longrightarrow CO_2\left(X1\Sigma_g^+\right).$$

The mechanism of the latter reaction can be represented as follows

$$O(^3P) + CO\left(X^1\Sigma_g^+\right) \rightleftharpoons CO_2(^3B_2)$$

$$CO(^3B_2) \rightleftharpoons CO_2(^1B_2)$$

$$CO_2(^1B_2) \longrightarrow CO_2\left(X^1\Sigma_g^+\right) + h\nu$$

$$CO_2(^1B_2) + M \longrightarrow C_2\left(X^1\Sigma_g^+\right) + M.$$

The production of electronically-excited states, that is, nonadiabatic transitions, is a characteristic feature of homogeneous combination processes observed in almost all the cases studied. Under low pressures, those transitions caused by the reverse of the predissociation process lead to radiative combination.

Combination occurring from levels lying below the dissociation boundary was found to bring about a negative temperature dependence for the rate constant.

(xviii) Heterogeneous combination

$$N + N \xrightarrow{\text{wall}} N_2$$

The following elementary processes result in electronic excitation and ionization.

(1) Collisional processes:

$$A(n) + e^- \rightleftharpoons A(k) + e^-$$
$$(n < k)$$

Excitation of molecule from its ground or excited state by electron impact

Quenching of excited state by electron impact

$$A(n) + e^- \rightleftharpoons A^+(n') + e^- + e^-$$

Ionization by electron impact

Three-body recombination

$$A(n) + B(m) \rightleftharpoons A(k) + B(l)$$

Excitation by heavy particle (atom, ion) impact

Quenching of excited state by heavy particle impact

$$A(n) + B(m) \rightleftharpoons A^+(n') + B(l) + e^-$$

Ionization by heavy particle (atom, ion) impact

Three-body recombination on heavy particles

$$A^+(n) + B(m) \rightleftharpoons A(n) + B^+(m)$$
$$A^+(n) + B(m) \rightleftharpoons A(n) + B^{2+}(m) + e^-$$

Charge transfer

$$A(n) + e^- \rightleftharpoons A^-(n)$$
$$A(n) + e^- \rightleftharpoons A^-(k) + h\nu$$

Electron capture

Detachment of excess electron

(2) Radiative processes

$$A(n) + h\nu \rightleftharpoons A(k)$$

Photoexcitation

Light emission (spontaneous and stimulated radiations)

$$A(n) + h\nu \rightleftharpoons A^+(n') + e^-$$

Photoionization

Photocapture and photorecombination to the ground state

Here $A(n)$ is an atom or molecule in its ground state ($n = 1$) or in and excited ($n > 1$) state. The process of radiationless recombination or deactivation is called quenching or an impact of the second kind.

As shown in the preceding sections of this chapter, there are a wide variety of intermediate species, and they can take part in chemical reactions of different types. However, the complete set of reactions known for the principal transients (ions, electrons, radicals, and excited species) mainly responsible for the ultimate formation of the chemical products in HEC processes can be represented as two groups of reactions: first-order and second-order reactions. This classification is illustrated in Table 2.25.

It is seen that first-order reactions include degradation and rearrangements of transient species. Collisional deactivation and radiative deactivation are considered separately as typical only of excited species. Second-order reactions include these of addition, charge transfer, excitation transfer, or transfer of an unpaired electron. The reactions of (re)combination are termination processes for ions and radicals (there are no other modes of decay for these species). Combination is less typical for excited molecules. As far as transient species of mixed nature (radical ions, excited ions, and excited radicals) are concerned, they enter all those types of reactions which are characteristic for both of the corresponding constituents.

Table 2.25. Reactions of transient species

Reaction type		Positive ions	Negative ions	Excess electrons	Electronically excited states	Radicals
				Transient species		
	1st-order reactions					
Degradation		$AB^+ \rightarrow A + B^+$	$AB^- \rightarrow A + B^-$	$e_s^- \rightarrow e^-$	$ABCD^* \begin{cases} \rightarrow A + BCD \\ \rightarrow AB + CD \end{cases}$	$\dot{R}AB \rightarrow \dot{R} + AB$
Rearrangement		$ABC^+ \rightarrow ACB^+$	$ABC^- \rightarrow ACB^-$	—	$ABC^* \rightarrow ACB$ $M^* \rightarrow M + Q$ (collisional deactivation) $M^* \rightarrow M + h\nu$ (luminescence)	$\dot{R}AB \rightarrow \dot{R}BA$
	2nd-order reactions					
Addition		$A^+ + M \rightarrow AM^{+(*)}$ (associative ionization) $A^+ + BC \rightarrow$ $AB^{+(*)} + C^{(*)}$ (dissociative ionization)	$A^- + M \rightarrow AM^{-(*)}$ (associative ionization) $A^- + BC \rightarrow$ $AB^* + C^{-(*)}$ (dissociative ionization)	$e^- + M \rightarrow M^{-(*)}$ $e^- + AB \rightarrow$ $A^{-(*)} + B^{(*)}$	$M^+ + M \rightarrow MM^*$ (excimer formation) $M^* + A \rightarrow AM^*$ (exciplex formation)	$\dot{R} + M \rightarrow \dot{R}M^{(*)}$

Table 2.25. (continued)

Reaction type	Transient species				
	Positive ions	Negative ions	Excess electrons	Electronically excited states	Radicals
Solvation	$A^+ + nM \rightarrow A_s^+$	$A^- + nM \rightarrow A_s^-$	$e^- + nM \rightarrow e_s^-$	—	$\dot{R} + nM \rightarrow \dot{R}_s$
Transfer (transformation)	$A^+ + B \rightarrow$ $B^{+(*)} + A^{(*)}$ (charge transfer)	$A^- + B \rightarrow$ $B^{-(*)} + A^{(*)}$ (charge transfer)	—	$A^* + B \rightarrow B^* + A$ (excitation transfer)	$\dot{R} + R'H \rightarrow RH + \dot{R}'$ (H-atom abstraction)
Recombination and disproportionation	$A^+ + B^- \rightarrow A^{(*)} + B^{(*)}$ (neutralization)		$e^- + M \rightarrow M^*$ (neutralization)	$T_1 + T_1 \overline{}\begin{array}{l}\rightarrow S_0 + S_1 \\ \rightarrow T_2 + S_0\end{array}$ $S_1 + S_2 \rightarrow S_0 + S_2$	$\dot{R} + \dot{R}' \rightarrow \dot{R}R'^{(*)}$ (combination) $\dot{R} + \dot{R}' - M + M'$ (disproportionation)

Note: Subscript s refer to solvated state; e^- mean quasi-free electron; asterisk in parentheses indicates the possibility of formation of excited state in this type of reaction.

3

Kinetics of processes in high-energy chemistry

The rate of a chemical reaction, which is the subject matter of chemical kinetics, depends on many factors, the main contribution under thermodynamic equilibrium (or near-equilibrium) being played by the thermodynamic parameters p and T (for systems of a constant volume comprising a large number of molecules). The simplest case is when the rate of a chemical reaction is determined by its dependence on concentration and temperature, that is, on the energy introduced into the system. The so-called reaction rate constant is described by the simple Arrhenius law $k = k_o \exp(-E_a/kT)$. As seen from this equation, the term 'constant' does not seem too felicitous since k is a strongly temperature-dependent function, although its dependence on other quantities is weaker. In accordance with this law (assuming that deviations from equilibrium are always small), three regions $E_a \gg kT$, $E_a \approx kT$, and $E_a < kT$ can be defined.

The first case predominates decisively in conventional chemical kinetics (hence the application of the Arrhenius equation with sufficient accuracy is always legitimate), but high-energy chemistry is concerned mostly with the second and third regions (note that even near $E_a \approx kT$ ($E_a/kT \cong 1$), the exponential law already does not make sense). Moreover, with $E_a < kT$ the principal assumption of a small departure from thermodynamic equilibrium is untrue. The single (Maxwell–Boltzmann) temperature of a chemically reacting system disappears, while the temperatures (or, more accurately, the energies) of translational motion and of the quantum degrees of freedom (rotational, vibrational, and electronic levels of molecules) come into action for molecules participating in the reaction in a different way and interacting with each other, and so reactions of excited species are brought about and new particles, such as ions, electrons, etc., emerge.

Processes of this type can be described only with the use of nonequilibrium chemical kinetics. The latter covers a wider variety of reactions (for example electron impact-induced reactions) and includes the possibility of appearance of new reaction pathways (a multichannel character is a principal concern in the statement of the problem). The definition of the reaction rate constant itself changes because of the change in definition of temperature (and the emergence of many 'temperatures') and a more complex but more realistic concept of the *rate coefficient* of a chemical reaction appears, so that the *rate constant* is a particular (limiting) case of this coefficient for $kT < E_a$. In accordance with this, the physical model of the kinetics of chemical reaction also changes, thus causing a corresponding change in the corresponding mathematical model developed for computations. The physical model takes into account not only chemical reactions by themselves and their probabilities (cross-sections) for different channels of interaction, but also fluctuations in the parameters of a reacting system and various physicochemical processes such as mixing of the components, diffusion transfer (in particular, diffusion-controlled reactions), etc. The reacting system turns out to be nonequilibrium not only energetically but also spatially (with inhomogeneity in the spatial distributions of reactants, products, and transient species). This especially affects (homogeneously) heterogeneous reactions and is displayed in the presence of interfacial boundaries whose active centres take part in chemical reactions and, in particular, facilitate the formation of products of different morphology and structure (ultradisperse powders, surface films, etc.).

3.1 PRINCIPAL QUANTITIES, CONCEPTS, AND ASSUMPTIONS IN CHEMICAL KINETICS

To assist understanding of the material of this chapter, we consider it appropriate to give more detailed definitions of the basic quantities (variables and parameters) and concepts. A comprehensive presentation of these can be also found in earlier publications [88–96].

(i) The fundamental difference between *equilibrium* and *nonequilibrium* systems with constant mass and energy flows is their behaviour on reversal of time. In equilibrium systems, by definition, each flow in one direction is counterbalanced by a flow in the opposite direction, so the system is invariant with respect to reversal of time. This symmetry can be violated by flows across the system which will cause departure of the latter from its equilibrium state. Near equilibrium, the reacting system is stable and any perturbations applied decrease with time [88–97].

(ii) The *equations of chemical kinetics* (both equilibrium and nonequilibrium kinetics) establish the relation $dc_i/dt \equiv \dot{c}_i = f_i(c, k)$ between $\dot{c}_i$, c, and t where $c_i = c(t)$, the concentration of the i-th species, which is continuous and is at least twice-differentiable scalar statistical quantity; f_i are continous functions containing certain coefficients k_i (see section 3.4) and having continuous derivatives but not including time in an explicit form.

Let us define a range for the existence and variation of c_i:

$$0 \leq c_i \leq \infty,$$

that is, c_i has an essentially non-negative value. If considering c_i as a generalized coordinate of a system, it differs from the generalized coordinates of mechanics q_i insofar as for the latter we have

$$-\infty \leq q_i \leq +\infty.$$

(iii) *Threshold energy and activation energy.* For a chemical reaction to become possible, the system of colliding particles must possess a certain minimum energy (and appropriate orientation). This energy will be called the threshold energy, or E_{th}. It is a parameter which characterizes a given chemical reaction and is independent of the state of the reacting particles.

This value is dynamic in its nature (though this is usually ignored): it cannot be obtained directly from traditional kinetic experiments [98].

The activation energy of a chemical reaction E_a, or the difference in the mean energy of the reacting particles and the average energy of all particles, is a statistical quantity[†] which can be found experimentally. This definition of activation energy was mainly developed by Tolman [99]. The deduction itself requires the assumption of the Maxwell–Boltzmann distributon. Differentiating the logarithmic form of the equation for $k(T)$, we find, with this assumption,

$$E_a = \langle \tilde{E} \rangle - \langle E \rangle, \tag{3.1}$$

where $\langle \tilde{E} \rangle$ is the average energy of collisions resulting in chemical reaction; $\langle E \rangle = 3/2kT$ is the mean energy of all collisions in the system whether or not they lead to chemical change.

From Eq. (3.1) it follows that, because of the rapid decrease in the number of particles in the high-energy region of the Maxwell–Boltzmann distribution with increasing energy, only those collisions of energy close to the threshold energy of a given reaction will be the main contributors to reaction. The energy $\langle \tilde{E} \rangle$ increases less strongly than $\langle E \rangle$ with increasing temperture, therefore E_a is temperature-dependent. This dependence on temperature is weak because of the value of E_{th} or $\langle \tilde{E} \rangle$ is large compared with kT. The latter situation shows why the Arrhenius scheme (where $E_a \neq E_a(T)$) is used successfully in so many cases [101]. There are other cases, however, where the function $\langle \tilde{E} \rangle = f(T)$ cannot be neglected and the Arrhenius kinetics does not seem to be sufficiently accurage even for quasi-equilibrium systems.

[†] Note that although in Arrhenius' equation for $k(T)$ it is assumed that $E_a \neq E_a(T)$, this was not the case in his own experiments. This has been analyzed carefully by Moelwyn-Hughes [100].

When considering chemical reactions in nonequilibrium systems, it is more convenient to use E_{th} since the introduction of E_a through Eq. (3.1) will meet problems with non-Maxwellian distributions [102]. The relation between E_{th} and E_a can be described by a particular mathematical expression only for a definite type of distribution function, insofar as the activation energy depends on this function [103].

(iv) *Temperature.* Unlike systems described at least approximately with the Arrhenius equation, systems studied with nonequilibrium physical and chemical kinetics generally have no uniform temperature, which is the parameter of the Maxwell–Boltzmann distribution. Thus it is often appropriate to introduce and to rely on the concepts of

(a)   rotational temperature

$$\frac{n_{j'}^{(v)}}{n_{j''}^{(v)}} = \frac{g_{j'}^{(v)}}{g_{j''}^{(v)}} \exp\left(-\frac{E_{j'}^{(v)} - E_{j''}^{(v)}}{kT_{rot}}\right), \tag{3.2}$$

(b)   vibrational temperature

$$\frac{n_{v'}}{n_{v''}} = \exp\left(-\frac{E_{v'} - E_{v''}}{kT_{vib}}\right), \tag{3.3}$$

(c)   distribution temperature

$$\frac{n_i}{n_k} = \frac{g_i}{g_k} \exp\left(\frac{E_i - E_k}{kT_{dist}}\right), \tag{3.4}$$

(d) k-th level population temperature

$$n_k = n_i \frac{g_k}{Z_{col}} \exp\left(-\frac{E_k}{kT_{pop}}\right), \quad \text{and} \tag{3.5}$$

(e)   ionization temperature

$$\frac{n_e n_{i,z+1}}{n_{i,z}} = \frac{2(2\pi m_e kT_{ion})^{3/2}}{h^3} \frac{Z_{col}^{z+1}}{Z_{col}^z T_{ion}} \exp\left(\frac{E_{ion}^{z+1}}{kT_{ion}}\right). \tag{3.6}$$

Here $n_{j'}^{(v)}, n_{j''}^{(v)}, g_i^{(v)}$, and $g_j^{(v)}$ are the concentrations and the degeneracies (statistical weights) of particles with excited vibrational (v) and rotational (j' and j'') levels respectively. $E_{j'}^{(v)}, E_{j''}^{(v)}$, etc. are the energies of the corresponding levels; $n_{v'}$ and $n_{v''}$ are the populations of the vibrational levels v', and v'' and n_i and n_k are the populations of the ground and k-th electronic states, respectively;

$$Z_{col} = \sum^{n_o} g_k \exp\left(-E_k/kT_{pop}\right)$$

is the electronic partition function (n_o is the last discrete energy level occupied upon collisional interaction); g_k and E_k are respectively the degeneracy and the energy of the k-th state counting from the ground state; n_i^{z+1} is the observed concentration of ions with a multiplicity of $z + 1$ (for $z = 0$, $n_{io} = n_{at}$ is the concentration of neutral atoms); Z_{col}^{z+1} and E_{ion}^{z+1} are the partition function and the ionization energy, respectively, for an ion of multiplicity $z + 1$.

The physical meaning of the quantities defined with Eqs (3.2)–(3.6) is as follows. T_{rot} and T_{vib} are the parameters of the corresponding (Boltzmann) energy distributions of the particles in their rotational and vibrational levels, respectively. These concepts make sense only in the cases when T_{rot} and T_{vib} are unified parameters for the entire set of rotational and vibrational energy levels, so Eqs (3.2) and (3.3) are true for any two energy levels of the considered totality having identical parameters T_{rot} or T_{vib}, respectively. T_{dist} is the parameter of the Boltzmann energy distribution of particles in their excited electronic levels; in this case only that distribution may be taken as Boltzmann-like which also includes the ground states of the atoms, that is, which satisfies Eq. (3.4). The physical meaning of T_{ion} is clear from Eq. (3.6).

(v) *Reaction rate order and molecularity.* In its general form the kinetic equation for the decrease in concentration c_i of a substance will be

$$- \dot{c}_i = k c_1^a c_2^b \dots c_N^m. \tag{3.7}$$

The overall order of reaction is defined as the sum of the powers for all of the reactant concentrations. Generally speaking, it is quite unnecessary for the reaction rate order to be an integer; it is taken as the best value fitting the experimental data. Similarly, there is no relation between the stoichiometry of a chemical reaction and its reaction order. For example, the reaction

$$2N_2O_5 \rightarrow 4NO_2 + O_2$$

is a first-order reaction, but

$$2NO_2 \rightarrow 2NO + O_2$$

is a second-order reaction.

A reaction usually occurs through a series of stages, which separate the initial reactants from the final products, and the reaction rate is generally determined by the slowest stage. Each of these stages is termed elementary. The molecularity of a reaction indicates how many reactant molecules take part in the elementary reaction. Unimolecular reactions are reactions of first order, bimolecular reaction are of the second order, etc., but the opposite is not necessarily true. The concept of reaction order is applied to the empirical rate law, while that of molecularity refers to the molecular mechanism of an elementary stage.

(vi) *Steady states.* Very important among the possible states of a reacting system is the stationary (steady) state when none of the thermodynamic properties of the system changes with time. The properties can vary in space, and the intensive parameters of the system may be discontinuous at its boundary, across which mass and energy transfer between the system and the surroundings are realized. If a system is in its stationary state, the corresponding flows of mass and energy are constant in time [104, 105].

In a stationary system some parameters, especially those describing the state of the surroundings (temperature T, pressure p, and chemical potentials μ) remain constant or nearly constant under the influence of the state of the system. The difference between the system and its surroundings requires us to make the assumption that the latter affects the former but not vice versa.

When considering the stability of a stationary state, it is obligatory to take into account which type of perturbation is included: infinitesimal, finite, spatially uniform (independent of x, y, z), or nonuniform. The stability of a steady state is usually expressed with a special parameter.

(vii) *Equilibrium state.* Equilibrium is defined as that stationary state in which the intensive properties of a system remain continuous upon transition across the boundary. In other words, net mass and/or energy flows are equal to zero at the boundary.

(viii) *Rate constant of a chemical reaction.* Let us consider a multicomponent system in which a chemical reaction takes place and the components $c_j (j = 1, \ldots, n)$ transform into each other in accordance with the law of mass action

$$\sum_{j=1}^{n} \bar{v}_{\lambda j} c_j \; \underset{k_{-1}}{\overset{k}{\rightleftharpoons}} \; \sum_{j=1}^{n} \bar{\bar{v}}_{\lambda j} c_j \qquad (\lambda = 1, \ldots, l). \tag{3.8}$$

Introducing

$$\vec{R} = k_\lambda c_1^{v_{\lambda 1}} \ldots c_n^{v_{\lambda n}} \quad \text{and} \quad \bar{R} = k_{-\lambda} c_1^{v_{\lambda 1}} \ldots c_n^{v_{\lambda n}} \tag{3.9}$$

we write the kinetic equation in its conventional form

$$\dot{c}_j = \sum_{\lambda=1}^{1} (\bar{v}_{\lambda j} - \bar{\bar{v}}_{\lambda j})(\vec{R}_\lambda - \bar{R}_\lambda) \tag{3.10}$$

where $\bar{v}_{\lambda j}$ and $\bar{\bar{v}}_{\lambda j}$ are the stoichiometric coefficients and $\vec{R}_\lambda$ and $\bar{R}_\lambda$ the rates of change in concentration of a component R for the forward and reverse reactions, respectively. This deterministic equation is fundamental. For reactions desribed with Eq. (3.10) the laws of conservation of energy and of total mass are fulfilled.

It is necessary to note that the traditional experimental procedures of chemical kinetics cannot provide complete information on a reaction mechanism, in particular, when concerning data on the concentration and behaviour of transient species such as radical ions, ions, free radicals, excited molecules, etc. Experiments of this conventional type enable us to determine the more-or-less complete set of stable final products of reaction and to observe some components emerging or disappearing during the process.

Assume the stoichiometric equation of a reaction has the form of Eq. (3.8). Then, by definition, the reaction rate W_i of the forward reaction is

$$W_i = \left| \frac{1}{\vec{v}_{\lambda j} - \vec{v}_{\lambda j}} \right| \frac{dc_j}{dt} \frac{1}{V} \tag{3.11}$$

At the same time, W_i, is proportional to the product of the reactant concentrations, that is, for Eq. (3.8) we get

$$\vec{W}_i = k_\lambda \prod_j c_j^{\vec{v}_{\lambda j}} \tag{3.12}$$

and

$$\vec{W}_i = k_{-\lambda} \prod_j c_j^{\vec{v}_{\lambda j}}. \tag{3.13}$$

Adding, we obtain W, or the overall rate for the i-th reaction

$$W = \left| \vec{W}_i - \vec{W}_i \right| = \left| k_\lambda \prod_j c_j^{\vec{v}_{\lambda j}} - k_{-\lambda} \prod_j c_j^{\vec{v}_{\lambda j}} \right|. \tag{3.14}$$

At equililbrium, $W = 0$, and then

$$\frac{k_\lambda}{k_{-\lambda}} = \frac{\prod_j \bar{c}_j^{-v_{\lambda j}}}{\prod_j \bar{c}_j^{-v_{\lambda j}}} \tag{3.15}$$

where $\bar{c}_j$ is the equilibrium value of c_j.

The right-hand part of Eq. (3.15) depends on the equilibrium parameters of the reaction under consideration, and it is termed the equilibrium constant

$$K = k_\lambda / k_{-\lambda}. \tag{3.16}$$

Naturally, this relationship, which is important for equilibrium chemical kinetics, is not fulfilled for nonequilibrium kinetics, except for the case when the characteristic time of the chemical reaction greatly exceeds the period of relaxation by the internal degrees of freedom of the reacting molecules. However, this would be a case of a near quasi-equilibrium system.

On the basis of the above considerations it is easy to get the fundamental equation of equilibrium chemical kinetics. From the well-known equation

$$\frac{d \ln K}{d(1/T)} = -\frac{\Delta H^o}{R} \tag{3.17}$$

where ΔH^o is the change in enthalpy for the conversion of 1 mole of a reactant to a product (provided that all the reactants and products are in their standard states), we obtain from Eqs (3.16) and (3.17)

$$\ln K = \ln k_\lambda - \ln k_{-\lambda}$$

and

$$\frac{d \ln K}{d(1/T)} = -\frac{E_a}{R},$$

where $E_{forward} - E_{back.rxn} = \Delta H^{\,o}$.
In the integral form it will look like

$$k = A \exp(-E_a/kT) \tag{3.18}$$

This is the fundamental equation for the reaction rate constant in equilibrium chemical kinetics, that is, the Arrhenius equation.

The rate of a chemical reaction, according to the principal hypothesis of collision theory, is proportional to the number of collisions between molecules

$$W = Z \Delta n \, c_j c_i \tag{3.19}$$

where Z is the number of encounters of molecules A_i with A_j per second and Δn is the fraction of reactants possessing an energy sufficient for chemical reaction to occur [106].

Assume further that chemical reactions of interest to us satisfy all the requirements of valency theory. A chemical transformation occurring within a period less than the time between two collisions will be considered as an elementary event of a chemical reaction. Thus a chemical reaction is a sequence of elementary events

$$\sum_i M_i \rightarrow \sum_j M_j' \tag{3.20}$$

where M_i and M_j are different types of molecule, the molecule being taken as a system of nuclei and electrons with a single minimum potential energy.[†]

Here the chemical reaction is defined as a process resulting in the appearance of 'new' molecules as species with a different potential energy minimum. Note that any type of excitation of a molecule keeps the value of its minimum potential energy unchanged. This enables a distinction to be made between chemical kinetics proper and the closely related physical kinetics.

Let us now consider the two fundamental quantities of chemical kinetics which are the concentration c_i of a substance and time t. The latter is normal physical time (the continuity and the unidirectionality of time are assumed), and the former is defined as the number of particles n_i of a given component of mass m_i in a given energy state ε_i in some unit volume (the volume is assumed to contain particles in numbers sufficiently large for the macroscopic averaging of elementary events). Thus c_i is an essentially statistical quantity.

† Reaction (3.20) can take place only in the case of the generation of some particular configuration of a system with a characteristic type of interaction usually described with one or other model.

Let us postulate that the law of conservation of mass is fulfilled[†] in a system where chemical reactions take place (this law is actually the first postulate of chemical kinetics):

$$\sum_i m_i n_i = \text{const} \qquad (i = 1, \ldots, s). \tag{3.21}$$

Since $n_i/V = c_i$ (where V is the volume of the system) then

$$V \sum_i m_i c_i = \text{const} \tag{3.22}$$

and for a constant volume

$$\sum_i m_i c_i = \text{const} \tag{3.23}$$

Turning to the number of atoms N_j of the j-th type in the system, we write

$$\sum_i N_j = \text{const} \tag{3.24}$$

which is the law of conservation of the numbers of atoms of each type in a chemically reacting system. Having obtained Eq. (3.24) we get the discrete conservation law formally similar to those in quantum physics (and in the theory of elementary particles). This situation is still far from receiving proper attention in its use in the analysis of the fundamental aspects of chemical kinetics.

The rate of the chemical reaction is defined as

$$W = \frac{1}{V} \lim_{t \to 0} \frac{\Delta n_i}{\Delta t} \tag{3.25}$$

where Δn_i includes only those changes in n_i which are caused by the reaction. If the volume of the system is constant, then $W = \dot{c}$. We define $\dot{c}_i$ as

$$\dot{c}_i = f_i(c_1, c_2, \ldots, c_n), \tag{3.26}$$

the functions f_i being themselves continuous, having continuous derivatives, and containing some coefficients k_i (the rate constants of chemical reaction) but not including time in an explicit form. Systems of this type are known in probability theory as Markov systems [107–110].

Eq. (3.26) establishes the basic relation between $\dot{c}_i$, t, and c_i in chemical kinetics. To impart to this relation a clear physical meaning, it is necessary to accept some preliminary axiomatic statements:

---

† The law of conservation of mass (Eq. (3.21)) is needed principally for the sound definition of c_i. As with laws of conservation, it is fulfilled at any moment of time in a closed reacting system. In the case of an open system, generalization of the law can be carried out without difficulty if including the flow terms.

(i) Together with the law of mass conservation, the principles of conservation of energy, of momentum, and of moment are fulfilled[†] in chemical reactions.

(ii) For chemical reaction to occur, the encounter of at least two particles is needed.

(iii) Species participating in a chemical reaction are statistically independent.

(iv) A chemical reaction is one of the pathways of energy redistribution in a system leading the system ultimately to a state with minimum potential energy.[‡]

(v) A system of colliding particles must possess a certain minimum energy (called the threshold energy) and an appropriate orientation for the chemical process to become possible.[§]

(vi) There exists a composition

$$\sum_j c_j$$

that corresponds to the equilibrium state invariant with time so that all $\dot{c}_i = 0$ and all $c_i' = dc_i/dq_j = 0$ ($j = 1, 2, 3, \ldots$) [102].

Later, for the sake of simplicity and clarity of presentation, we shall consider a chemical reaction in the gas phase (assuming that the gas deviates negligibly from ideal behaviour).

Suppose some reaction

$$A(i) + B(j) \rightleftharpoons C(l) + D(m) \tag{3.27}$$

where i, j, l, and m characterize the energy states of the molecules.

(Generalization to more sophisticated chemical reactions is not fundamental and does not seem to be difficult.)

From Eq. (3.26) and postulates (i) to (iii) (based on consideration of the energy states for the reactant and product molecules), we find that

$$\dot{c}_A = -k_{AB} c_A c_B = k_{CD} c_C c_D \tag{3.28}$$

where k_{AB} and k_{CD} are the rate constants for chemical reactions independent of concentration and (in explicit form) of time.[§§]

Similar equations may be written for $\dot{c}_B$, $\dot{c}_C$, and $\dot{c}_D$. The solution of the set of equations thus obtained gives $c_B(t)$, $c_C(t)$, and $c_D(t)$ with the prescribed initial condition $c_i|_{t=0} = c_i(0)$.

---

† Possible complications related to the principle of conservation of momentum during inelastic collisions are not considered here. This problem is discussed elsewhere [111].

‡ 'Pumping' of energy by electromagnetic, electric, and other fields proceeds via creation of transient reactive species (for example electrons).

§ If the first four postulates define the conditions necessary for the occurence of chemical reaction, then the fifth principle is one of the sufficient conditions considered together with postulate (iv).

§§ Furthermore, the factor k involved in the equation of chemical kinetics will be called the *rate coefficient* of chemical reaction in the case of nonequilibrium reactions and the *rate constant* in the case of equilibrium and quasi-equilibrium reactions (the rate constant is not actually constant but a function of temperature, $k(T)$).

Eq. (3.28) is a trivial representation of the so-called law of mass action, so that the implementation of some of the principles introduced above might seem to be unnecessary. However, the law of mass action is based on thermodynamics and its statistical interpretation, which lead finally to Eq. (3.28). The law itself was stated as early as 1854–1867 by Guldberg and Waage from a generalization of their experimental data. Later, it achieved statistical thermodynamical substantiation [105–112]. This substantiation is strictly valid for the conditions of chemical equilibrium or for states close to equilibrium.

The equations of chemical kinetics for isothermal homogeneous gas-phase reactions can be written in a general form as follows

$$\dot{c}_i = \sum_{m=i}^{n} \sum_{j \leq m} k_{imp} c_m c_p + \sum_{p=1}^{m} k_{ip} I(t) \qquad \begin{array}{l} t > t_0 \\ i = 1, 2, \ldots, n \end{array} \qquad (3.29)$$

where c_i is the dimensionless concentration related to the dimensional factors $T_0/V_0 T$, $I(t)$ is the intensity of flow of particles, k_{imp} and k_{ip} are the rate constants for second-order and first-order reactions respectively, t_0 is the time of the beginning of reaction, $0 \leq c_i \leq 1$, k_{imp} is symmetrical with respect to permutation of the indices.

Let us develop these principles. Simplifying Eq. (3.28), assuming that the reverse reactions do not occur, we have for the reaction $A + B \rightarrow$ Products

$$\dot{c}_i = k_{AB} c_A c_B. \qquad (3.30)$$

This example is convenient for making comments on postulate (ii). First of all, note that molecules, atoms, radicals, ions, and species like the neutron, electron, a-particle, etc. as well as photons or particles with a rest mass of zero, can take part in the collisions. In all of these cases the corresponding kinetic equation includes a flux density for a particular type of species (or, sometimes, radiant power, dose rate, etc.) instead of its concentration. It is the acceptance of the second postulate which enables a kinetic equation to be written as Eq. (3.28) in collision theory. However, only the acceptance of postulate (iii) enables us to state that increasing, for example twice the concentration of one of the components, results in doubling the reaction rate.

Mathematical expressions of the collision theory involve a concentration-independent factor z which is incorporated into the rate constant k together with some other factors. Since an inelastic encounter does not necessarily result in chemical reaction and, besides, inelastic collisions in a system are usually accompanied by elastic collisions, there is a need for postulate (iv).

We shall not dwell on the details of the process of collision and related phenomena (in fact, these problems are outside the scope of chemical kinetics). Nevertheless, some comments are inevitable.

On its encounter with an atom or molecule, a photon can be either scattered or absorbed. It is the latter case which prompts a chemical reaction to take place.

The specific feature of this process is that one of the participants ceases to exist. Note that in this case, as in encounters of heavy particles, all the conservation laws (postulate (i)) are wholly fulfilled.[†] Thus the vanished (absorbed) photon transfers the molecule (atom) to an excited state whose total energy differs from the initial one by the value of hv. The total energy of a molecule is composed of the energies of its internal degrees of freedom and the kinetic energy

$$E_{tot} = E_{tran} + E_{el} + E_{vib} + E_{rot}.$$ (3.31)

The redistribution of energy during collision can take place, that is, via translational-vibrational $(v - T)$, vibrational-rotational $(v - R)$, etc. transitions. This situation determines the multichannel occurrence of the process which is expressed exactly by postulate (iv).

To complete the picture it should be noted that the passage of a chemical reaction is not indifferent to the mutual orientation of the colliding particles, although the equation for translational energy[‡] includes the square of the relative velocity of the particles. Thus the mutual orientation may be neglected only in the approximate model of 'point-sized' particles. Otherwise, the concepts of 'steric factor', 'steric hindrance', etc. usually appears in chemistry.

One additional note should be made here. The fulfilment of the basic principles does not by itself provide a practically measurable reaction rate based on the given level of experimental techniques.

Postulate (vi) is never strictly fulfilled for a system where chemical reactions take place, since any process occurring at a finite rate disturbs the state of equilibrium even if the system was initially in equilibrium.

3.2 PRINCIPLES OF MICROSCOPIC REVERSIBILITY AND DETAILED EQUILIBRIUM

The principle of microscopic reversibility follows directly from the symmetry of the Schrödinger equation (or classical Liouville equation) with respect to reversal of time. This principle relates the cross-sections of the forward and reverse reactions. The principle of detailed equilibrium establishes a statistical relationship between the rate constants of the forward and reverse processes at equilibrium. The principle of detailed equilibrium for the rate constants of forward and reverse reactions can be deduced as follows from the equality of the rates for the direct and reverse reactions at equilibrium as well as from the

† With the 'wide' interpretation of the nature of collision of particles (M + M, M + R, M + hv, etc.) presented here, the second postulate defines particular conditions providing the fulfillment of the first postulate and, first of all, the law of conservation of energy.
‡ The relative energy in the centre-of-mass of the colliding particles, but not in the laboratory coordinate system, is meant. We do not consider here the problem of coordinate systems for the other terms of Eq. (3.31).

relationships of microscopic reversibility with the use of the Maxwell-Boltzmann equilibrium distribution in terms of velocity and internal energy.

Omitting mathematical operations we write the following general equations.

(i) The principle of microscopic reversibility in the case (a) when molecules do not possess an internal moment (spin, orbital, or rotational moment):

$$P_{in}^2 \vec{\sigma}[ij/ml; v_{in}, \Omega] = P_{fin}^2 \vec{\sigma}[ml/ij; v_{fin}, \Omega] \tag{3.32}$$

and (b) when molecules have an internal moment in the absence of external electric and magnetic fields:

$$P_{in}^2 \vec{\sigma}[ij/ml; v_{in}, \Omega] = P_{fin}^2 \vec{\sigma}[m'l'/i'j'; v_{fin}, \Omega], \tag{3.33}$$

where P_{in} and P_{fin} are the initial and final relative momenta, respectively, the arrows over σ indicate the forward ($\rightarrow$) and reverse ($\leftarrow$) reactions respectively, ij and ml are the initial and final states respectively, v_{in} and v_{fin} are the initial and final velocities, respectively, and Ω is the solid angle of scattering of the particle. The primed quantum states differ from the non-primed by the sign of projection of the moment on the 'z'-axis.

Note that the quantum mechanical probabilities of transitions

$$\omega[ij/ml; v, \Omega] = \omega'[ml/ij; v', \Omega'] \tag{3.34}$$

are referred to particular initial and final quantum states, while the cross-sections define the fraction of the initial flux of particles scattered within a solid angle $\Delta\Omega$. Hence, in order to find the cross-sections from the transition probabilities, it is necessary to divide ω by the initial flux and to multiply the quotient by the number of final states in a given energy range and a given $\Delta\Omega$ interval.

In the analysis of experimental data, a relationship between the averaged cross-sections describing the transition from one set of degenerate levels to another is used

$$P_{in}^2 g_i g_j \overline{\sigma}[\overline{ij}/(\overline{ml}); \overline{v}_{in}, \Omega] = P_{fin}^2 g_m g_l \overline{\sigma}[\overline{ml}/(\overline{ij}); \overline{v}_{fin}; \Omega'] \tag{3.35}$$

which is applicable only in the absence of external fields when the sets of states $\overline{i}\,\overline{j}$ are degenerate.

(ii) The principle of detailed equilibrium:

$$\vec{k}(T)/\overleftarrow{k}(T) = K \tag{3.36}$$

where K is the equilibrium constant for a reaction of type A + B $\rightarrow$ C + D.

The principle of detailed equilibrium is a macroscopic representation of the principle of microscopic reversibility.

These principles make it possible (i) to find the cross-section $\overleftarrow{\sigma}$ of the reverse reaction on the basis of the principle of microscopic reversibility provided that the cross-section $\vec{\sigma}$ is known, (ii) to calculate $\overleftarrow{k}(T)$ for a given $\vec{\sigma}$ if the transitional and internal degrees-of-freedom distribution are known to be equilibrium,

(iii) to find $\bar{k}(T)$ with the principle of detailed equilibrium once $\vec{k}(T)$ has been calculated.

Thus, the principle of microscopic reversibility may be stated as follows: for every elementary chemical process there exists a reverse process. These two processes as a whole make up a reversible chemical reaction which realizes chemical equilibrium between reactants and products. The chemical reaction at equilibrium has to balance itself so that the mean rate of the forward process is equal to that of the reverse process.

Microscopic reversibility is the principal feature of elementary chemical reactions at the molecular level. At the same time, the microscopic description of the same processes must include irreversibility when a large number of molecules are involved simultaneously. The sign of 't' has no importance for an elementary chemical process, but it is very important in describing the evolution of the system towards equilibrium. Time is a vector indicating the direction to equilibrium for an isolated multiparticle reacting system [88].

3.3 DISTRIBUTION FUNCTIONS

For simplicity consider a certain amount of a gas in a rigid vessel with completely impermeable walls. For a sufficiently long period of observation, the gas is found to be uniformly distributed throughout the vessel (neglecting, of course, changes in density resulted from gravitation forces and slight variations in density near the walls due to attractive or repulsive forces). This system will be characterized by its equilibrium state, that is, by a certain energy and by identical pressure and temperature throughout the vessel. From the molecular point of view, the pressure is a result of random collisions of molecules with the walls, while the energy of the system is simply equal to the sum of the energies of the separate molecules. Therefore, to describe the system it is first of all necessary to know the number of molecules having a given velocity or kinetic energy (neglecting the internal energy of the molecules and the intermolecular forces, which is true for an ideal monoatomic gas). Information on the number of molecules with a given velocity is represented as a velocity (energy) distribution. For the above case, such information directly provides a value for the total kinetic energy

$$E_{\text{kin}} = \frac{1}{2} m \sum_{i=1}^{N} v_1^2 n(v_i) \tag{3.37}$$

where $n(v_i)$ is the number of molecules having velocity v_i. This sum can be replaced by an integral provided that $n(v_i)$ is given as a continuous function of v:

$$E_{\text{kin}} = \int_0^{\infty} v^2 n(v) \, dv. \tag{3.38}$$

To determine the properties of an ideal gas in its equilibrium state, we have to know the velocity distribution function. With non-ideal gases or liquids, the spatial distribution function is also required, while nonequilibrium systems changing in time need to be treated with time-dependent velocity distribution functions and time-dependent spatial distribution functions.

For an equilibrium system, the function of the velocity distribution of molecules is named the Maxwell–Boltzmann distribution function and has the form:

$$f_M(v) = \left(\frac{m}{2\pi kT}\right)^{3/2} \exp\left(-\frac{mv^2}{2kT}\right). \tag{3.39}$$

This function defines the probability of finding a molecule with a velocity v in a gas under equilibrium conditions. The Maxwell–Boltzmann distribution has an important property, that is, it does not depend on a particular type of interaction between the molecules if there is any interaction. Because of this, the distribution becomes universal. Any other type of velocity (or translational energy) distribution becomes in time the equilibrium distribution (unless this process is artificially retarded), and this is one of the features of distribution in nonequilibrium systems.

Some energy distribution functions (EDFs) met in HEC are shown in Fig. 3.1 for molecules of a binary gas mixture. The case 'b', for example, is realized in a glow discharge or microwave discharge, the case 'e' for a gas stream injected into a volume occupied by an equilibrium gas, and the case 'f' in reactions occurring in monoenergetic fluxes and beams.

Fig. 3.1a illustrates an example from conventional chemistry when both components of the isothermal reaction of type A + B have the same Maxwellian distribution: Fig. 3.1b shows the case when the two components have different temperatures (mean energies) in reaction with each other. Example 'e' is rather interesting insofar as one of the components has a monoenergetic distribution.

Let us examine the influence of EDF on k. Fig. 3.2 shows the dependence of log k on E_A/E_B (E_A and E_B are the energies of species A and B, respectively, averaged over their corresponding distribution) for a reacting system composed of two Maxwellian gases with different mean energies. Since E_A is taken as constant for simplicity and $E_B = kT$, then $E_A/E_B \approx 1/T_B$, and, consequently, the plot in Fig. 3.2 is Arrhenius in character. Instead of the Arrhenius rectilinear dependence we have a curve. This curve can be divided roughly into 3 or 4 linear segments for conditions resembling those of routine chemical experiments. Breaks on a plot like this are usually seen as referring to a change in mechanism, although such a conclusion seems rather ambiguous.

The Arrhenius treatment of experimental data will be incorrect for highly exothermic reactions or branched chain reactions occurring via excited molecules, even if in the reacting system the equivalence $T_{tran} = T_{vib}$ is maintained to a high degree of approximation (Fig. 3.3).

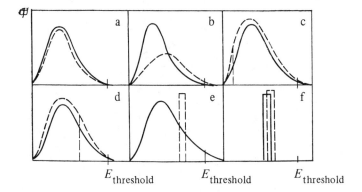

Fig. 3.1. Types of distribution function for components A (solid line) and B (dashed line) of a gaseous system. In the figures a–e F_A is the Maxwell function with $E_A > 0.15\, E_{th}$: a–$F_A = F_B$; b–$E_B \neq E_A$; c–F_B is cut off in the low-energy region; d–F_B is cut off in the high-energy region; e–F_A is a monoenergetic δ-distribution; f– F_A and F_B are idential monoenergetic distributions.

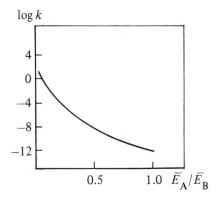

Fig. 3.2. log k vs $E_A/E_B \approx 1/T$.

The time-dependence of the distribution function is shown in Fig. 3.4. Methane is introduced as an additive to a thermal bath, the energy (temperature) of the methane being less than that of the bath. As time passes, the energy distribution function of the methane molecules changes both the position of its maximum and its shape in its approach through a series of intermediate stages to the Maxwell distribution function of the gas thermostat. This process is known as Maxwellization of the energy distribution function.

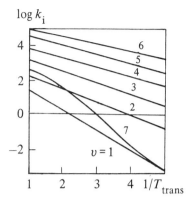

Fig. 3.3. log k as a function of $1/T_{trans}$ for dissociation of O_2 for δ-distributions of the vibrational energy of molecules (1–6) and at $T_{trans} = T_{vib}$ (7).

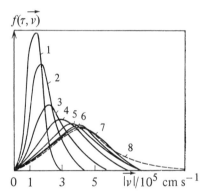

Fig. 3.4. Velocity distribution function for molecules of methane in a gas heat bath. 1—$t = 0$; 2—1×10^{-10} s; 3—2×10^{-10} s; 4—4×10^{-10} s; 5—5×10^{-10} s; 6—6×10^{-10} s; 7—1×10^{-9} s; and 8—Maxwell function at the heat bath temperature.

3.4 THE RATE COEFFICIENT OF A CHEMICAL REACTION AND ITS PHYSICAL MEANING

3.4.1 Definition of rate coefficient

Let us define statistical coefficients (constants) k for the rate of a chemical reaction as the average rates (with respect to unit density of the reacting components) by all of the available (dynamically and energetically) pathways of collisions. The averaging has to be carried out over the velocities and internal quantum states of the reactants. This means that the equation for k must include a distribution function of the reactants and products in an explicit form which allows k to be determined for nonequilibrium reactions [88, 113–115].

In the simplest case of the dissociation of a diatomic molecule introduced as a minor additive into an inert gas heat bath, when taking into account only the vibrational degrees of freedom, we find for the i-th vibrational level [114]:

$$k_i = \int_{E_{th}}^{\infty} \sigma_i(E)\sqrt{E}f_i(E)\,dE \tag{3.40}$$

where σ_i is the cross-section of reaction and $f_i(E)$ is the energy distribution function.

Let k_i be called the level rate coefficient of chemical reaction to distinguish it from k to $k(T)$ [88].

The total rate coefficient k of chemical reaction giving the summarizing characteristics of dissociation from *all* of the levels, both ground and excited, is as follows

$$k = \sum_i a_i k_i \tag{3.41}$$

where a_i is the relative population of the corresponding quantum level and $a_i k_i$ the contribution of the i-th level to the total rate coefficient of chemical reaction.

So long as the elementary stage (elementary event) of a chemical reaction is a process involving a given molecule in a given energy state, k_i is the rate coefficient for the elementary event. The factor k determined experimentally in chemical kinetics with the equation $k = k_0 \exp(-E_a/kT)$ is always a complex function, depending not only on $\sigma(E)$ and $f(E)$ but also on a_i, and it coincides with k_i only in very specific cases. This can be realized at a relatively low temperature and in slow (preferably thermoneutral) reactions. As to the most general cases, k is a function of the population of individual levels or of the relative concentration of molecules in definite energy states. These concentrations vary over very wide limits, which is why conclusions about a change in reaction mechanism are considered incorrect if they are drawn only from the appearance of breaks in the Arrhenius dependence [88].

The cross-sections involved in the equations for k and k_i can be determined in many cases from molecular beam experiments, while the distribution functions and the level populations can be found from data obtained with spectral and contact diagnostics methods. If the reactions concerned are slow, the cross-sections may be assumed to be Maxwellian or Boltzmann-type. Then, knowing the threshold energy for a reaction, it is possible to calculate k and thus to compare the quantity obtained with the value of k found experimentally.

The chemical reactions possible for a molecule and its reaction rates are determined by the structure of this molecule (as well as by the structure of the other participants of reaction in the case of non-unimolecular reactions), and by the interaction potentials. Molecular structure may be conceptualized in different ways. For our purposes, it is better to state it as follows: the structure of a molecule composed of several atoms is the system of its quantum levels and the spatial distribution of the constituent particles.

One of the most important problems in classical chemical kinetics, as in all chemistry, is the elucidation of the relation between the reactivity and structure of a molecule. However, within the limits of the molecular model used in kinetics, this problem cannot be solved. It is solved automatically in non-equilibrium chemical kinetics because the manner of description of reactions requires it. Of course, this will be the case if the vague term 'reactivity' is replaced by $\tilde{k}$ for a particular reaction. The development of instrumental techniques and computation in addition to the emergence of new branches of mathematical physics has made it possible to study the kinetics of nonequilibrium chemical reactions, taking into consideration the real complex energetic and spatial structures of reacting molecules and the dynamics of inelastic collisions.

Familiar in classical chemical kinetics is the equation of the type

$$\frac{d[c_i]}{dt} = \sum_{j=1}^{n} k_{ij}[c_i][c_j] - k_{ml}[c_m][c_l] \tag{3.42}$$

where k_{ij} and k_{ml} are the reaction rate constants and $[c_i]$, $[c_j]$, $[c_m]$, and $[c_l]$ are the concentrations of the corresponding components.

To simplify this equation we omit the terms describing nonbinary processes, which changes nothing in principle. Eq. (3.39) can be derived as a natural consequence of the Boltzmann kinetic equation. Under these conditions it acquires a more universal character and embraces the phenomena both of physical (population of molecular quantum levels, ionization, etc.) and chemical kinetics.

Let us write the Boltzmann equation in the form[†]

$$\frac{df_i}{dt} = \sum_{j,m,l} I_{ij}^{ml} \tag{3.43}$$

where f_i represents the velocity distribution functions for the particles and I_{ij}^{ml} the integrals of binary (elastic and inelastic) collisions. The indices ij, etc. indicate both the type and state of the particles.

Generalization of the collision integrals to large multiplicities will not affect the nature of the following operations. If the collision integrals are written as

$$I_{ij}^{ml} = \iint_{4\pi} \iint \int_{-\infty}^{+\infty} (f_m f_l - f_i f_j) V_{ij} \sigma_{ij}^{ml} \, d\Omega \, d\vec{P}_j, \tag{3.44}$$

$$\vec{V}_{ij} = \left| \frac{1}{m_i} \vec{P}_i - \frac{1}{m_j} \vec{P}_j \right|$$

where the function f are normalized by numerical densities $n(t)$, and f_i is substituted by $n_i F_i$ where

---

† The absence of the flux term does not restrict the general approach.

$$\int \int \int_{-}^{+} F_i \, d\vec{P}_i = 1,$$

then the integration of Eq. (3.44) by momenta $\vec{P}_i$ results in the disappearance of all of the integrals of elastic collision so that we obtain the following set of equations

$$\frac{dc_i}{dt} = \sum_{i,m,l} \left(W_{ij}^{ml} c_m c_l - W_{ij}^{ml} c_i c_j \right), \tag{3.45}$$

in which

$$W_{ij}^{ml} = \int\int_{4\pi} \int \int \int_{-\infty}^{+\infty} \int \int \int_{-\infty}^{+\infty} \vec{V}_{ij} \sigma_{ij}^{ml} F_i F_{-j} \, d\Omega d\vec{P}_i d\vec{P}_j \tag{3.46}$$

where σ_{ij}^{ml} is the differential cross-section for a collision of the i-th particle with the j-th one producing the m-th and l-th particles, $d\Omega \equiv \sin v \, dv \, d\varphi$ where φ and φ are the scattering angles and $d\vec{P}_i = dP_{i1} dP_{i2} dP_{i3}$, and $d\vec{P}_j = dP_{j1} dP_{j2} dP_{j3}$.

More detailed consideration of this type of equation can be found in [111, 116–119].

Thus, the quantities W_{ij}^{ml} are the distribution-averaged probabilities of the transition, in particular, the rate coefficients (constants) of the chemical reaction k_i determined for molecules in particular quantum states.

Suppose that the reactants A and B are in different quantum states and an irreversible gas-phase reaction of Type (3.27) is considered. To avoid a complex equation we simplify the expression for Reaction (3.27) as follows:

$$A(i) + B(j) \rightarrow \text{Products.} \tag{3.47}$$

Then we obtain

$$k_i = \int_{V_A} \int_{V_B} \sigma \left[\frac{ij}{ml}, \vec{V}_{AB} \right] |V_{AB}| F_{A(i)}(\vec{V}_{A(i)}) F_{B(j)}(\vec{V}_{B(j)}) \, d\vec{V}_{A(i)} d\vec{V}_{B(j)} \tag{3.48}$$

for the rate coefficient of chemical reaction from the i-th level. The total reaction rate coefficient k will be expressed as

$$\tilde{k} = \sum_{ij} \alpha_i \beta_j \int_{V_A} \int_{V_B} \sigma \left[\frac{ij}{ml}, \vec{V}_{AB} \right] |V_{AB}| F_{A(i)}(\vec{V}_{A(i)}) F_{B(j)}(\vec{V}_{B(j)}) \, d\vec{V}_A d\vec{V}_B \tag{3.49}$$

where α_i and β_j are the populations of the corresponding quantum levels normalized to unity.

However, if only one of the components possesses internal degrees of freedom, then

$$k = \sum_i \alpha_i \int_{V_A} \int_{V_B} \sigma(i, \vec{V}_{AB}) |V_{AB}| F_{A(i)}(\vec{V}_{A(i)}) F_{B(j)}(\vec{V}_{B(j)}) \, d\vec{V}_A d\vec{V}_B$$

$$(3.50)$$

Eq. (3.49) in it short form is represented by Eq. (3.41).

We have considered the molecules to have only vibrational levels although a more correct approach requires rotational levels also to be taken into account.

The theoretical rate coefficient for a chemical reaction k_i is a function of the following type:

$$k_i \sim k_i(V, f) \sim k(\sigma, f)$$

where V is the intermolecular interaction potential.

The nature of k_i is rather similar to that of transfer coefficients, and many good results can be obtained on this basis. The problem as to how close is the analogy between k_i and the transfer coefficients deserves special attention and will be the subject of separate consideration.

3.4.2 Expressions of rate coefficients for simple individual cases

Maxwell–Boltzmann distribution. It has been shown [120, 121] that k is the Laplace transform of $E\sigma(E)$ for this case; $f_{A(i)}(\vec{v}_A)$ and $f_{B(j)}(\vec{v}_{B(j)})$ have the Maxwell form and are independent of internal energy states. Thus

$$f_A(v_A, T) = \left(\frac{m_A}{2\pi kT}\right)^{3/2} \exp\left(-\frac{m_A v_A^2}{kT}\right) \qquad (3.51)$$

and

$$k(T) = \iint v\sigma(v) f_A(\vec{v}_A) f_B(\vec{v}_B) \, d\vec{v}_A d\vec{v}_B. \qquad (3.52)$$

Assuming that $T_A = T_B = T$, the transition from the velocities of the laboratory coordinate system to a relative velocity and the velocities of the centre-of-mass coordinate system is carried out in accordance with equations

$$\vec{v} = \vec{v}_A - \vec{v}_B \qquad \text{and} \qquad (m_A + m_B)\vec{V} = m_A \vec{v}_A + m_B \vec{v}_B. \qquad (3.53)$$

Here $\vec{v}$ is the relative velocity and $\vec{V}$ is the velocity of the centre of mass

$$d\vec{v}_A d\vec{v}_B = d\vec{V} d\vec{v}, \qquad (3.54)$$

$$\tfrac{1}{2} m_A v_A^2 + \tfrac{1}{2} m_B v_B^2 = \tfrac{1}{2}\mu v^2 + \tfrac{1}{2}(m_A + m_B)V^2,$$

where

$$\mu = m_A m_B / (m_A + m_B)$$

is the reduced mass.

From Eqs (3.51)–(3.54) and integrating $\vec{V}$, we obtain

$$k(T) = \left(\frac{\mu}{2\pi kT}\right)^{3/2} \int \sigma(v) \exp\left(-\frac{\mu v^2}{2kT}\right) dv. \tag{3.55}$$

Taking account the relative kinetic energy $E = \frac{1}{2}\mu v^2$, we have in polar coordinates

$$k(T) = \frac{1}{kT}\left(\frac{8}{\pi\mu kT}\right)^{1/2} \int_0^\infty E\sigma(E) \exp(-E/kT)\, dE \tag{3.56}$$

Let us now introduce the hard sphere interaction model accepted in Arrhenius kinetics. For this model

$$\sigma(E) = 0 \text{ for } E < E_{th}$$

and

$$\sigma(E) = \pi d^2 (1 - E_{th}/E) \text{ for } E \geq E_{th}. \tag{3.57}$$

Substituting $\sigma(E)$ for Eq. (3.57) in Eq. (3.56) we have

$$k(T) = \frac{1}{kT}\left(\frac{8}{\pi\mu kT}\right)^{1/2} \int_{E_{th}}^\infty \pi d^2 (1 - E_{th}/E) \exp(-E/kT) E\, dE \tag{3.58}$$

or

$$k(T) = \pi d^2 \left(\frac{8}{\pi\mu}\right)^{1/2} \exp(-E_{th}/kT). \tag{3.59}$$

This is the Arrhenius equation where the pre-exponential factor is expressed as [115]

$$\pi d^2 \left(\frac{8kT}{\pi\mu}\right)^{1/2} \sim (T/\mu)^{1/2}.$$

Chemical reactions at reactant temperatures of T_1 and T_2.
The equilibrium velocity distributions of reacting particles in this case are expressed as

$$f_1(v) = \left(\frac{m_1}{2\pi kT_1}\right)^{3/2} \exp\left(-\frac{m_1 v_1^2}{2kT_1}\right)$$

and

$$f_2(v) = \left(\frac{m_2}{2\pi kT_2}\right)^{3/2} \exp\left(-\frac{m_2 v_2^2}{2kT_2}\right) \tag{3.60}$$

For reaction from the same ground quantum level we obtain

$$k = \iint v\sigma(v)f(\vec{v}_1)f(\vec{v}_2)\,d\vec{v}_1 d\vec{v}_2 \tag{3.61}$$

where $v = |v_1 - v_2|$ is the relative velocity.

After some transformation and integration we find

$$k(T_1, T_2) = 4\pi \left(\frac{1}{2\pi k\left(\dfrac{T_1}{m_1} + \dfrac{T_2}{m_2}\right)} \right)^{3/2} \times \int_0^\infty v^3 \sigma(v) \exp\left(-\frac{v^2}{2\pi k\left(\dfrac{T_1}{m_1} + \dfrac{T_2}{m_2}\right)} \right) dv$$

$$\tag{3.62}$$

On introducing the reduced mass

$$\mu = \frac{m_1 m_2}{m_1 + m_2} \tag{3.63}$$

and the 'effective temperature'

$$T_{\text{eff}} = \frac{m_1 m_2 + m_2 T_1}{m_1 + m_2}, \tag{3.64}$$

an equation similar to that for the single-temperature system is obtained where the actual equilibrium temperature T is substituted by T_{eff}:

$$k(T_{\text{eff}}) = 4\pi \left(\frac{\mu}{2\pi k T_{\text{eff}}} \right)^{3/2} \int_0^\infty v^3 \sigma(v) \exp\left(-\frac{\mu v^2}{2k T_{\text{eff}}} \right) dv. \tag{3.65}$$

A disadvantage of Eqs (3.62) and (3.65) is that they do not take into consideration mixing of the reagents at the molecular level as required for the reaction. The mixing of gases at different temperatures obviously proceeds with T_1 and T_2 changing with time; that is, it is necessary to take into account, depending on the particular conditions, that during reaction, processes occurring at various T_1 and T_2 and different T_1/T_2, contribute to it. This takes place with sufficiently close values of m_1 and m_2 and T_1 and T_2. If this is not the case, the consideration of the temperature of only a high-energy component will be a rather good approximation; this approximation provides quite satisfactory results for $m_1 > m_2$ (for example in problems on electron impact-induced chemical reactions when energy exchange is inhibited).

Ionization of atoms by electron impact. In the case of a low-temperature plasma we have T_e and T_n as the temperatures of electrons and neutral atoms, respectively, T_e being greater than T_n and the ionization potential I being much greater than kT_e. Taking the cross-section for the first ionization potential in the threshold form as

$$\sigma(v) = 0 \quad \text{for} \quad \mu v^2/2 < I$$

and

$$\sigma(v) = a\left(\frac{\mu v^2}{2} - I\right) \quad \text{for} \quad \mu v^2 \geq I \tag{3.66}$$

where $\mu = m_e m_n/(m_e + m_n) \approx m_e$ and a is the cross-section parameter, in $10^3 \ \text{m}^2 \ \text{J}^{-1}$, we obtain from Eq. (3.62)

$$k(T_e) = 2a\left(\frac{2kT_e}{m_e}\right)^{1/2} \exp(-I/kT_e). \tag{3.67}$$

The value for k is usually small owing to the exponential term.

Ion–molecule reactions. Let us take the long-range potential for the radial motion of the ion and molecule in the form [122, 123]:

$$V(r) = -\frac{\alpha e^2}{2r^4} + \frac{Eb^2}{r^2} \tag{3.68}$$

where α is the polarizability of the neutral particle, e is the electronic charge, E is the initial relative kinetic energy, b is the collision parameter, and r is the distance between the ion and the neutral molecule. Then

$$\sigma(v) = 2\pi \int_0^{b_{max}} gb \, db = \pi g b_{max}^2 = \frac{2\pi g}{v}\left(\frac{\alpha e^2}{\mu}\right)^{1/2} \tag{3.69}$$

where

$$b_{max} = \left(\frac{4\alpha e^2}{\mu v^2}\right)^{1/4}$$

and the factor g is equal to unity for exothermic reactions and to zero for endothermic reactions.

According to Eq. (3.69), $v\sigma(v)$ is a constant, hence k is independent of temperature:

$$k = 4\pi\left(\frac{\mu}{2\pi kT}\right)^{3/2} \int_0^\infty v^3 \sigma(v) \exp(-\mu/2kT) \, dv = 2\pi\left(\frac{\alpha e^2}{\mu}\right)^{1/2} \tag{3.70}$$

Thus, ion–molecule reactions occur without an activation energy, and k does not depend on temperature. Usually, these reactions are characterized by large cross-sections ($\geq 10^{-14} \ \text{cm}^2$).

Vibrational-transitional energy exchange. The simplest version of the theory of vibrational-translational energy interchange considers a harmonic oscillator which undergoes a time-dependent perturbation caused by collision with an atom or molecule. In the first-approximation perturbation theory, the probability of exciting an oscillator from its ground state to its first vibrational level is given by

$$P_{0 \to 1}(v) = \frac{1}{h^2} \left| \int_{-\infty}^{+\infty} < \varphi_1^* |V(x,t)| \varphi_0 > \exp(-i\omega_{01}t) \, dt \right|^2 \tag{3.71}$$

where $\omega_{01} = (E_1 - E_0)/h$; $\varphi_1(x)$ is the wave function for the i-th level of the harmonic oscillator, and $V(x, t)$ is the velocity- and time-dependent interaction potential.

Considering only the linear terms on the harmonic oscillator coordinates and being restricted only to the one-dimensional classical trajectory defined by the repulsive exponential potential [124], we obtain

$$\sigma_{0 \to 1}(v) \approx v^2 \exp(-4\omega L/v) \tag{3.72}$$

where L is the parameter of the potential and v the relative velocity of the colliding particles.

For the three-dimensional Maxwell distribution, $k_{0 \to 1}$ appears as follows

$$k_{0 \to 1}(T) \approx (4\pi LkT/\mu)^{3/2} (\pi kT/6\mu)^{1/2} \exp\left[-\frac{3}{2}\left(\frac{16\mu\omega_{01}^2 L^2}{kT} \right)^{1/3} \right] \tag{3.73}$$

or

$$k_{0 \to 1}(T) \sim AT^2 \exp(-B/T^{1/3}). \tag{3.74}$$

The influence of temperature on $k_{0 \to 1}$ is determined mainly by the exponential term; the exponential dependence on $T^{-1/3}$ over a wide temperature range is well-established experimentally.

The effect of different types of nonequilibrium distribution on $k_{0 \to 1}$ is described elsewhere [125, 126]. As has been frequently emphasized, the nonequilibrium occurrence of chemical reactions can create energy-rich and, accordingly, unusually reactive (excited or 'hot') molecules; it also enables control of the kinetics of reactions to an appreciable extent (in particular, in chemical and gas lasers).

Experimental determination of a reaction rate coefficient for dissociation from the i-th level. The rate coefficient $k_{i,diss}$ can be determined either by means of theoretical calculations or on the basis of experimental data.

Let us consider the second possibility in detail. There are several ways of doing this, depending on the completeness of the experimental data. If experimental operations have provided

(i) the time-dependence for the population of all levels, then it is possible to find all values of $k_{i,diss}$ by using, for example, the 'ravine' method [127];

(ii) information on the kinetics of population of some (not all) of the levels, then under these conditions one can determine only a part of the $k_{i,diss}$ value, corresponding to the transitions from the levels specified;

(iii) only one (general) k_{exp} kinetic curve by the disappearance of an active component, then it is necessary to define somehow a type of dependence of $k_{i,diss}$ on the temperature of a thermal bath, $k_{i,diss} = k_{i,diss}(T_i, \theta)$, where θ is a parameter. This makes it possible to find the parameter θ for $k_{diss} = k_{diss}(T_i, \theta)$, using a value for the overall rate constant known from the experiment.

The form of this function is determined from the dependence of the dissociation cross-section for the level i on energy and from the temperature of the thermal bath, and it can be found by using Eq. (3.56).

To obtain values for the cross-section (except for direct experiments like molecular beam studies of reactions) the following ways are suggested.

(i) If there is a theoretically-based form of the function $\sigma = \sigma(E, \Phi)$, where Φ is an adjustable variable, the latter variable can be selected on the basis of an experimentally-derived $\sigma - E$ relationship.

(ii) If the potential of intermolecular interaction is known, then a cross-section–energy relationship can be obtained by solving the equations of motion for a system of particles (a reactive component plus a thermal bath) with regard to different initial conditions (relative velocities of motion and collision angle) with subsequent averaging over the set of trajectories.

(iii) If the potential is unknown, one may try to find a reasonable form of $U(r, \theta)$ by solving the inverse problem provided that the cross-section has been determined experimentally. Then θ should be prescribed and the dynamic equation solved to get σ. If it does not fit experimental data, θ needs to be changed and the iterations continued. The iteration procedure can be realized by different techniques (see, for example, [94]).

Thus, a combination of computational mathematics and experimental data makes it possible to find numerical values for k_{diss}, at least in simple cases, without a huge body of complicated, extremely difficult, and in most cases, not unambiguous experiments.

3.5 EQUATIONS OF NONEQUILIBRIUM CHEMICAL KINETICS

3.5.1 Models of chemical reaction

Models for chemical reactions occurring in space can be distinguished by the following two features: (i) as providing a global description (that is, a spatially uniform or 'well-mixed' case without diffusion) and a local description (which includes diffusion or a spatially nonuniform case), and (ii) providing a deterministic description (macroscopic, phenomenological, or in terms of concentrations) and a stochastic description (at the level of the number of particles with regard to internal fluctuations).

Let us compare the resulting four possible types of description of chemical reaction, restricting ourselves, for the sake of clarity, to the case of a reactant $x = x(t)$ in one-dimensional space participating in the following reaction scheme [128]

$$\sum_{j\text{-}1}^{r} a_{ij}A_j + \beta_i x \underset{k_i'}{\overset{k_i}{\rightleftharpoons}} \sum_{j=1}^{r} a_{ij}'A_j + (\beta_i + 1)x \tag{3.75}$$

where a_{ij}, a_{ij}', and β_i are the elements of $N = \{0, 1, 2, 3, \ldots\}$, k_i and $k_i' \geq 0$, and $i = 1, 2, 3, \ldots, s$.

(i) The global deterministic model

$$\frac{dc(t)}{dt} = f[c(t)], \qquad c_0 = c(0) = 0 \tag{3.76}$$

where $c = c(t)$ is the concentration of reactant x.

(ii) The local deterministic model

$$\frac{\partial c(r,t)}{\partial t} = f[c(r,t)] + D\Delta c(r,t) \tag{3.77}$$

where $c = c(r, t)$ and $D > 0$ is the diffusion coefficient.

(iii) The global stochastic model

$$\dot{P}_k(t) = \lambda_{k-1}P_{k-1}(t) + \mu_{k+1}P_{k+1}(t) - (\lambda_k + \mu_k)P_k(t), \tag{3.78}$$

$$\sum_k P_k^{(0)} = 1,$$

where λ_k and μ_k are the transition (formation and decay) rates and $P_k(t)$ the probabilities.

This is the case of a Markov spasmodic process described by the so-called master equation.

(iv) The local stochastic model

$$\dot{P}_k(t) = \sum_{j=1}^{n}\left[\lambda_{k_j-1}(t)P_{k-e_j}(t) + \mu_{k_j+1}P_{k+e_j}(t) - (\lambda_{k_j} + \mu_{k_j})P_k(t)\right]$$

$$+ \frac{D}{2}\sum_{j=2}^{N}\left[(k_j+1)P_{k+e_j-e_{j-1}}(t) - k_jP_k(t)\right]$$

$$+ \frac{D}{2}\sum_{j=1}^{N-1}\left[(k_j+1)P_{k+e_j-e_{j+1}}(t) - k_jP_k(t)\right] \tag{3.79}$$

where λ_{k_j} and μ_{k_j} are the boundary rates of formation and decay, respectively, and D is the diffusion coefficient; the boundary conditions are that the flows at the boundaries are zero, that is, the Markov spasmodic process [128].

Some additional comments are needed here. It is well-known that the evolution of a chemically reacting system can be described by a deterministic differential equation

$$\dot{c}_i = f(c_i, \lambda) \tag{3.80}$$

where λ is the totality of the parameters of the process.

The macroscopic variable characterizes the average state of a large system and is obtained via averaging over many independent subsystems. The form of $f(c_i, \lambda)$ is defined by the mechanism of reaction under consideration. A stationary state is related to the condition $\dot{c}_i = 0$. If $f(c_i, \lambda)$ is nonlinear then the system may have many stationary states. The deterministic description can be made more realistic if fluctuations are taken into account. Concentrations fluctuate and are represented by the random variable $c(t)$. From this point of view, the equation for $\dot{c}(t)$ describes the average behaviour of a large number of statistical variables. A more correct description of the reacting system can be obtained by taking into consideration the reaction volume V and the whole-numbered set of random variables $x(t)$ which represents the number of molecules in volume V at time t. The mean of $x = x(t)/V$ corresponds to the deterministic concentration.

Fluctuations play an important role in the evolution of macroscopic systems obeying nonlinear kinetics far from equilibrium. This is observed, in particular, for systems close to the point where the property of absolute stability is lost and the evolution acquires a statistical nature. If macroscopic fluctuations of a decisive value arise in a system on a scale necessary for this system, then the chaotic stage comes to an end and a new stable state emerges.

It is the existence of fluctuations that reveals the approximate nature of classical thermodynamics which considers matter as a continuous medium while it actually consists of discrete particles.

Classical statistical mechanics allows us to calculate the mean square deviation of a thermodynamic value in the equilibrium state. However, time-dependent fluctuations (also called noise or Brownian motion in a general sense) have to be considered for our purpose, that is, it is necessary to find the way in which the deviations (fluctuations) occurring at different times correlate with each other [129].

So far as the stochastic description of the kinetics of chemical reactions is concerned, although this does not seem to be as customary for kineticists as the deterministic description, there are many reactions for which the latter description is inadequate and so stochastic models should be applied. The list of examples of the most important applications of stochastic techniques to chemical kinetics and to more complicated systems which actually require such a description is quite extensive, but some examples are the distribution of chain lengths and the distribution of copolymer composition in polymer kinetics, the kinetics of reactant isolation, polymerization kinetics or biological macromolecules on templates, the denaturation of polypeptides and proteins, and certain visual and blood-clotting mechanisms in which a few activated molecules initiate an avalanche-type reaction. Some nonbiological and nonpolymeric examples are diffusion-controlled chemical reactions, models of sterilization, chromatography, relaxation of vibrational nonequilibrium distributions in shock waves, the theory of absorption of gases onto solid surfaces, the theory of homogeneous and heterogeneous nucleation in vapours, the statistical process of aggregation, isotope exchange, etc. [130–138].

The possibility of the stochastic description of chemical reaction kinetics and relevant mathematical procedures are considered in detail in [139].

3.5.2 Pauli equation
The kinetics of both equilibrium and nonequilibrium chemical reactions may be described either through concentrations or by means of distribution functions. Correspondingly, either the Pauli equation or the Boltzmann equation is to be used.

Note that the Pauli equation for a particular case when physical processes (excitation, transitions between quantum levels, etc.) have been completed long before any observable chemical change, transforms to the normal Arrhenius kinetics [88, 140].

The equation derived by Pauli in 1928 is a type of 'master equation' [188, 129, 136, 141–145].

The master equation takes the form

$$\dot{P} = \tilde{W}P \tag{3.81}$$

where $\tilde{W}$ is the linear operator operating on functions of some independent variable q.

Equations of this sort are used widely in physics, for example equations describing diffusion and heat conduction, Schrödinger equation, Liouville equation, etc.

When considering cases, not uncommon in nature, of $x(t)$ being a Markov process, the equation for the evolution will be a master equation of the general form

$$\frac{\partial P(x,t)}{\partial t} = \int \left[W(|x')P(x't) - W(x'|t)P(x,t) \right] dx' \tag{3.82}$$

where

$$P(x,t) \equiv P\left(x,t \middle|_{x_0,t_0}\right)$$

is the probability of a transition between t_0 and t_1 and $W(x|x')\Delta t$ the same probability of the transition taken for a period Δt sufficiently small for P to change not too dramatically, but large enough for the assumption on the Markov character of the process to be justified. Then

$$P(x, t + \Delta t | x', t) = W(x|x')\Delta t + \delta(x - x')\left[1 - \Delta t \int W(x'|x)\, dx' + 0(\Delta t)\right]$$

(3.83)

The master equation describes the evolution as a whole including the fluctuations, the explicit term $W(x|x')$ reflecting the properties of a particular system. In many cases, for example in chemical reactions, the variable x takes only integral values, while in others, like Brownian motion, it varies continuously.

So far as the Pauli equation is concerned, it can be obtained in two ways: (i) on the basis of the general concepts of probability theory, and (ii) from the Liouville equation.

The more general approach implicit in the first of these gives the Pauli equation (or, more generally, a 'master' equation) a meaning going beyond the limits of classical and statistical mechanics.

The second approach relates the Pauli equation to the fundamental postulates of classical mechanics and makes it possible to apply the Hamiltonian and Lagrangian formalism when using the equation.

Although the description of the evolution of a reacting system requires, in general, the use of stochastic master equations allowing the fluctuation characteristics of the system to be incorporated, simple kinetic problems disregarding the fluctuations can be solved rather satisfactorily with the Pauli equation [146].

For a large-structure matrix of probability density $\rho(n_i, t)$ of weakly interacting particles to be found in a set Δn of closely-located states (such a set is the analogue of the phase space cell in classical mechanics), the Pauli equation will be written as [143]:

$$\frac{\partial \rho(n, t)}{\partial t} = \sum_{n'} v_n P_{nn'} \rho(n', t) - \sum_{n'} v_{n'} P_{n'n} \rho(n, t)$$

(3.84)

where v_n is the number of states in Δn, $P_{nn'}$ and $P_{n'n}$ the probability of transition from state n' to n, and vice versa, per unit time, respectively.

For the case of a unimolecular reaction occurring in a thermal bath of an inert gas (the gas concentration being taken as constant), the Pauli equation is written as

$$\frac{\partial n_i}{\partial t} = \sum_j \omega P_{ij} n_j(t) - \sum_j \omega P_{ji} n_i(t) - k_i n_i(t) + R_i(t)$$

(3.85)

where n_i is the concentration of reactant molecules in their i-th energy state at time t, P_{ij} is the probability (per collision) of the transition from the j-th to the i-th energy state in the collision of a reactant molecule with a molecule of the gas of the bath, P_{ij} is the probability for the transition from the i-th to the j-th state, k_i is the rate coefficient for a chemical reaction in the i-th energy state, R_i

is the rate of excitation ('pumping') of the i-th level, and ω is the collision frequency.

In addition to the purely mathematical advantages which the Pauli equation gives from the standpoint of computational techniques, the following points should be carefully noted. Strictly speaking, this is a balance equation, that is, if correctly written, it is always valid, as any balance relationship. It enables us to unify transitions between levels and chemical transformations proper [88]. The physical interpretation of terms on the right-hand part of Eq. (3.85) is obvious, that is, the first term expresses the increase in probability density caused by transitions from the cells $\Delta n' \neq \Delta n$ to the cell Δn, and the second term to the decrease in the density function due to transitions from Δn to $\Delta n'$. The time variable of the function $\rho(n, t + \Delta t)$ at time $t + \Delta t$ (Δt is much larger than the duration of one transition) is defined by the distribution of probabilities $\rho(n, t)$ at time t and does not depend on the values of $\rho(n, t')$ for $t' < t$. This form of evolution of a system is termed Markovian. Unlike the Liouville equation Eq. (3.85) describes the irreversible course of a system approaching monotonically its equilibrium state [113, 142, 147, 148].

Disregarding any chemical reaction ('run-off') and 'pumping' in the kinetics of population of the vibrational energy levels we obtain

$$\dot{n}_k = Z\left(\sum_j n_j P_{kj} - \sum_j n_k P_{jk} \right) \tag{3.86}$$

where n_k and n_j are the populations of the k-th and j-th vibrational levels, respectively, P_{kj} is the probability per collision of a transition from the j-th to k-th level, Z is the collision frequency, which is assumed to be independent of the vibrational state of the molecule (generally speaking, this assumption is incorrect since σ and, consequently Z do vary with the vibrational state but these variations are small and may be neglected).

The system of Eq. (3.86) is written with two assumptions: (i) the collision may be considered in the classical way (for calculations of Z), and (ii) the behaviour of the vibrational degrees of freedom may be considered independent of the state of the translational, rotational, and electronic degrees of freedom.

In the ranges of temperature and density of molecules relevant to chemical kinetics, the correction related to the finite value of the de Broglie wavelength is very small (say, even at $T = 300$ K, for air $\lambda = \hbar/(\mu v) = 5 \times 10^{-10}$ cm whereas the radius of intermolecular interaction is $ca.$ 5×10^{-9} cm). The second assumption is justified only when the relaxation times of the various degrees of freedom differ from each other strongly enough: $\tau_{trans} \approx \tau_{rot} \ll \tau_{vib} \ll \tau_{diss} < \tau_{ion}$, that is, the degrees of freedom of dissociation and recombination behave as if 'frozen'.

Implied here is the assumption that the Maxwell distribution is unperturbed by chemical reaction which is, strictly speaking, never fulfilled even in a system initially at equilibrium. Nevertheless, in many systems for which $E_a > kT$ (large activation energies and low temperatures) perturbations of the Maxwell–Boltzmann distributions are small, at least, because they concern a relatively small group of particles. By contrast, the distribution is violated when $E_a \approx kT$ [125, 149, 150].

It is very important also to note that Eq. (3.85) is, in a certain way, a direct generalization of the equations of chemical kinetics (and of the kinetics of level populations, etc.). On disregarding the transitions between levels, it reduces to normal kinetic equations.

It is obvious that many of the methods developed in classical chemical kinetics (in particular, the steady-state concentration method) will find wide use in generalized nonequilibrium chemical kinetics. It is to be borne in mind that the rate of a chemical reaction may be considered from the classical point of view over certain finite increments of change in temperature (energy) and in the concentrations of reactants and products (taking into account experimental error).

We wish to emphasize that the Pauli equation is valid for an ensemble of many weakly interacting particles (for example rarefied gases, weakly ionized plasma, etc.). So, in the strict sense, the equation is inapplicable to systems featuring strong interactions, for example chemical reactions at solid surfaces, strongly non-ideal gases and plasmas, highly ionized plasmas, etc. These types of situation require us to investigate the limits of applicability of the Pauli equation for each individual case [88].

All these types of argument include the assumption of the statistical independence of molecules before collision. If a is the radius of interaction, λ the mean free path, and N the density of molecules, then the condition of statistical independence may be written as

$$a/\lambda \approx N/a^3 < 1 \tag{3.87}$$

The usual substantiation of the Pauli equation, first made by Pauli himself [143], implies that the approach to equilibrium is caused by a perturbation term H_1 in the Hamiltonian $H = H_0 + H_1$ of a system, H_1 being so small that the transition probabilities P_{ij} can be calculated to a first approximation by means of nonstationary perturbation theory. Under these conditions the derivation of the Pauli equation is based on the statistically random distribution of the wave function phases belonging to differing eigenvalues of H_1, that is, the density matrix is considered to be diagonal in the unperturbed Hamiltonian representation. The random phase hypothesis applies not only to the initial state but is used many times after every time interval for which the unperturbed energy H is conserved upon a transition. A similar (and deeply unsatisfactory) situation arises with the random walk approximation in the derivation of the Boltzmann kinetic equation. This problem is related to the requirement to achieve time irreversibility while the original equations of dynamics are reversible [143].

Note that the Pauli equation includes a description at a level intermediate between the microscopic and macroscopic levels. It is not invariant with respect to time inversion, and its solution tends to a certain fixed equilibrium distribution. This equation is the expression of the probability of the distribution over different states. The evolution of a system is described by the equation as a stochastic process. Consequently, the process is considered to be Markovian, that is, the Pauli equation determines the probabilities at any moment $t > 0$ if they are known at $t = 0$ [151–156].

3.5.3 Boltzmann equation

The description of chemical reactions using concentrations as average macroscopic quantities observable in a kinetic experiment implies the loss of information on many essential features of a chemical process.

It is to be hoped that distribution functions, which are the fundamental characteristics of the evolution of a statistical ensemble will become measurable in the coming decades. A crucial parameter is the distribution function in terms of the coordinates $\vec{r}$ and momenta $\vec{p}$ for a separate molecule, that is, the probability of its being in a state with values of the coordinate and momentum between $\vec{r}$ and $\vec{r} + d\vec{r}$ and $\vec{p}$ and $\vec{p} + d\vec{p}$ is equal to $f(\vec{r}, \vec{p}) \, d\vec{r} \, d\vec{p}$ for the molecule taken out of the whole ensemble. This distribution is defined in a six-dimensional space which is often called the μ-space.

An equation describing the development in time of such a distribution function is known as the Boltzmann equation

$$\frac{\partial f}{\partial t} + \vec{v}\frac{\partial f}{\partial r} + \vec{F}\frac{\partial f}{\partial p} = \left(\frac{\partial f}{\partial t}\right)_{col} \qquad (3.88)$$

where $\vec{v}$ is the velocity and $\vec{F}$ the external force acting on the particle.

Equation (3.38) may also be written in the form:

$$\frac{\partial f}{\partial t} = \left(\frac{\partial f}{\partial t}\right)_{col} + \left(\frac{\partial f}{\partial t}\right)_{drift}, \qquad (3.89)$$

$$\left(\frac{\partial f}{\partial t}\right)_{drift} \equiv -\vec{v}\frac{\partial f}{\partial r} - \vec{F}\frac{\partial f}{\partial p} \qquad (3.90)$$

where $(\partial f/\partial t)_{drift}$ is a term describing the change in distribution function caused by the collisionless motion (drift) of particles.

For a time dt, such a movement leads to the following changes in $\vec{r}$ and $\vec{p}$:

$$\vec{p} \to \vec{r}' = \vec{r} + \vec{v}dt \quad \text{and} \quad \vec{p} \to \vec{p} + \dot{\vec{p}}dt = \vec{p} + \vec{F}dt$$

which define a flux in the μ-space.

A change in f must satisfy the continuity equation

$$f(\vec{r}, \vec{p}, t + dt) = f(\vec{r}, \vec{p}, t) \qquad (3.91)$$

or

$$f(\vec{r} + \vec{v}dt, \vec{p} + \vec{F}dt, t + dt) = f(\vec{r}, \vec{p}, t) \qquad (3.92)$$

The collisional term in Eq. (3.88) incorporates the effect of collisions of particles, with each other or with scattering centres, on the distribution function. In elementary theory, this term is defined intuitively by admitting that the number of collisions in a period t is equal to the probability of finding the particle in unit space multiplied by the number of scattering centres. The random walk approximation is used here, which means that the dynamic connections between consecutive collisions are quickly lost owing to the large and random distribution of the scattering centres, as well as to the assumption that the collisions are binary.

The ordinary Boltzmann equation describes the evolution of the distribution function in the phase space of a single particle. The equation includes two terms: the flux term describing the motion of molecules by trajectories in the phase space and represented by the differential operator, and the collisional term describing the change in velocity caused by collisions and represented by the integral operator. The Boltzmann equation is thus an integrodifferential equation, the collisional term being nonlinear. This nonlinearity is the major obstacle to elaboration of the methods of its solution, the more so as the collision integral is closely related to the law of intermolecular interaction, with rather incomplete and often contradictory information being available on this law.

The solution of the Boltzmann equation as a nonlinear integrodifferential equation has so far been obtained for only a few particular cases. Nonetheless, if the deviation of φ from equilibrium is small then the Boltzmann equation can be linearized after getting

$$f(\vec{r}, \vec{p}, t) = f_0(1 + \varphi), \qquad \varphi \ll 1. \qquad (3.93)$$

Sometimes the collision term is substituted by a simple approximate expression

$$(\partial f / \partial t)_{col} = (f - f_0)/\tau = -\varphi/t \qquad (3.94)$$

where τ is a constant called the relaxation time.

Generalization of the Boltzmann equation to a mixture of several simple gases does not, in principle, seem to be difficult. However, generalization of ternary (rather than binary) collisions has not yet been performed in a form amenable to calculation. This means that reverse reactions cannot be included into the scheme of the Boltzmann equation, while in most cases it is impossible to describe a process without taking into account these reactions (for example three-body recombination, etc.) [157]

Any chemical reaction represents one of the pathways of redistribution of the mass and energy of particles in the course of their interaction. Like other pathways, it has a statistical nature and can be described by its differential cross-section which enables us to use, for the description of its results, a system of Boltzmann-type equations:

$$\left(\frac{\partial}{\partial t} + \vec{v}\frac{\partial}{\partial \vec{r}} + \frac{a}{m}\frac{\partial}{\partial \vec{v}}\right)f = \int\int|\vec{v} - \vec{v}'|\sigma(\vec{v})(f\tilde{f}' - f\tilde{f})\,\mathrm{d}\Omega\mathrm{d}v' \qquad (3.95)$$

The notations here are conventional. The function f to be found signifies the probability density of finding a given number of particles in the unit volume $\mathrm{d}r\,\mathrm{d}v$ of the single-particle phase space. Thus, knowing the function, it is easy to calculate such properties of the medium as concentration, flux density of particles, etc.

The set of Eqs (3.95) allows reactions of the 'two-to-two' type:

$$A + B \rightarrow C + D \qquad (3.96)$$

to be described. To deal with reactions

$$A + B \underset{M}{\rightleftharpoons} C + D \qquad (3.97)$$

we need an equation which contains the integral of ternary collisions. Equations of Type (3.95) are classical or at least quasi-classical (if quantum differential cross-sections are used) [158–171].

For various mathematical reasons, the solution of Eq. (3.95) (not to mention more sophisticated equations) encounters great difficulties. Most frequently the solution is carried out by means of the Chapman–Enskog[†] [104, 156] or Grad [162] methods developed explicitly for Eq. (3.95). The former method is applicable in some 'small' vicinity of the equilibrium distribution (as we have seen, a chemical reaction can disturb it strongly); the latter is perhaps more general, but neither makes it possible to achieve good approximation because of the enormous body of calculations.

Moreover, the Chapman–Enskog method involves no criterion allowing us to determine how far it lets us depart from equilibrium. Therefore, in solving a given problem, an experimental trial is needed to prove the fit of calculated and experimental quantities, which is a clear disadvantage. The procedures of linearization also sometimes turn out to be inapplicable since the interaction of identical particles is a nonlinear phenomenon.

---

† The idea of the Chapman–Enskog method is as follows. The distribution function is split into two parts: the first, Maxwellian part $f_M(r, v, t)$, gives values for the local concentration, velocity, and energy density in a gas, while the second defines the fluxes of heat and momentum. These two parts of the distribution function are related to each other by the linearized collision operator in such a manner that determination of the heat conductivity and friction is reduced to a solution of a linear nonuniform second-order integral equation.

Naturally, all of these considerations lead to the idea of using powerful computers which could facilitate the derivation of one or other type of relationship. But the equations turn out to be so complex that the creation of a satisfactory scheme for calculation does not seem to be a simple procedure although numerous but somewhat eccentric efforts have been made in this direction [96, 112, 139].

The collision integral in the Boltzmann equation has a complex nonlinear structure. Two approaches are therefore used in the solution of this equation: (i) linearization of the equation or (ii) use of a model Boltzmann equation.

To linearize the Boltzmann equation, we make use of Eq. (3.93). Substituting it into Eq. (3.95), assuming $\vec{F} = 0$, and neglecting the squared and higher exponents of φ, we obtain the equation:

$$\frac{d\varphi}{dt} = \frac{\partial \varphi}{\partial t} + \vec{v}\frac{\partial \varphi}{\partial r} = \int f_{M_1}(\varphi' + \varphi_1' - \varphi - \varphi_1)gb\, d\vec{v}\, d\vec{r} \qquad (3.98)$$

where g is the degeneracy and b the impact parameter, which can be fruitfully applied in the cases of small perturbations of an equilibrium state. We shall not consider the model equation here but point out that it leads to the relaxation equation (3.94), from which it follows that the rate of approach to equilibrium for states close to equilibrium is proportional to their departure from equilibrium.

To conclude, we make one important observation concerning the development and generalization of the Boltzmann equation. It is easily seen that the characteristic linear dimension for the Boltzmann equation is the mean free path λ, and the characteristic time is the mean period τ between collisions of molecules. This distinguishes the Boltzmann equation from almost any other equation of mathematical physics describing irreversible behaviour of the medium on distances which must be longer than λ and within time intervals which must be large with respect to τ. This situation is also manifested by the fact that traditional thermodynamics of irreversible processes deals with small (linear) departures from equilibrium, whereas the Boltzmann equation copes with large (nonlinear) deviations. Thus, it is necessary to discriminate between the nonlinearity of equations of hydrodynamics and the nonlinearity of the mechanism of irreversibility (for example the proportionality of a heat flux to a temperature gradient) [140, 163–166].

Let us pose some crucial questions arising in connection with the Boltzmann equation, (3.95).

(1) Is it possible to abandon the restriction to a monoatomic gas, that is, to extend the Boltzmann equation to particles possessing quantal internal degrees of freedom?

(2) It is possible to revise the assumption on binary collisions?

(3) Can the Boltzmann equation describe fluctuations of gas properties? What is the physical meaning of hydrodynamic equations of higher order?

(4) Can the Boltzmann equation give an approximation to equilibrium in the region of the condensed state for a Boson gas of solid balls?

(5) Is it possible to generalize the Boltzmann equation to the case of relativistic particles with a prescribed rest mass?

3.5.4 Langevin equation

The requirement to consider fluctuations (noise) in a system caused by an external[†] source, such as an applied hydrodynamic and/or electrodynamic field, has led to the formulation of the Langevin equation. In its general form the equation is

$$\dot{v} = F(v, \lambda) + g(v)L(t) \tag{3.99}$$

where F and g are defined functions, and $L(t)$ is a random function, a stochastic or fluctuating force whose stochastic properties are postulated; the function $g(v)$ defines the magnitude of the fluctuations (it can be found on the basis of physical approaches to the nature of the source of noise).

Initially no probability distribution is introduced for $L(t)$, but only an assumption on correlation functions. The postulate on the stochastic properties of $L(t)$ is an essential element in the Langevin approximation (usually $L(t)$ is postulated to be Gaussian white noise). Suppose that a system is autonomous, that is, F and g do not include t explicitly. Then for the case of a single variable x (if g is independent of x) Eq. (3.99) takes the form:

$$\dot{x} = F(x) + gL(t) \tag{3.100}$$

Eq. (3.99) defines unambiguously the stochastic process $x(t)$, $t \geq 0$. This is a Markov process, and the probability of transition $P(x, t | x_0, t_0)$ from x_0 at t_0 to the interval $x + dx$ at t) obeys the Fokker–Planck equation

$$\frac{\partial P}{\partial t} = -\frac{\partial}{\partial x} F(x)P + \frac{g^2}{2} \frac{\partial^2 P}{\partial x^2}. \tag{3.101}$$

If g is x-dependent, then Eq. (3.100) loses its meaning. Indeed, according to this equation every fluctuation in $L(t)$ results in a fluctuation in x and, consequently, a surge in x. Because of the above, the value of x involved in $g(x)$ is indefinite (and, hence, so is the amplitude of the surge) [167–170].

To investigate in detail the character of the Langevin equation (its appearance and the properties of its solutions) we consider Brownian motion, since the equation was first suggested as describing this motion. One must note that this

---

† The term 'external' refers to the two conditions applied on the interacting system + surroundings: (i) the system itself does not affect the source of noise, and (ii) there exists a parameter enabling, in principle, the noise to be eliminated.

equation has the same form for other processes (in particular, for chemical reactions), so any conclusions drawn from considering Brownian motion may be extended to any stochastic process of similar type [171].

The Langevin equation for the description of the motion of a Brownian particle as a random-walk process looks like

$$\dot{v} = -\gamma v + A(t) \tag{3.102}$$

where γ is the frictional coefficient for unit mass.

The term $A(t)$ is the fluctuating force independent of v, and, in addition, satisfies the condition that its time-averaged value is zero:

$$\langle A(t) \rangle = 0 \tag{3.103}$$

In writing the Langevin equation in the form of Eq. (3.102), we make an implicit but very important assumption, namely, we have assumed that the phenomenon described can be divided into two parts, the discontinuity of events being essential in one of them but not so in the other. In spite of the fact that this assumption is rather intuitive, it is justified *a posteriori* by its successful applications. This important circumstance is clearly evident, for example, in consideration of the problems of stellar dynamics.

3.6 INTERDEPENDENCE OF CHEMICAL REACTIONS AND THE ENERGY DISTRIBUTION FUNCTIONS OF PARTICLES

One problem which excites attention is as follows: how do chemical reactions disturb the equilibrium distribution function, and conversely, how do perturbations of the distribution function affect the rate of chemical reaction? This problem can be resolved by using Monte Carlo methods.

Consider the velocity distribution of the molecules in the reaction $CH_4 \rightarrow \dot{C}H_3 + H\cdot$ where the reaction cross-section is assumed to be zero for $E_{kin} < D$ and constant $E_{kin} \geq D$ (D is the dissociation energy) [172]. The lifetime in a state with $E_{kin} \geq D$ is distributed exponentially as shown by the Monte Carlo method [166]. If the lifetime of a molecule [166] as 'gambled' by a computer turned out to be less than its mean free path, it was assumed that chemical reaction had taken place. Otherwise, an elastic collision of this molecule with an inert atom of the methane thermal bath was deemed to have occurred. Taking $D = 361$ kJ mol^{-1} and an argon concentration of 10^{18} molecules cm^{-3}, calculations were carried out for $T = 5000, 7500, 10\,000, 12\,500,$ and $15\,000$ K (where T is the heat bath temperature and the initial temperature of methane). For each of the T's, the value for the mean lifetime of the molecule in a state with $E_{min} \geq D$ was also varied, which naturally led to a change in reaction order. The number of molecules was taken as 5×10^4, so the standard error did not exceed ±1%. Output data were taken from the computer memory at regular intervals.

The results thus obtained characterized both the kinetics of the transition of the system from its initial to the equilibrium state, and the equilibrium state itself.

Let us consider the thermal decomposition of a low methane concentration $CH_4 \rightarrow \dot{C}H_3 + H$ in an inert gas heat bath. The velocity distribution functions in terms of methane molecules are shown in Fig. 3.4, the average time of the mean free path of a molecule t_{av} being approximately equal to 0.5×10^{-9} s. For $\tau \ll t_{av}$, the high-energy wing of the distribution function almost disappears because the rate of chemical reaction significantly exceeds the rate of activation of molecules in collisions with the atoms of the heat bath. Thus the thermal decomposition is a second-order reaction.

For $\tau \gg t_{av}$ the velocity distribution functions of the molecules differ slightly from their initial functions. The decrease in the number of molecules with $E_{kin} \gg D$ ($D = 361$ kJ mol^{-1}) due to reaction is recouped almost instantaneously via the collisional activation of further molecules.

For $\tau \approx t_{av}$ both of the processes (the activation and dissociation of molecules) are in a state of dynamic equilibrium, and the distribution function has a discontinuity at the point where $E_{kin} = D$.

The universality of the results obtained does not depend on the particular concentration of the heat bath atoms, the dissociation energies, and the mean lifetime of the molecules as used in the calculations. Indeed, a change to other concentrations is equivalent to the introduction of a constant factor to the physical time (the dimensionless quantity τ/t_s and the determinant parameter D/RT are actually used in calculations) [126, 172].

Calculations show that the pre-exponential factor in the Arrhenius equation is temperature-dependent. This situation is probably due to the decrease in the distribution function in the region $\langle E_{kin} \rangle < E_{kin} < D$ (where $\langle E_{kin} \rangle$ is the average kinetic energy of the molecules in the equilibrium state) compared with the equilibrium distribution (see [170] for details).

In the treatment of experimental data, the temperature of the gaseous heat bath (a monoatomic gas) is taken as the temperature involved in the Arrhenius equation. The relations between $\ln \Delta N/N$ and reciprocal temperature $1/T$ for three different lifetimes τ of a molecule in a state where $E_{min} \geq D$ are shown in Fig. 3.5. The rate constant for each of these values of τ obeys the Arrhenius equation.

Since the nonequilibrium distribution functions of molecules differ significantly from the Maxwell distribution, the concept of temperature for a system of molecules does not make sense. Nevertheless, we formally define $T^* = \frac{2}{3}\langle E_{kin} \rangle/R$. It is seen from Fig. 3.5 that the relation between the rate constant and $1/T^*$ is described by the Arrhenius equation for τ equal to 0.5×10^{-8} or 0.5×10^{-9} s; that is, unlike the case when $\tau = 0.2 \times 10^{-13}$ s. The reason for this discrepancy lies in the difference in the kinetic orders of the reaction. In the first two cases, the reaction rate is largely determined by the stationary velocity distribution function of the molecules, whereas in the third it is determined by

$-\ln(\Delta N/N)$

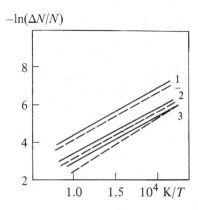

Fig. 3.5. Dependence of the equilibrium values of ln $\Delta N/N$ on $1/T$ (solid lines) and $1/T^*$ (dashed lines).1—$\tau = 0.5 \times 10^{-8}$ s; 2—$\tau = 0.5 \times 10^{-9}$ s; 3—$\tau = 0.2 \times 10^{-13}$ s.

the frequency of collision of the molecules with the atoms of the heat bath, resulting in activation of the molecules. The activation itself depends mainly on the form of the high-energy wing of the distribution function for the heat-bath atoms, which is supposed to be Maxwellian within the limits of the model considered.

In the nonisothermal case, the only quantity which can be substituted for the temperature in the Arrhenius equation is T^*. Hence, reaction rates for the second-order thermal dissociation of molecules cannot be described with a Arrhenius-type relationship. These nonisothermal conditions cause the greatest deviations from the Maxwell distribution.

Thus, it is evident that the Monte Carlo method can also be successfully applied to the investigation of the processes of vibrational relaxation of molecules with regard to the dissociation process. These aspects are considered in detail elsewhere [172]. In the absence of multiquantum transitions, the distribution function conserves its Boltzmann form. Further on, the dissociation reduces the population of the higher levels. In taking into account multiquantum transitions, there is no common vibrational temperature even at a moment when dissociation does not occur to any appreciable extent. This is due to the stronger dependence of the transition probability on the vibrational energy, compared, for example, with that for the simple harmonic oscillator.

One of the potential applications of the Monte Carlo method is the so-called 'classical trajectory method', enabling the rate constants and cross-sections of chemical reactions to be determined.

The influence of the vibrational excitation of molecules on the reaction rate constant is shown in Fig. 3.6. A semi-empirical method for calculating the effect of vibrational excitation on the rate of chemical reaction will be discussed in section 3.11.

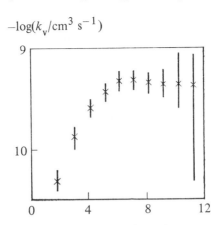

Fig. 3.6. Dependence of k_v on the vibrational level number for the reaction
$$O^+ + N_2 \rightarrow NO^+ + N.$$

3.7 STEADY-STATE CONCENTRATION METHOD

This technique does not depend on whether a system is at equilibrium or not, and it is widely used in chemical kinetics. We consider the method by taking two simple and rather typical examples, first of all, however, implying a definition of the conditions for its application [88, 173].

Example 1. Consider a sequence of reactions

$$A \xrightarrow{k_1} 2M_{(1)}, \quad M_{(1)} + C \xrightarrow{k_2} P + M_{(2)},$$

$$M_{(2)} + A \xrightarrow{k_3} P + M_{(1)}, \quad M_{(1)} + M_{(1)} \xrightarrow{k_4} A \tag{3.104}$$

Their sum gives the stoichiometric equation

$$A + C \rightarrow 2P \tag{3.105}$$

where A and C are the reactants, P the molecular products, and $M_{(1)}$ and $M_{(2)}$ the reactive intermediates.

Now the following kinetic equations can be written:

$$- d[A]\big/dt = k_1[A] + k_2[M_{(2)}][A] - k_4[M_{(1)}]^2,$$

$$- d[C]\big/dt = k_2[M_{(1)}][C],$$

$$d[P]\big/dt = k_2[M_{(1)}][C] + k_3[M_{(2)}][A],$$

$$d[M_{(1)}]\big/dt = 2k_1[A] + k_3[M_{(2)}][A] - k_2[M_{(1)}][C] - 2k_4[M_{(1)}]^2,$$

and

$$d[M_{(2)}]\big/dt = k_2[M_{(1)}][C] - k_3[M_{(2)}][A], \tag{3.106}$$

the number of moles consumed being $[A_0] - [A] = [P]/2 + [M_{(1)}]/2$ for A, and $[C_0] - [C] = [P]/2 + [M_{(2)}]/2$ for C. At $t = 0$ $[P]$, $[M_{(1)}]$ and $[M_{(2)}]$ are equal to zero.

Let $[M_{(1)}]$ and $[M_{(2)}] \ll [A]$, $[C]$, and $[P]$ for any value of t. Then

$$[A_0] - [A] = [P]/2 \quad \text{and} \quad [C_0] - [C] = [P]/2, \tag{3.107}$$

or

$$-d[A]/dt = \frac{1}{2}d[P]/dt \quad \text{and} \quad d[C]/dt = \frac{1}{2}d[P]/dt \tag{3.108}$$

Eqs (3.108) mean that there is no delay (or, at least, it is very small) between the disappearance of reactants A and C and the emergence of the product P. In this case $[M_{(1)}]$ and $[M_{(2)}]$ turn out to be practically equal to zero, so that the relevant equations in Set (3.106) appear to become algebraic and solvable with respect to $[M_{(1)}]$ and $[M_{(2)}]$.

The last equation in the system of Eqs (3.106) gives

$$k_2[M_{(1)}][C] = k_3[M_{(2)}][A].$$

Then the penultimate equation gives

$$[M_{(1)}] = (k_1/k_4)^{1/2}[A]^{1/2} \quad \text{and}$$

$$[M_{(2)}] = (k_2/k_3)(k_1/k_4)^{1/2}[C]/[A]^{1/2}$$

or

$$\frac{[M_{(1)}]}{[A]} = \left(\frac{k_1}{k_4}[A]\right)^{1/2} \quad \text{and} \quad \frac{[M_{(2)}]}{[C]} = \frac{k_2}{k_3}\left(\frac{k_1}{k_4}[A]\right)^{1/2}. \tag{3.109}$$

The latter equations specify the quantitative conditions required for the steady-state method to be applied, namely $[M_{(1)}]$, $[M_{(2)}] \ll [A]$, $[C]$. If the error in the concentration must be, for example, less than 1%, it is necessary that $[M_{(1)}]/[A] \leq 0.01$ and $[M_{(2)}]/[C] \leq 0.01$. For $[M_{(1)}]/[A]$, this takes place only when $k \leq 10^{-4} k_4[A]$ and the inequality $k_2 < k_3$ is true for $[M_{(2)}]/[C] - [M_{(1)}]/[A] \leq 0.01$. These inequalities enable the supposed mechanism to be verified.

Example 2. Under conditions of continuous energy supply (a so-called stationary action regime), highly reactive intermediates enter reactions with other species as the former are produced. With an increasing concentration of intermediate species, the rate of these reactions increases and gradually approaches the rate of formation; that is, the concentration of intermediate R reaches its stationary value during the process provided that $[R] \ll [A]$. Suppose that the transient R emerges at a constant rate $v(R)$ and it reacts with a species $A_{(1)}$ in accordance with

$$R + A_{(1)} \xrightarrow{\; k_1 \;} RA_{(1)} \tag{3.110}$$

with the rate constant k_1 to form a stable product.

The kinetics are described by the differential equations

$$d[R]/dt = v(R) - k_1[R][A_{(1)}], \tag{3.111}$$

$$- d[A_{(1)}]/dt = k_1[R][A_{(1)}].$$

With the initial conditions $t = 0$ and $[R] = 0$, the solution of System (3.111) will be

$$[R] = \frac{v(R)}{k_1[A_{(1)}]} \left\{ 1 - \exp(-k_1[A_{(1)}]t \right\} \tag{3.112}$$

For a sufficiently large value of t, the exponential term becomes small and the concentration $[R]$ takes almost a constant value. Since the experimental error in determination of the concentration of any species is worse than 5%, the stationarity is defined with the same accuracy. Thus, the stationary (steady) value of the concentration is believed to be attained when $\exp(- k_1[A_{(1)}]t) < 0.05$, hence the time required to attain the steady-state concentration is $t_{st} = 3/k_1[A_{(1)}]$. So long as the concentration of solute $[A_{(1)}]$ changes in the process, the steady-state concentration of transient species will continuously adjust itself to the value corresponding to the given concentration of solute.

This principle of stationary (steady-state) concentrations lies at the basis of the kinetic procedure for the treatment of experimental data known as the steady-state method (SSM) or the quasi-steady state concentrations method suggested by Bodenstein and developed further by Semenov. A simplified version of this method was derived in photochemistry a little earlier under the name of the Stern–Volmer method or Stern–Volmer formalism. The principal feature of this method is the change from differential equations describing the behaviour of transient species to algebraic equations. The above example will be treated then as

$$v(R) - k_1[A_{(1)}][R] = 0 \tag{3.113}$$

In general, for every transient species, the following steady-state equation may be written

$$d[R]\big/ dt = \sum v_{form} - \sum v_{dec} = 0 \tag{3.114}$$

or

$$\sum v_{form} = \sum v_{dec}, \tag{3.115}$$

that is, the sum of all formation rates v_{form} for a given species R is equal to the sum of all its decay rates v_{dec}. So, the treatment of kinetics under steady-state conditions is reduced to the solution of a system of algebraic equations which do not involve time, hence the possibility of finding values for reaction rate constants is excluded. The question arises as to what might be gained by this method? Consider two principal cases.

(1) Each transient species enters only one type of decay process. Thus we get Eq. (3.113) for each of the species, and a reaction rate constant can be determined only if the concentration $[R]_{st}$ is measured experimentally and the rate of formation $v(R)$ is known. In most processes of HEC occurring under conditions of continuous irradiation (action), the concentration of transients is less than the experimentally detectable limit. Thus the solution of the system of algebraic equations gives only an equation relating the rate of conversion of solute with the formation rate of transient species, which does not contain the reaction rate constants.

(2) At least one of the transient species takes part in two reactions, for example with two different solutes $A_{(1)}$ and $A_{(2)}$

$$R + A_{(1)} \xrightarrow{\ k_1\ } RA_{(1)} \tag{3.116}$$

$$R + A_{(2)} \xrightarrow{\ k_2\ } RA_{(2)}. \tag{3.117}$$

Thus, there are two reactions competing for the same transient species in the system. Then the set of equations will be as follows

$$d[R]/dt = v(R) - k_1[R][A_{(1)}] - k_2[R][A_{(2)}] = 0 \tag{3.118}$$

$$v(-A_{(1)}) = -d[A_{(1)}]/dt = v(RA_{(1)})$$
$$= d[RA_{(1)}]/dt = k_1[R][A_{(1)}] \tag{3.119}$$

$$v(-A_{(2)}) = -d[A_{(2)}]/dt = v(RA_{(2)})$$
$$= d[RA_{(2)}]/dt = k_2[R][A_{(2)}], \tag{3.120}$$

assuming that $[R]$ has attained its steady-state value.

This set of three equations contains three unknowns namely $[R]$, k_1, and k_2, but the system is degenerate (with a rank of two) and it is impossible to find the numerical values of all three quantities. Meanwhile, the rate constant ratio k_1/k_2 is easily determined. We find $[R]$ from Eq. (3.118) and substitute in, for example, Eq. (3.119). Thus

$$[R] = \frac{v(R)}{k_1[A_{(1)}] + k_2[A_{(2)}]} \tag{3.121}$$

$$v(-A_{(1)}) = \frac{k_1[A_{(1)}]}{k_1[A_{(1)}] + k_2[A_{(2)}]} \times v(R) \tag{3.122}$$

$$\frac{1}{v(R)} + \frac{1}{v(R)} \times \frac{k_2[A_{(2)}]}{k_1[A_{(1)}]} = \frac{1}{v(-A_{(1)})}. \tag{3.123}$$

When solved graphically, an equation of type (3.123) makes it possible to find two values at once: the rate of formation of a given species, which can be determined from the intercept on the ordinate, and the reaction rate constant ratio k_1/k_2 as a component of the slope. Similar relationships can be obtained for $v(-A_{(2)})$, $v(-RA_{(1)})$, or $v(RA_{(2)})$. The factor before $v(R)$ in Eq. (3.122) defines the role the given reaction plays in the total decay of transient species, and it may be generalized to cover any number of solutes

$$k_1[A_i] \bigg/ \sum_{j=1}^{j=n} k_j[A_j].$$

The ratio k_1/k_2 can also be found in another way, namely by a search for such a ratio of reaction rates in a system that the final expression does not include the concentrations of transient species. For the system under consideration this ratio is easily found as

$$\frac{v(-A_{(1)})}{v(-A_{(2)})} = \frac{k_1[R][A_{(1)}]}{k_2[R][A_{(2)}]} = \frac{k_1[A_{(1)}]}{k_2[A_{(2)}]}. \tag{3.124}$$

Consequently, the reaction rate constant ratio will be

$$\frac{k_1}{k_2} = \frac{v(-A_{(1)})}{v(-A_{(2)})} \times \frac{[A_{(2)}]}{[A_{(1)}]}. \tag{3.125}$$

If only one of the rates $v(-A_{(1)})$ is measured experimentally, the rate of another reaction can be found from the steady-state equation (3.118):

$$v(-A_{(2)}) = v(R) - v(-A_{(1)}) \tag{3.126}$$

3.8 PULSE TECHNIQUES IN STUDIES ON THE KINETICS OF HOMO-GENEOUS PROCESSES

Transient species in HEC processes carried out under continuous irradiation are generally found in very low concentrations inaccessible to direct observation, so various pulse techniques have been developed. The basis of these techniques is that, within a time appreciably less than the lifetime of the species under investigation, a system is supplied with some energy which creates a concentration of the transient species of a size that can be measured experimentally by means of a fast detection method, for example with absorption spectroscopy, luminescence, Raman spectroscopy, voltamperometry, conductimetry, ESR, etc. Relevant combinations of the above methods applied under conditions chosen to carry out the process make it possible to determine physicochemical quantities describing the transients such as their molar extinction coefficient, energetic and quantum yields, lifetime and reaction rate constant, equilibrium constant, redox potential, mobility in an electric field, sign and magnitude of electric charge, etc. The techniques of pulsed excitation by UV and visible light [174–176], of pulsed ionizing radiation [177, 178], and of pulsed electric discharge [179, 180] have been covered thoroughly. Let us consider some procedures for the determination of rate constants for reactions occurring after pulse excitation. It should be noted that for a complex reaction scheme, different methods of mathematical simulation to obtain the reaction rate constants (see sections 3.10 and 3.12) are desirable.

3.8.1 Irreversible processes

Pulse methods are used extensively in studies on the mechanism of HEC processes, in investigations of the nature of excited species and other transient products, and in determination of the lifetimes and reaction rate constants of transients [174–176, 181, 182]. When the duration of the exciting pulse (or, more correctly, the resolving time of the apparatus) is rather less than the lifetime of the species under study, there is usually no serious problem in the analysis of the resulting kinetics. The general form of the differential equation for a change in concentration of the species may be written as

$$d[N]/dt = v(t) - \sum_i k_i [N]^{m_i} [X]^{n_i} \qquad (3.127)$$

where $v(t)$ is the formation rate of N, k_i is the rate constant for the decay of N on interaction with X_i (or in a unimolecular process), and m_i and n_i are the reaction orders in N and X_i, respectively.

Take $v(t) = 0$ and $[N] = [N]_0$ for $t = 0$ for the above conditions. As regards excited molecules, they normally decay via processes of light emission, internal conversion and intersystem crossing, radiationless energy transfer, and quenching, that is, interaction (either physical or chemical) with other molecules, resulting in deactivation of the excited species or in their transformation to reaction products. Most decay processes for excited molecules are first-order and pseudo-first-order reactions, the decay kinetics in these cases being described with the simple exponential law

$$[N] = [N]_0 e^{-t/\tau} \qquad (3.128)$$

where

$$\tau = 1 \bigg/ \sum_i k_i$$

is the lifetime of the excited molecule.

If only the spontaneous processes of decay (emission, internal conversion, or intersystem crossing) take place, the quantity τ is called the *natural lifetime* and is denoted τ_0.

In the presence of foreign additives (quenchers), new processes come about:

$$N + X \rightarrow \ldots, \tag{3.129}$$

being also first-order in molecule N, so that their decay kinetics remains exponential but the lifetime decreases

$$\tau_0 / \tau = 1 + k_i \tau_0 [X_i] \tag{3.130}$$

where τ_0 and τ are the lifetimes in the absence and presence of X, respectively, and k_i is the rate constant for interaction of N and X_i.

Eq. (3.130) is called the Stern–Volmer equation. It describes the relation between the lifetime and the concentration of a quencher. Using relationships of this type, it is easy to determine the rate constant for the interaction of N with X_i.

There can also be bimolecular decay processes for excited molecules, for example the so-called triplet–triplet annihilation

$$^3M + {}^3M \rightarrow {}^1M + {}^1M^* \tag{3.131}$$

which is second-order in the excited molecule. The decay kinetics for M in this case are given by a hyperbolic relationship

$$[^3M] = \frac{1}{1/[^3M] + 2kt}, \tag{3.132}$$

provided that Reaction (3.131) dominates all other processes. With comparable rates, a mixed order in decay of N (^3M) is observed:

$$d[N]/dt = -k_1[N] - k_2[N]^2. \tag{3.133}$$

Let us turn now to the analysis of kinetic data. Determination of the rate constant for simple first-order and second-order rate laws is usually carried out graphically, using semilogarithmic or reciprocal coordinates, respectively (Fig. 3.7).

$$\log([N]/[N]_0) = -2.303 \, k_1 t \tag{3.134}$$

$$1/[N] = 1/[N]_0 + 2k_2 t \tag{3.135}$$

where $k_{1,2}$ are the overall rate constants and N and N_0 the current and initial concentrations of the species under investigation, respectively (the factor 2 is due to the simultaneous decay of *two* particles in one event of the second-order reaction).

Since pulse techniques frequently include absorption and emission spectroscopy measurements as detection methods, the optical absorbance A or the intensity of emitted light I, respectively, are substituted for the concentration of transients. This does not complicate the first-order processes, and the semilogarithmic plot $\log(\Delta A)$ or $\log I$ *vs* t directly provides a value for the rate

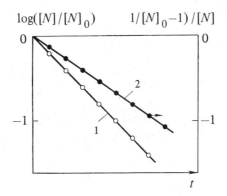

Fig. 3.7. Linear transformations of first-order (1) and second-order (2) rate laws for the decay of transient species.

constant which has the units of s^{-1}. However, to find the rate constant for a second-order process it is necessary to know the absolute concentration of N at the initial (or any other) instant. If this concentration is not measurable, then the rate constant obtained is uncertain by some factor, for example if measuring optical absorption, the value obtained is k/ε as determined from the plot in accordance with

$$1/A = 1/A_0 + 2kt/\varepsilon l \tag{3.136}$$

where ε is the extinction coefficient for N (other substances are assumed to be nonabsorbing in the relevant spectral region) and l the optical pathlength.

To find k in this case, it is necessary to know ε. The task becomes largely that of determination of the extinction coefficient of a transient.

The determination of molar extinction coefficients ε (or, similarly, of emission rate constants k_f and k_p) for transient short-lived species, unlike stable compounds, presents some difficulties related to the necessity to measure their concentration. There are some procedures used for the purpose [183] which can fall into four different groups.

(i) A transient species is formed from its precursor with a known yield ϕ_T. If this precursor is the ground electronic state, then by measuring the optical density at two wavelengths, so that the precursor absorbs at one of them (λ_1), and either both, or only the transient species, at the other (λ_2), it is possible to determine the extinction coefficient of the transient form ε', provided that the extinction coefficient ε of the precursor and the change in optical density are known for both of these wavelengths:

$$\varepsilon_2' = (\varepsilon_2 - \varepsilon_1 \Delta A_2 / \Delta A_1)/\phi_T \tag{3.137}$$

This procedure is inapplicable when it is impossible to find a wavelength region where only the substrate absorbs light, or if some other products interfere in the same optical region as the transient species we are interested in. Then a saturation technique can be used.

The absorbance of the transient species increases nonlinearly with increasing energy of the exciting pulse

$$[T] = c_0 \{1 - \exp(-2303\varepsilon\phi_T W_i)\}, \qquad (3.138)$$

where ε is the extinction coefficient of the precursor at the excitation wavelength and W_i is the number of incident photons per unit area (in einstein cm^{-1}), and ε' can be determined either from the limiting absorbance value ΔA_{max} (corresponding to complete transformation of the precursor to the transient species):

$$\varepsilon' = \Delta A_{max} / c_0 l \qquad (3.139)$$

where c_0 is the initial concentration of the precursor and l is the optical pathlength, or from the dependence of A on the excitation energy W_i in the nonlinear region:

$$\Delta A = c_0 \varepsilon' l (1 - \exp(-2303\varepsilon\phi_T W_i)) \qquad (3.140)$$

This procedure can be applied only in those cases when the precursor and the transient product are the only types of species present in the system and there are no by-products formed to an appreciable extent.

A widely-used procedure is actinometry, in which the number of photons N_{abs} absorbed in an analysed volume V is measured (with the use of actinometers such as the ferrioxalate, Aberchrome-540, etc.) in a separate experimental run, and the concentration of the transient species formed [T] is calculated as follows:

$$[T] = \phi_T N_{abs} / V. \qquad (3.141)$$

This method is convenient in the case of monochromatic laser excitation, but, when dealing with lamp excitation, there can be some difficulties caused by differences in the absorption spectra of the precursor and the actinometer. It is necessary to know the quantum yield of formation of the transient species and to take into account the inhomogeneity of the light absorption in the sample as well as any interfering absorption by the transient species and other substances present in the system.

In some cases (especially for triplet species) it is very convenient to use a method based on radiationless energy transfer from a sensitizer. Here, two experiments are carried out, the absorbance of the sensitizer ΔA in the absence of the compound studied being measured in one of them, and the absorbance of the transient species $\Delta A'$ (at any wavelength) for such a concentration of precursor that the excited states of the sensitizer are completely quenched in the other. Then

$$\varepsilon' = \varepsilon \Delta A' / \phi_T \Delta A \qquad (3.142)$$

(ii) Methods of conversion to a product. These methods consist of measurements of the concentration of a product formed from the transient species spontaneously or with the use of specially-added agents either directly (in the case of a stable product) or by the change in optical density (if the extinction coefficients are known). With triplet species, energy transfer to a standard compound can be used.

(iii) Kinetic methods are based on calculations of the concentration of the transient species, and they require knowledge of the lifetime τ_T for these species under the conditions of a particular experiment. In the simplest cases, the steady-state concentration $[\overline{T}]$ can be expressed via the number of photons absorbed per unit volume in unit time W_a:

$$[\overline{T}] = \phi_T \tau_T W_a \tag{3.143}$$

Under a constant photon flux, the steady-state concentration of the transient can be expressed as a function of the rise time τ_r and decay time τ_T:

$$[\overline{T}] = c_0 (1 - \tau_r / \tau_T). \tag{3.144}$$

(iv) In some cases, it is possible to measure in the same experiment, or under identical excitation conditions, both the absorption of the transient and its concentration by an independent technique (ESR, calorimetry, etc.).

When the decay of the species occurs via several pathways simultaneously and the process is characterized by a mixed order (for example in the case of triplet–triplet annihilation and radical combination), an integral method is more suitable for the treatment of the kinetic curves. Integrating Eq. (3.133) we obtain

$$\ln[N]_0 - \ln[N] = k_1 t + k_2 \int_0^t [N] \, dt \tag{3.145}$$

and the graphical representation of Eq. (3.145) is

$$\frac{2.303}{t} \ln \frac{[N]_0}{[N]} = k_1 + k_2 - \int_0^t \frac{[N] \, dt}{t} \tag{3.146}$$

gives k_1 from the intercept with the ordinate and k_2 from the slope of the line (Fig. 3.8) (k_1 is obtained directly, but k_2 contains an uncertainty factor related to the absolute concentration of N).

If the lifetime of a species is comparable or less than the duration of the exciting flash, which is rather common, then the implementation of pulse techniques in such cases has some specific features. Deconvolution methods must be used to extract the true decay parameter or true decay function $f(t)$ from the experimentally-observed kinetic curve $F(t)$ and experimental response function $E(t)$ (a combination of the excitation pulse and the photomultiplier response). These methods are very important for fluorescence decay analysis and are also used in flash photolysis and pulse radiolysis. The observed kinetic curve is related to $f(t)$ and $E(t)$ as follows:

$$[\int A(t)\mathrm{d}t]/t$$

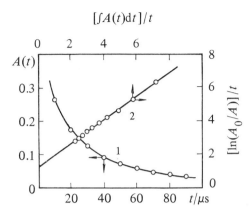

Fig. 3.8. Mixed-order decay kinetics for excited molecules of anthracene (1) and the linear integral transformations (2) of the rate law (Eq. 3.146).

$$F(t) = \int_0^t E(t')f(t-t')\,\mathrm{d}t' \qquad (3.147)$$

There are many mathematical methods used to solve the above equation [184, 185], some being based on Laplace or Fourier transforms. The most widely applied method of deconvolution is curve fitting employing the nonlinear least squares technique. Since this method is a statistical fitting procedure, large data sets must be used to ensure its validity. Programs in Basic and Fortran which can be used to achieve simple least squares deconvolution are given in [184, 185]. Software for such data reduction is also available.

The second most popular deconvolution method is the so-called method of moments [184. Eq. (3.147) can be solved in terms of the statistical moments of the functions $f(t)$ and $E(t)$. Other deconvolution methods have been described [184, 185].

There are very simple methods for the calculation of $\tau(k)$ in the case of single exponential decay. The simplest one is recommended by Birks & Munro [186] for comparable decay times of the fluorescence and of the exciting spark:

$$\tau^2 = \tau_F^2 - \tau_E^2 \qquad (3.148)$$

or

$$1/k^2 = 1/k_F^2 - 1/k_E^2 \qquad (3.149)$$

where $k = 1/\tau$ is the true decay rate constant, $k_F = 1/\tau_F$ is the apparent decay rate constant, and $k_E = 1/\tau_E$ is the decay rate constant of the exciting flash.

Another procedure [187], based on integration of the corresponding differential equation (phase plane method), enables the lifetime to be determined more accurately

$$d[N]/dt = v(t) - k[N] \tag{3.150}$$

where $v(t) + bE(t) + \varphi(1 - 10^{-A})E(t)/V$ is the rate of excitation, A is the absorbance (assumed to be almost independent of time), V is the sample volume, and $E(t)$ is the intensity of the exciting light.

Integrating Eq. (3.150), we obtain

$$[N] - [N]_0 = \int_0^t v(t)\,dt - k\int_0^t [N]\,dt. \tag{3.151}$$

Note that the directly measurable quantities are usually the optical density (absorbance) in the absorption band of N, $A_\lambda = \varepsilon_\lambda l[N]$, or the intensity of emission, $I = a[N]$. Let us define

$$W(t) = \int_0^t E(t)\,dt, \quad R(t) = \int_0^t A(t)\,dt, \quad \text{and} \quad V(t) = \int_0^t I(t)\,dt$$

Then

$$A_\lambda(t)/W(t) = \varepsilon l b - kR(t)/W(t) \tag{3.152}$$

and

$$I(t)/W(t) = ab - kR(t)/W(t). \tag{3.153}$$

The integrals $W(t)$, $R(t)$, and $V(t)$ are calculated by numerical integration from the kinetic curves for $E(t)$, $A(t)$, and $I(t)$, respectively; then, plotting $A(t)/W(t)$ vs $R(t)/W(t)$ or $I(t)/W(t)$ vs $V(t)/W(t)$, it is easy to find k from the slope of the linear function thus obtained (Fig. 3.9).

Simultaneously, it is possible with this procedure to determine the quantum yield of formation of the product under investigation (there is no need even to extrapolate the data to an uncertain moment of initial time). The intercept of the plot of Eq. (3.152) is defined by $\varphi \varepsilon_\lambda l(1 - 10^{-A})/V$. Hence, knowing the total dose absorbed by the sample $Q_a = (1 - 10^{-A})W(\infty)/V$ and values for ε and l, it is no problem to find φ. It is also convenient to use for the same purpose a simplified version of this procedure (Fig. 3.10), provided that the duration of the exciting flash is shorter than the lifetime of N. Then, beginning from a certain moment, the value of $W(t)$ becomes approximately constant and Eq. (3.152) takes the form

$$A_\lambda(t) \approx \varphi \varepsilon l Q_a - k\int_0^t A(t)\,dt. \tag{3.154}$$

This procedure appears to be no more laborious for determination of $k = 1/\tau$ than that based on semilogarithmic plots. An advantage of this method is the sensitivity of the plots to nonexponentiality.

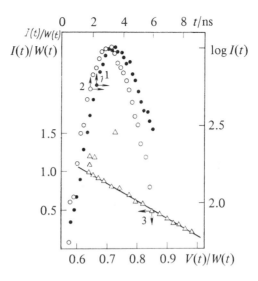

Fig. 3.9. Fluorescence kinetics of 4-phenylphosphenyl-3′-methoxystilbene (1). 2—Exciting pulse shape; 3—Linear transformation of Eq. (3.153).

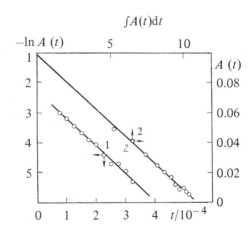

Fig. 3.10. Decay kinetics for 9,10-dicyanoanthracene triplets: semilogarithmic (1) and integral (2) plots.

3.8.2 Reversible processes

In some cases, molecules in their excited state can take part in reversible processes, producing other excited molecules [188, 189]:

$$M^* + X \rightleftharpoons P^* + Y \qquad (3.155)$$

These processes may be, for instance, excimer or exiplex formation and photo-induced proton transfer. Similar reversible reactions are also possible for other transient species (for example acid–base equilibria for radicals). The kinetics of such reversible reactions are described by a system of differential equations [188]:

$$d[M^*]/dt = bE(t) - \{1/\tau_0 + (k_1 + k_2)[X]\}[M] + k_{-1}[Y][P^*] \qquad (3.156)$$

$$d[P^*]/dt = k_1[X][M^*] - \{1/\tau_0' + (k_{-1} + k_{-2})[Y]\}[P^*] \qquad (3.157)$$

where $bE(t)$ is the rate of excitation, τ_0 and τ_0' are the lifetimes of M^* and P^*, respectively, in the absence of chemical reaction (of reactants X and Y), k_1 and k_{-1} are the rate constants of the forward and backward reactions, 3.155, and k_2 and k_{-2} are the rate constants for the competing reactions of M^* with X and P^* with Y, respectively, giving correspondingly no P^* and M^* (for example deactivation).

If the excitation pulse is sufficiently short, it may be assumed that $E(t) = 0$, $[M^*]_0 = m_0$, and $[P^*]_0 = 0$ at $t = 0$. Defining

$$\mu = 1/\tau_0 + (k_1 + k_2)[X], \qquad (3.158)$$

$$\mu' = 1/\tau_0' + (k_{-1} + k_{-2}([Y], \qquad (3.159)$$

and

$$\beta = k_1 i_{-1}[X][Y], \qquad (3.160)$$

after some transformation, we obtain the system of Eqs. (3.156) and (3.157) in the form:

$$d^2[M^*]/dt^2 + (\mu + \mu')\,d[M^*]/dt + (\mu\mu' - \beta)[M^*] = 0 \qquad (3.161)$$

$$d^2[P^*]/dt + (\mu + \mu')\,d[P^*]/dt + (\mu\mu' - \beta)[P^*] = 0. \qquad (3.162)$$

The solution of this system of homogeneous differential equations is as follows:

$$[M^*] = c(\theta e^{-\vartheta_1 t} + e^{-\vartheta_2 t}), \qquad (3.163)$$

$$[P^*] = c'(e^{-\vartheta_2 t} - e^{-\vartheta_1 t}), \qquad (3.164)$$

where

$$\vartheta_{1,2} = \frac{1}{2}\left\{\mu + \mu' \pm \sqrt{[(\mu - \mu')^2 + 4\beta]}\right\} \qquad (3.165)$$

and

$$\theta = \frac{\mu - \vartheta_2}{\vartheta_1 - \mu} = \frac{\mu - \mu' + \sqrt{[(\mu - \mu')^2 + 4\beta]}}{-\mu + \mu' + \sqrt{[(\mu - \mu')^2 + 4\beta]}}. \qquad (3.166)$$

Kinetics of this kind are typical for reactions of complex formation and of proton transfer in excited singlet states; it is usually studied by means of fluorescence measurements. The fluorescence kinetics of M^* and P^* are also described by the sum and difference of two exponentials, respectively,

$$f(t) = f_0(\theta e^{-\vartheta_1 t} + e^{-\vartheta_2 t}) \qquad (3.167)$$

and

$$f'(t) = f_0'(e^{-\vartheta_2 t} - e^{-\vartheta_1 t}). \qquad (3.168)$$

The decay of fluorescence of M^* is represented by the sum of two exponentials, each of them contributing in accordance with the ratio of the reaction rate constants in the system. It is then convenient to present experimental data by plotting the logarithm of $F(t)$ vs t (Fig. 3.11). There can be cases when both components are observed (two linear segments with a different slope) or only one of them is actually seen. The fluorescence decay of the product P^* is described by the exponential with the decay constant ϑ_2, that is, the smaller of the two decay rate constants. The increase in fluorescence with time is linearized as follows:

$$\log[e^{-\vartheta_2 t} - F'(t)/F_0'] = -0.434\,\vartheta_1 t. \qquad (3.169)$$

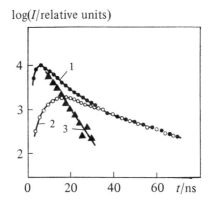

Fig. 3.11. Semilogarithmic plot of fluorescence kinetics of naphthalene (1) and its exciplex with triethylamine in ethyl acetate (2). 3—log $(I - I')$.

It should be noted that the situation in which the fluorescence decays of M^* and P^* are described with identical exponentials does not necessarily correspond to equilibrium between the species. Generally, this state is supposed to be steady, and the concentration ratio of M^* to P^* is given by

$$\frac{[Y][P^*]}{[M^*][X]} = \frac{k_1}{\vartheta_1 - \mu} = \frac{2k_2}{-\mu + \mu' + \sqrt{[(\mu - \mu')^2 + 4\beta]}}, \qquad (3.170)$$

which coincides with the equation for an equilibrium constant only in the particular case when $k_1 \gg k_2$ and $|k_1[X] - k_{-1}[Y]| \gg |1/\tau_0 - 1/\tau_0'|$.

To find ϑ_1 it is appropriate to normalize $F'(t)$ so that it coincides with $F(t)$ for large t:

$$aF'(t) = F(t). \tag{3.171}$$

Then for small t

$$F(t) = aF'(t) = F_0(1 + \theta)\exp(-\vartheta_1 t), \tag{3.172}$$

and, plotting the logarithm of $\{F(t) - aF'(t)\}$ vs time, it is possible to determine ϑ_1 and θ. Making use of the fact that

$$\vartheta_1 + \vartheta_2 = \mu + \mu' = 1/\tau_0 + \mu' + (k_1 + k_2)[X] \tag{3.173}$$

and plotting $\vartheta_1 + \vartheta_2$ vs $[X]$, it is easy to obtain $k_0 = k_1 + k_2$ from the slope and μ' from the intercept so long as the value of τ_0 is known. An essential disadvantage of this procedure is that the major contribution to the sum comes from ϑ_1, the value of which is usually much higher but is determined with less accuracy than ϑ_2.

Many deconvolution methods have been used for multiexponential decay analysis [184, 185] and some software is commercially available. The use of any deconvolution methods for more-than-three component functions is discouraged. While statistical parameters (χ^2, etc.) are important criteria for judging the goodness-of-fit, the plots of weighted residuals and autocorrelation of residuals should be strongly recommended for the visual inspection of data and their fit. It should also be mentioned that the parameters obtained for multiexponential functions can be mutually correlated. Usually, the fit is very sensitive to variation of any parameter taken separately, but it may be affected only slightly if a correlated change in two or more parameters takes place simultaneously. This is a general property of the inverse Laplace transform (in contrast to the Fourier transform).

The form of fluorescence kinetics curves is specified by three parameters: by the coefficients ϑ_1 and ϑ_2 involved in the exponential power, and by their relative contribution θ (Eqs (3.165) and (3.166)). These three parameters, in turn, depend nonlinearly on the combination of rate constants and quencher concentration. The general appearance of Eqs (3.165) and (3.166) is such that it shows the nature of the dependence of ϑ_1, ϑ_2, and θ on the concentration of quencher to be rather complex.

The general analysis of nonlinear relationships (3.165) and (3.166) is a laborious procedure which does not seem to produce convincing results. To present ϑ_1, ϑ_3, and θ as a function of quencher concentration in a more explicit form, it is convenient to use a graphical approach applying the following procedure [175].

First, we define dimensionless variables x and y as linear functions of the concentration of quencher X

$$x = \beta/\mu'^2 = \frac{k_1 k_{-1}[X][Y]}{\mu'^2} \qquad (3.174)$$

$$y = \mu/\mu' = \frac{1}{\tau_0 \mu'} + \frac{k_1 + k_2}{\mu'}[X]. \qquad (3.175)$$

With changing quencher concentration, a figurative point in the graph will move monotonically in the x, y plane along the line

$$y = 1/\tau\mu' + \frac{(k_1 + k_2)\mu'}{k_1 k_{-1}[Y]}x \quad \text{or} \quad y = p + qx \qquad (3.176)$$

where

$$p = 1/\tau_0 \mu' \qquad (3.177)$$

and

$$q = \frac{(k_1 + k_2)\mu'}{k_1 k_{-1}[Y]}. \qquad (3.178)$$

It should be noted at once that the tangent of this line cannot be less than unity so long as $k_1 + k_2 \geq k_1$ and $\mu' = 1/\tau_0 + k_{-1}[Y] \geq k_{-1}[Y]$. Now Eq. (3.165) is expressed as

$$\begin{aligned}
\vartheta_{1,2} &= \mu' + \frac{\mu'}{2}\left[-1 + \frac{\mu}{\mu'} \pm \sqrt{\left(1 - \frac{\mu}{\mu'}\right)^2 + \frac{4\beta}{\mu'^2}}\right] \\
&= \mu' + \frac{\mu'}{2}\left[-1 + y \pm \sqrt{\{(1-y)^2 + 4x^2\}}\right].
\end{aligned} \qquad (3.179)$$

Defining $r_{1,2} = \vartheta_{1,2}/\mu'$, we obtain a system of two linear equations[†] connecting the variables x and y for the same relative value of the coefficients to an exponential power:

$$y = r_1 + x/(1 - r_1) \qquad (3.180)$$

$$y = r_2 + x/(1 - r_2). \qquad (3.181)$$

---

† This equation is also linear with respect to the concentration of X so long as x is proportional to the concentration. The equation could be written in an explicit form as a function of [X], but this would require different scales in the axes for different systems, depending on the values of the rate constants. Therefore it is more convenient to keep x as a variable.

The real state of the system must be relevant to both equations; that is, it corresponds to the intersection of the lines described by Eqs (3.180) and (3.181). Under these conditions, values of $y < x$ have no physical meaning, since $r_{1,2} \geq 0$ and $1/(1 - r_{1,2}) \geq 1$. The corresponding diagram for different values[†] of the dimensionless variable r taken at intervals of 0.2 is shown in Fig. 3.12a. Every point in the plane (that is, every pair of values of x and y) corresponds to two values of r equal to $r_1 = \vartheta_1/\mu'$ and $r_2 = \vartheta_2/\mu'$, and conversely, each pair of values of ϑ_1/μ' and ϑ_2/μ' corresponds to only one pair of x and y.

Similarly, it is possible to obtain an expression for θ:

$$y = 1 + (\theta - 1)\sqrt{(x/\theta)}, \qquad (3.182)$$

the conditions of $\theta > 1$ for $y > 1$ and $\theta < 1$ for $y < 1$ being fulfilled. The corresponding diagram is shown in Fig. 3.12b.

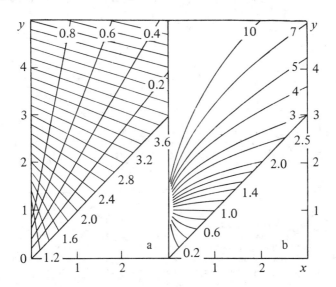

Fig. 3.12. Diagram of the dimensionless parameters x and y for the exponents $r_1 = \vartheta_1/\mu'$ and $r_2 = \vartheta_2/\mu'$ (a), and for the factor θ (b).

Using the diagrams presented above, an analysis of the dependences of ϑ_1, ϑ_2, and θ on solute concentration can be carried out, and, secondly, rate constants for the elementary reactions can be determined from experimental data on ϑ_1 and ϑ_2.

‡ In spite of the fact that the slope of $y(x)$ tends to infinity at $r = 1$, it should be noted that there is no specific point in this case. Each of the exponentials $\exp(-\vartheta_1 t)$ and $\exp(-\vartheta_2 t)$ does not refer to any real physical phenomenon separately. The behaviour of molecules is described by the sum and difference of these exponentials. For $r = 1$, $\vartheta_1 = \vartheta_2$, and it looks as though the degrees of the exponential power exchange places on passing this point.

The concentration-dependent functions ϑ_1, ϑ_2, and θ (see Eqs (3.163) and (3.164)) may behave in a different manner.

(i) If $1/\tau_0 > \mu'$ (that is, $p > 1$), then ϑ_1 increases almost linearly with increasing concentration of quencher, while ϑ_2 decreases, approaching a limiting value equal to

$$\vartheta_{2,\text{lim}} = \frac{1}{\tau'_0} + \frac{k_{-1}k_2[\text{Y}]}{k_0}.$$

Under these conditions θ will decrease rapidly (from an infinitely large value at $[X] = 0$) at first and then it will slowly increase (Fig. 3.13)). For large θ, the fluorescence decay rate curves of M^* and P^* will have a single-exponential character with a different decay time. A double-exponential mode of fluorescence decay of M^* can be observed only in the region of θ values from *ca.* 10 to 10^2.

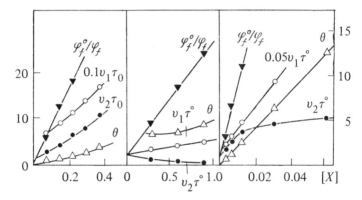

Fig. 3.13. Dependence of the rate parameters $\vartheta_1\tau_0$ and $\vartheta_2\tau_0$, the θ-factor, and fluorescence quenching parameter φ_0/φ on amine concentration in different systems: a—pyrene + triethylamine in heptane at −24°C; b—perylene + diethylaniline in heptane at +20°C; c—pyrene + diethylaniline in heptane at +20°C (ordinate axis is to the left for (a) and (b) and to the right for (c)).

(ii) If $1/\tau_0 < \mu'$ (that is, for $p < 1$), ϑ_1 will increase almost linearly with increasing concentration and θ will also increase (beginning from zero). In practice, it is very difficult to observe two exponentials in the fluorescence decay curve of M^*. If the concentration of solute increases, the contribution from the exponential of ϑ_1 increases, becoming more significant. However, ϑ_1 itself increases under these conditions, and then $1/\vartheta_1$ becomes comparable to the flash duration so that observation of the component relevant to the exponential ϑ_1 turns out to be complicated. In any case, there is always a region of optimum concentration of quencher where the measurement of the double-exponential fluorescence decay of M^* is possible. The behaviour of ϑ_2 is determined by the ratio of the parameters p and q.

Three possibilities may be realized here:

(i) $k_0 k_1 > k_{-1}[Y]/(\mu' - 1/\tau_0)$, that is $(1 - p) > 1/q$. Thus ϑ_2 will increase with increasing concentration of quencher approaching the limiting value $\vartheta_{2,\lim} = 1/\tau_0 + k_{-1}k_2[Y]/k_0$.

(ii) $k_0/k_1 = k_{-1}[Y]/(\mu' - 1/\tau_0)$, that is $(1 - p) = 1/q$. Varying the concentration of solute, ϑ_2 will remain constant and equal to $\vartheta_2 = 1/\tau_0 = 1/\tau_0' + k_{-1}k_2[Y]/k_0$.

(iii) $k_0/k_1 < k_{-1}[Y]/(\mu' - 1/\tau_0)$, that is $(1 - p) < 1/q$. The value of ϑ_2 will decrease on increasing the solute concentration, approaching the same limit $\vartheta_{2,\lim} = 1/\tau_0' + k_{-1}k_2[Y]/k_0$.

(iii) If $1/\tau_0 = \mu'$, that is $p = 1$, it is unreasonable to introduce the dimensionless variables. Equation (3.165) can be immediately simplified to produce:

$$\vartheta_{1,2} = 1/\tau_0 \pm \sqrt{(k_1 k_{-1}[X][Y])}; \qquad \theta = 1.$$

Generally, the quantum yields φ and φ' of fluorescence of M^* and P^* depend on the concentration of quencher solute in the following way[†]

$$\frac{\varphi_0}{\varphi} = 1 + \kappa[X] = 1 + \left(k_2 + \frac{k_1}{1 + \tau_0' k_{-1}[Y]} \right) \tau_0[X] = 1 + \frac{q-1}{p}x, \qquad (3.183)$$

$$\frac{\varphi'}{\varphi} = \frac{k_f'}{k_f} \frac{k_1 \tau_0'}{1 + \tau_0' k_{-1}[Y]}[X] = \frac{k_f' k_1}{k_f \mu'}[X], \qquad (3.184)$$

where k_f and k_f' are the rate constants of fluorescence emission for M and P, respectively.

Eqs (3.165) and (3.166) can be represented as a function of the dimensionless parameters and the variable x:

$$\vartheta_{1,2}\tau_0 = \frac{1}{2p}\left(1 + p + qx \pm \sqrt{[(p - 1 + qx)^2 + 4x]} \right), \qquad (3.185)$$

$$\theta = \frac{1 \pm \sqrt{[1 + 4x/(p - 1 + qx)^2]}}{-1 \pm \sqrt{[1 + 4x/(p - 1 + qx)^2]}}. \qquad (3.186)$$

Eq. (3.185) is an analogue of Eq. (3.183), the Stern–Volmer relationship for pulsed photoexcitation. Comparing Eqs (3.183) and (3.185), one can see that the fluorescence quenching φ_0/φ and the rate constants ϑ_1 and ϑ_2, in general,

---

[†] For the steady-state case, Eqs (3.183) and (3.184) are easily obtained from the system of Eqs (3.156) and (3.157), assuming the derivatives are equal to zero.

depend on solute concentration in a different way. Therefore, the values of φ_0/φ and τ_0/τ usually do not coincide with each other when plotted *vs* concentration of solute, even in the absence of static or nonstationary quenching. Experimentally, values for τ are usually found from the slope of the fluorescence decay rate curve[†] in the semilogarithmic coordinates log I *vs* t. Then the value of $1/\tau$ will coincide either with ϑ_1 (for $\geq 10^2$) or ϑ_2 (for $\theta \leq 1$), depending on the value of θ. If $k_{-1}[Y] < 1/\tau_0'$, the values of φ_0/φ and τ_0/τ must coincide.

Use of the graphical method makes it possible to predict the change in form of the fluorescence decay curves; this also helps us to choose optimum experimental conditions for determination of the kinetic parameters of elementary processes.

3.8.3 Diffusion gradients, solvent relaxation, and energy-transfer effects

Even in the absence of any reversible reaction of excited molecules, there may occur a nonexponential decay of the excited states caused by nonstationary diffusion gradients in the presence of a quencher [190]. Usually, rate equations are based on the assumption of stationary diffusion in a liquid medium. In solution, excited molecules are formed by light absorption for a static distribution of quencher molecules. Extension to cover nonstationary diffusion results in the following expression for fluorescence decay [191]:

$$I(t) = I_0 \exp(-t/\tau_0 - A\sqrt{t}) \tag{3.187}$$

where A is a function of the diffusion coefficient.

This type of deviation from the exponential law is observed over small (up to 10 ns) post-flash periods.

Even in the absence of solute, nonstationary phenomena related to solvent relaxation may also be observed. Molecules in their excited state often have both a different geometrical structure and dipole moment compared with the ground state. The transition to the excited state occurs almost instantaneously, while the solvent needs time to take up the energetically most favourable configuration. In experimental observations this phenomenon is manifested as follows: the longer the time after the exciting flash, the further is the red-shift of the centre of the emission band; for example the shift approaches 50 nm and the relaxation time several tens of ns at $-70°C$ for 4-aminophthalimide in propanol-1. The lifetimes measured at different wavelengths differ more than two-fold [192]. If the lifetime is not too short, the relaxation process follows an approximately exponential law. The smaller values for the lifetime obtained in measurements of emission at shorter wavelengths do not, of course, determine the real decay rate of the excited molecules involved in the rate equations (provided that the reaction takes place, as usual, from a thermalized fluorescent state).

[†] It should be remembered that the decay of fluorescence of M^* is not monoexponential in the case of reversible reactions of molecules in the excited state, so the term 'lifetime' is inapplicable. All that has been stated in section 3.8.2 is true unless P fluoresces.

The kinetics of fluorescence accompanied by reabsorption are, usually rather complicated. The concentration of excited molecules is changing with time and with sample depth, the excitation gradually propagating inwards. The reabsorption leads to an increase in fluorescence decay time, to deviation from an exponential dependence, and to a dependence of the decay mode on the fluorescence wavelength examined. The current treatment of reabsorption enabling the description of fluorescence with integral decay times and the estimation of deviations from exponential decay from the differences in times seems rather intricate [193]. However, there is a somewhat simpler method of analysing fluorescence kinetics [181], if we consider the total quantity $N(t)$ of excited molecules in the sample. For short-pulse excitation, the change in $N(t)$ after the end of the exciting pulse may be described by the equation[†]

$$\mathrm{d}N(t)/\mathrm{d}t = -N(t)/\tau_0 + \varphi\alpha N(t)/\tau_0 = -(1-\varphi\alpha)N(t)/\tau_0 \qquad (3.188)$$

where τ_0 is the true lifetime of the excited molecules, φ is the fluorescence quantum yield, and α is the integral probability of reabsorption of fluorescence in the sample depending on the overlap of absorption and fluorescence spectra, on the luminophore concentration and on the sample geometry.

If α is not time-dependent, the fluorescence decay time must increase because of reabsorption, the kinetics having to remain exponential. However, the integral probability of reabsorption is generally time-dependent because of the change in distribution of the excited molecules in the sample volume, so fluorescence occurring with reabsorption may decay nonexponentially.

Approximate evaluation of α_0 at zero time for a flat sample is based on the assumption that one half of the emitted fluorescence comes out through the front fact without being reabsorbed. Taking the optical pathlength to be equal to the sample thickness l we obtain:

$$\alpha_0 = \frac{\int f(v)(1-10^{-\varepsilon_v cl})\,\mathrm{d}v}{2\int f(v)\,\mathrm{d}v} \qquad (3.189)$$

where $f(v)$ is the fluorescence spectrum, ε is the absorption coefficient, and c is the luminophore concentration.

This estimation rather underestimates α. Taking into account the propagation of fluorescence in all directions with equal probability and that the actual pathlength is longer than the sample thickness, we have for a flat sample:

$$\alpha = \frac{\int f(v)\int_0^1 (1-10^{-A/(1-a)})\,\mathrm{d}a\,\mathrm{d}v}{2\int f(v)\,\mathrm{d}v} \qquad (3.190)$$

where A is the optical absorption in the sample at a given wavelength.

---

† The possibility of the yield of excited states to be different from unity on absorption of light is not explicitly considered here. It can be included in φ.

Let us consider in more detail the effects of reabsorption and re-emission on the fluorescence kinetics. When observing fluorescence in a wavelength region of strong reabsorption, the radiation detected is that coming mainly from the sample surface, which is emitted by primary-excited fluorophore molecules themselves, hence the kinetics of fluorescence must be near-exponential and the decay time close to τ_0.

In a region of weak reabsorption, the intensity $F(t)$ of fluorescence emitted by the entire sample is proportional to the total amount of excited molecules, and it is described by equation

$$d \ln F(t)/dt = d \ln N(t)/dt = (1 - \varphi\alpha)/v_0.$$ (3.191)

On excitation by weakly-absorbed light, the distribution of excited molecules in the sample is almost uniform, and α changes only slightly with time. Under these conditions the luminescence decay is near-exponential, but the decay time may be considerably longer than τ_0 (for strong reabsorption and a large fluorescence quantum yield).

With strong absorption of the exciting light, excited molecules are initially produced in a thin front-facing layer of sample. With a high probability of reabsorption ($\alpha \geq 0.3$), reabsorption is increased and α increases, but less than twice as the excitation propagates inside the sample. Then the apparent time of fluorescence decay

$$\tau = \tau_0/(1 - \alpha\varphi)$$ (3.192)

will gradually increase and the decay will be nonexponential, so that the smaller the difference $(1 - \alpha\varphi)$, the greater the deviation from exponential decay, that is, nonexponentiality is enhanced by high values of the probability of reabsorption and of the fluorescence quantum yield. This effect is significant only for luminophores with a sufficiently high fluorescence quantum yield.

If adding a solute which shows static fluorescence quenching to a system featuring reabsorption and re-emission, the integral probability of reabsorption remains unchanged but the quantum yield of fluorescence is reduced. As a result, the value of $\alpha\varphi$ decreases, which must lead to a decrease in non-exponentiality (if it is caused by reabsorption) and to shortening of the decay time. Thus, an increase in solute concentration will result in enhancement of static quenching causing a gradual reduction, with the occurrence of reabsorption and re-emission, of the apparent fluorescence decay time and thus simulating dynamic quenching.

The 'long-range' dipole–dipole (otherwise known as inductive-resonance) radiationless transfer of electronic excitation energy in viscous media also results in the nonexponential decay of fluorescence:

$$I(t) = I_0 \exp\left(-t/\tau_0 - 2q\sqrt{[t/\tau_0]}\right)$$ (3.193)

where q is a function of r_0 which is the distance at which the probability of donor-to-acceptor energy-transfer is 0.5. It is not surprising that Eq. (3.193) is analogous to Eq. (3.181) since the mathematical diffusion to a sphere of some radius r is considered. However, the radius is defined in a different way: in the first case it is determined by r_0, and in the second case it is the radius of diffusion. In low-viscosity solvents, stationary diffusion gradients are established faster and quenching by the Stern–Volmer mechanism with abnormally large cross-sections is usually observed.

Then, if $\ln vt \gg 1$, we find after integration

$$I(t) \approx I_0 \exp(-t/\tau_0 - A[\ln vt]^3) \qquad (3.196)$$

where $A = \pi a^3 N/6$ and $N = [X] \times 6.02 \times 10^{20}$.

Direct linearization of Eq. (3.196) is impossible since the value for I_0 is not known sufficiently accurately (because of the finite duration of the exciting flash), but its approximate form can be linearized:

$$\frac{I(t)}{I(t_1)} = \frac{t^{-\frac{1}{2}\pi a^3 N[\ln vt_1]^2}}{t_1} \times e^{-(t-t_1)/\tau_0} \qquad (3.197)$$

where t_1 is an arbitrary instant.

Eq. (3.191) may be appropriately represented for the treatment of experimental data as:

$$F(I) = -\beta \ln t/t_1 \qquad (3.198)$$

where

$$F(I) = \ln \frac{I(t)}{I(t_1)} - \frac{t - t_1}{\tau_0} \qquad (3.199)$$

and

$$\beta = \frac{\pi a^3}{2}[\ln vt_1]^2. \qquad (3.200)$$

A value for β can be easily found from experimental data, whereas it is rather difficult to determine a and v separately. Rate constants for reactions occurring via a tunnelling mechanism barely depend on temperature, and the kinetics of luminescence decay are practically the same at all temperatures.

3.8.4 Phosphorescence and delayed fluorescence kinetics

Phosphorescence is usually studied in the solid state since the rate constants for phosphorescence emission are usually small (10^{-1} to 10^3 s^{-1}) and impurities inevitable in fluid solutions quench it appreciably. Exceptions to this are such compounds as biacetyl and dibenzyl which also exhibit rather strong phosphorescence in liquid media. The phosphorescence of other compounds in liquid solutions can be detected only with the use of single-photon counting. Studies on triplet states in liquids can be carried out successfully by means of flash photolysis.

The quantum yield of phosphorescence is expressed as

$$\varphi_p = k_p \tau_T \varphi_T = k_p \tau_T k_{isc} \tau_0 \tag{3.201}$$

where k_p and k_{isc} are the rate constants for phosphorescence emission and intersystem crossing $S_1 \rightarrow T_1$, respectively, τ_T and τ_0 are the lifetimes of the excited triplet and singlet states, respectively, and φ_T is the triplet state quantum yield.

The decay of phosphorescence in the solid state usually occurs exponentially in the absence of additives:

$$I_p(t) = I_p(0) \exp(-t/\tau_0). \tag{3.202}$$

In the presence of large concentrations of a quencher, static quenching is observed which leads to a decrease in the quantum yield of phosphorescence without affecting the decay time (similarly to the static quenching of fluorescence). Certain compounds, especially substances containing heavy atoms which enhance intersystem crossing, form complexes with a phosphorescing material, thus changing the phosphorescence decay time and even increasing, in some cases, the quantum yield of phosphorescence owing to the increase in rate constant of phosphorescence emission by these complexes. Under these conditions the phosphorescence decays nonexponentially. The simplest decay kinetics are given by the sum of two exponentials, one of them corresponding to free molecules ^3M and the other to the complex ^3MX. The relative contribution of these two exponentials depends not only on the concentration of the complexing agent but on the duration of the exciting pulse, since the time needed to attain the steady state at a given intensity of the exciting light is different for species with different lifetimes.

The growth of luminescence intensity on excitation by light of constant intensity is described by:

$$I(t) = I(1 - e^{-t/\tau}) \tag{3.203}$$

where I is the stationary intensity and τ is the luminescence decay time.

Each of the components of the phosphorescence will grow at its own rate, depending on the value of τ. This provides the possibility of identifying more clearly one or another exponential simply by varying the duration of the exciting flash.

The emission observed from the $S_1 \rightarrow S_0$ transition with a decay time much longer than the lifetime of the excited singlet state S_1 is called 'delayed fluorescence'. There are three known types of delayed fluorescence: activation fluorescence, annihilation fluorescence, and recombination fluorescence. Having undertaken a detailed study on the processes of delayed fluorescence, Parker named the first two as E-(eosine) type and P-(pyrene) type delayed fluorescence, respectively. The activation-delayed fluorescence is observed in cases when triplet molecules can be brought back to the singlet excited states by thermal excitation: under these circumstances delayed fluorescence is possible when there is a small difference in energy between the triplet and singlet states, for example for solutions of fluoresceine in glassy boric acid and in thoroughly-deaerated solutions of eosine in glycerol or ethanol. The annihilation-delayed fluorescence arises through the interaction (annihilation) of two triplet states to produce one molecule in the excited singlet state and the other in the ground singlet state:

$$^3M + {}^3M \rightarrow {}^1M^* + {}^1M. \tag{3.204}$$

Annihilation-delayed fluorescence is typical of solutions of pyrene, anthracene, and phenanthrene, and of many other compounds. The recombination-delayed fluorescence is observed in media of extremely high viscosity where electrons produced in photoionization are trapped. Owing to diffusion, the electrons can recombine with the positive ions, yielding excited singlet molecules:

$$M + h\nu \rightarrow M^+ + e^- \tag{3.205}$$

$$e^- + M^+ \rightarrow {}^1M^* + h\nu'. \tag{3.206}$$

Delayed fluorescence of this type was identified in frozen solutions of aromatic amines, phenols, and some dyes. The recombination-delayed fluorescence is characterized by its comparatively long duration (up to several hundred seconds) and by a complicated nonexponential decay.

Let us consider the kinetics of the first two types of delayed fluorescence. The quantum yield of activation-delayed fluorescence is given by

$$\varphi_{af} = \varphi_f k_{exc} \tau_T \varphi_T. \tag{3.207}$$

The quantum yield ratio of delayed fluorescence φ_{af} to phosphorescence φ_p is

$$\frac{\varphi_{af}}{\varphi_p} = \varphi_f \frac{k_{exc}}{k_p} \tag{3.208}$$

where k_{exc} is the rate constant for thermal excitation of the $T_1 \rightarrow S_1$ transition, k_p is the rate constant of phosphorescence emission, φ_f is the quantum yield of normal fluorescence, and φ_T and τ_T are the quantum yield and the lifetime, respectively, of the triplet state.

The intensity of activation-delayed fluorescence is obviously dependent linearly on the intensity of the exciting light, and the decay occurs in parallel with the phosphorescence decay with the same rate constant. The rate constant for thermal activation may be expressed as

$$k_{exc} = k_{exc}^0 \exp(-\Delta E/RT) \tag{3.209}$$

where ΔE is the difference in energy between the S_1 and T_1 states.

Then for the quantum yield ratio of the delayed fluorescence to the phosphorescence, we obtain an Arrhenius-type equation

$$\frac{\varphi_{af}}{\varphi_p} = \frac{\varphi_f}{\varphi_p} k_{exc}^0 \exp(-\Delta E/RT). \tag{3.210}$$

The annihilation-delayed fluorescence is described by more sophisticated equations. The rate of emission is given by

$$v_{an.f} = \varphi_f k_{an} [^3M]^2 \tag{3.211}$$

where k_{an} is the rate constant of triplet–triplet annihilation.

If the rate of excitation is not too high, the decay of the triplet states usually occurs via a unimolecular process, and the steady-state triplet concentration is expressed as

$$[^3\overline{M}] \approx v_0 \tau_0 k_{isc} \tau_T. \tag{3.212}$$

The quantum yield of annihilation-delayed fluorescence

$$\varphi_{an.f} = v_{an.f}/v_0 = \varphi_f k_{an} v_0 (\tau_0 k_{isc} \tau_T)^2 \tag{3.213}$$

depends on the intensity of the exciting light. The intensity of this fluorescence is proportional to the square of the intensity of the exciting light. On pulsed excitation, the concentration of triplets in a solution decreases exponentially (for not very large initial concentrations of triplets when $k_{an}[^3M] \ll 1/\tau_T$):

$$[^3M] = [^3M]_0 \exp(-t/\tau_T), \tag{3.214}$$

and the intensity of the delayed fluorescence also decays exponentially:

$$I_{an.f}(t) = I_{an.f}(0) \exp(-2t/\tau_T). \tag{3.215}$$

The decay rate constant for annihilation-delayed fluorescence turns out to be twice as large as the rate constant of triplet state decay. In those cases where the decay of the triplet state is largely caused by triplet–triplet annihilation (for example, under flash photolysis conditions), the kinetics of delayed fluorescence depart from their exponential character.

3.9 SCAVENGING TECHNIQUES

Studies on the HEC processes of a pure material or of a system under closely defined experimental conditions cannot completely clarify the role of all intermediate stages or provide information on the whole set of transient species. Changing the physical conditions in carrying out the process (for example by freezing liquid or gaseous mixtures, by melting or vaporizing materials, or by applying an electric or magnetic field, etc.)., it is possible to obtain further information on the transient species. However, this information should be used with care since its quantitative and, sometimes, even qualitative aspects are not necessarily true for the same process carried out under the conditions we are interested in.

In response, a further method has been invented and is used extensively in HEC: the chemical technique of perturbing a system by doping it with a molecule (the 'scavenger'), the reactivity and transformations of which in the process under investigation are well known. Capable of capturing transient species of any type (excited molecules, transient ions, solvated or unsolvated electrons, radicals, and highly reactive molecules) this substance has acquired the name 'scavenger' (acceptor), and the method itself is called the 'scavenging technique' or 'scavenger competition' method.

A scavenger is selected in such a way that the product of its conversion differs from the products of transformation of the system under study. Scavengers are used both for pulsed and continuous irradiation. Introduction of a scavenger affects the occurrence of certain stages of a process, which is manifested by a change in composition and yield of the intermediate and final products. In addition, products of the conversion of the acceptor appear. By analysing the character of the observed changes and their kinetics, it is possible to deduce the chemical nature of the various transient species, the mechanisms of their formation, and their kinetic features. There is a tendency to use a specific solute which is highly reactive toward only one type of transient species. However, highly reactive solutes are often very efficient in reactions with different types of chemical species. So, well-founded conclusions can be obtained only by introducing different scavengers in turn, including, in particular, those with properties changing systematically (reactivity, redox potential, optical characteristics, etc.). If contrasting types of transient species are produced in a system, for example oxidizing and reducing species, then conjugate pairs of solutes should be used: one designed to react with the oxidizing intermediate, while the other with the reductant. 'Transmutating solutes' which convert one type of transient species into another, for example an oxidizing species into a reducing one or vice versa, can also be used.

Thus, intermediate stages of processes which are not amenable to direct observation in HEC can be elucidated by scavenging the transient species, using appropriate solutes. One might recall that the first information about the hydrated electron was obtained by using scavengers, and even the charge of hydrated electron was determined in this way [42]. In fact, this technique can also be used successfully in studies on thermal processes, provided that reactive transients are involved.

When a scavenger of transient species is added to a system, some reactions in the system are suppressed. Instead, we obtain information on just one or two reactions constituting the process (HEC processes generally include a large number of reactions, although they are very fast). Use of a sufficiently large range of different scavengers makes it possible, in principle, to identify all the intermediate steps of a process.

One procedure of the scavenging method is variation of the scavenger concentration. This procedure allows one to express the concentration of a certain reaction product as a function of scavenger concentration, to obtain the corresponding equation, to test its fit to experimental data, and to determine in the case of a good fit, the kinetic characteristics of the transient species, using the steady-state approximation (see section 3.7). By varying the scavenger concentration, it is possible to study processes of different duration including very short ones. Thus, the reactions of some radicals and excited molecules occurring over several ns have been studied in a 1 M solution of an effective scavenger. In the case of quasi-free electrons in polar and nonpolar systems, where the mobility of the electron is high, it is possible to interfere with processes occurring within a fraction of a ps by using scavengers. An example of this is the scavenging of the quasi-free electron before solvation (see Fig. 2.13). Thus the scavenger technique is a valuable supplement to other experimental methods, in particular, to pulse methods, provided that there is a rather wide selection of scavengers with well-defined kinetic characteristics of their reactions with a reasonable number of different transient species.

3.10 MATHEMATICAL SIMULATION OF PROCESSES IN HIGH-ENERGY CHEMISTRY

Mathematical simulation in chemical kinetics is based on the logic of processes stemming from their physicochemical nature [195]. The steps in a mathematical description are (i) that of the primary event, that is, the interaction of molecules, atoms, electrons, etc. leading, in particular, to a chemical reaction (definition of the cross-section of a process), (ii) the distribution function and population of quantum levels of the reactants and intermediates (averaging of the cross-sections by these distribution functions allows the rate coefficient or the rate constant as a macroscopic property in nonequilibrium or equilibrium chemical reactions, respectively, to be determined), and (iii) macroscopic representation in terms of concentration, using a necessary and sufficient (but not entire) set of elementary steps on the basis of the principle of microscopic reversibility as the final step of the description.

3.10.1 Primal kinetic problem: the stiffness

The formal expression of the primal kinetic problem is a system of ordinary (usually nonlinear) first-order differential equations with preset initial values. The solution of such systems is specific in the sense that the time-characteristics of the different variables differ from each other considerably. Kinetic systems of this type involve rapidly- and slowly changing variables. The fast variables are adjusted almost instantaneously to changes in the slow variables. This enables the steady-state approximation to be applied to solve such problems, that is, some differential equations of the system may be replaced by algebraic equations on taking the rates of change in the fast variables as equal to zero [196].

The existence of fast and slow subsystems specify the difficulties encountered in the numerical solution of the primal kinetic problem. To make the computation by fast variables sufficiently accurate, it is necessary to choose an integration step appreciably less than the total time of occurrence of the process as determined by the change in the slow variables.

The sets of differential equations describing the behaviour of both fast and slow subsystems are 'stiff'. This is the property of stiffness in the Cauchy problem

$$\vec{y} = \vec{f}(y), \quad t \text{ belongs to } [a, b], \vec{y}(0) = \vec{y}_0, \tag{3.216}$$

where $\vec{y} = [y_1(t), y_2(t), \ldots, y_n(t)]^T$ and $\vec{y}_0$ is preset, which is the principal cause of complications in the solution of such sets of equations.

In the case of the primal kinetic problem, the vector $\vec{f}$ always satisfies the Lipschitz condition over all n components of the vector $\vec{y}$, so it can be proved that a solution of the Cauchy problem exists and that it is unique [197].

The numerical solution of the Cauchy problem consists of the calculation of a series of approximations $\tilde{\vec{y}}_s = \tilde{\vec{y}}(t_s)$ in a series of discrete points $\{t_s\}$, $t_1 = a$, $t_s = b$, $t_{s+1} = t_s + h_s$, where h_s is the size of the step. The procedure for obtaining the numerical solution has to provide an approximation to the correct solution by numerical solutions at the discrete points taken.

A natural requirement for the procedure is the possibility of evaluation of the difference $\left|\vec{y}(t_s) - \tilde{\vec{y}}_s\right|$ at all the points taken. This difference must decrease as the step size of the net where the numerical solution is searched for, is reduced. Such a property of numerical methods is called the convergence. A procedure is said to be convergent if, for a step size tending to zero, the difference between the true and numerical solutions also tends to zero at all points of subdivision.

The second necessary property of the numerical method is its stability. This property means that small perturbations in the initial values or in the equation of difference caused by, for example, rounding-off or limited machine word length are suppressed or, at least, they do not increase. The same procedure can be either stable or unstable, depending on the step size. However, the stability of the method is determined not only by the numerical integration step but also by the properties of a particular Cauchy problem. Therefore, to describe the stability of a method of numerical integration regardless of a particular problem, the method is usually investigated for its stability in the solution of the model linear Cauchy problem

$$\dot{y} = \lambda y, \quad y(0) = y_0 \tag{3.217}$$

where λ is the complex constant. This is natural since the local behaviour of a solution of an initial problem is determined, to a first approximation, by the solution of the linear problem with a matrix which is the Jacobian of the original system. In theory, the concept of the absolute stability of the method is also used. A method is called absolutely stable if the global error $\left|\bar{\bar{y}} - \bar{y}(t_s)\right|$ for a prefixed step size of the integration remains limited at $s \to \infty$. Setting the problem in this way, for every method it is possible to find a region in the complex plane λh, where this method has the property of absolute stability.

Let us consider some general ideas [198] implemented in the numerical methods of solution of the Cauchy problem.

As has been pointed out, the stiffness is a property of the Cauchy problem exhibited upon the description of systems with essentially different time-characteristics of their processes. The stiffness of the problem can be identified in investigation of the local behaviour of the solution for a set of equations. To do this, the set of equations is linearized, that is, it is replaced by a linear system with a Jacobian matrix. If in the vicinity of the solution the Jacobian matrix changes insignificantly, then the linear system does describe the nonlinear set locally.

The Cauchy problem is said to be stiff [196] if, in a local region, the problem is stable, that is, the Jacobian eigenvalues have negative real parts $\mathrm{Re}(\lambda_i) < 0$ and $\mathrm{Re}(-\lambda_{max})/\mathrm{Re}(\lambda_{min}) \gg 1$.

The local stiffness in problems of kinetics can reach values of the order of 10^6 to 10^9. Difficulties encountered in the solution of stiff problems are determined by the fact that the step size of the integration in the numerical solution can be restricted by a requirement on the absolute stability of the method related to small perturbations arising in the process of realization of the method by means of a digital machine. Indeed, the step size of integration has to be selected in such a way that λ_{max} will be an element of the region of absolute stability of the method. Thus, the step size of integration is consistent with the characteristic time of a fast process $1/\mathrm{Re}(-\lambda_{max})$ whereas the characteristic time $1/\mathrm{Re}(-\lambda_{min})$ of a slow process is much greater, and the necessary number of integration steps will be comparable with $|\mathrm{Re}(\lambda_{max})/\mathrm{Re}(\lambda_{min})|$.

3.10.2 Sensitivity analysis of rate equations

A considerable aid in both the treatment of experimental kinetic data and the mathematical modelling of chemical reactions and processes can come from the so-called sensitivity analysis in regard to the equations of chemical kinetics. The task is the investigation of the sensitivity of the solution of the Cauchy problem to a change in the parameters involved in a set of ordinary differential equations:

$$\dot{X} = f(X, \vec{K}), \qquad x(0) = X_0$$

where $\vec{K}$ is the m-dimensional vector of the parameters and X is the n-dimensional solution vector.

The question to be answered in the sensitivity analysis is: if there is an uncertainty in setting the vector $\vec{K}$, what will be the uncertainty in values of the solution vector of Set $X(t)$?

As applied to the solution of a system of equations of chemical kinetics, the question may be reformulated thus: how will variation in a rate constant affect the form of the calculated rate curve? If the distribution densities $g(K)$ can be defined, then the problem of sensitivity analysis is reduced to that of calculating the distribution densities $g_i[X_i(t)]$.

Without dwelling upon the details of the calculation, we shall point out that the operation usually consists of determination of the first-order sensitivity coefficients which are the partial derivatives of the component concentrations for a chemically reacting system with respect to the rate constants. One should distinguish between local and global sensitivity analyses. The details of the method, and examples of its use in extracting information, can be found elsewhere [96].

3.10.3 Fitting the most probable mechanism to chemical reactions
Implementation of the methods of mathematical simulation in chemical kinetics gives the possibility of stating and, to a considerable (though not absolutely complete) extent, of formalizing in algorithmic language, the problem of finding the most probable mechanism for a set of reactions in a chemical process. To become more familiar with this productive approach, the reader is referred to monographs by Slovetskii [199] and Gagarin et al. [200].

3.10.4 Inverse kinetic problem. Stating the problem
The solution of the inverse kinetic problem consists of constructing a model describing the experimental data and in extracting the maximum information on the kinetic parameters of the mechanism from these data. Mathematically, the choice of such a model means constructing and writing the right-hand part of the set of differential equations of the primal kinetic problem. Thus, it is necessary to make assumptions on the set of components taking part in the chemical reaction and on the type of rate law (which can be the law of mass action, law of active surfaces, equations of nonequilibrium chemical kinetics, etc.). The next step in constructing the model is writing the proposed reaction scheme or several alternative mechanisms.

For a quantitative estimation of the agreement between the experimental and calculated data, a certain functional including parameters of the model is written (these parameters are usually the rate constants of chemical reactions). This functional has to include as much a priori available information as possible. It represents the mechanism of a process (chain, radical, etc.), semi-quantitative data on the concentrations of transient species, etc. [201]. Additional information of this sort provides the possibility of considerably reducing the area of research and controlling [98] the task of minimization.

The parameters of the model are varied to find the minimum of the chosen function. Then, using mathematical statistics, the hypothesis tested can be either dismissed or accepted with some definite degree of certainty, that is, the problem of the authenticity of the model can thus be solved   [201,   202]. The methods of mathematical statistics which are used in tests of authenticity are considered in detail elsewhere [202, 203]. If the hypothesis has to be abandoned, it is necessary to construct a new model.

Having tested a hypothesis for authenticity, one must find whether the corresponding model is unique. Let us consider the main reasons for the nonuniqueness of a solution of the inverse kinetic problem [202]. Since, in practice, experiments are carried out within a limited time, then, on one hand, this may turn out to be insufficiently long to determine rate constants for slow reactions (the procedure is asymptotic in terms of small constants) and, on the other hand, the time-resolution of the measurements may appear insufficient for determination of rate constants for fast reactions (the procedure is asymptotic in terms of large constants). There can also be a situation when both of these extreme cases are realized simultaneously. A sufficient condition for the existence of only one solution of the inverse kinetic problem is the possibility of measuring the concentrations of all components at any instant (usually measurable, in practice, are the concentrations of stable products). The relation between the uniqueness of the solution of the inverse kinetic problem with the properties of the matrix of the derivatives of experimentally-measurable quantities with respect to reaction rate constants is considered by Spivak & Gorskii [204].

One of the principal problems in the solution of the inverse kinetic problem is the definition of a set of parameters such that their values can be found from experimental data for a given kinetic scheme. The proposal of a complex mechanism, consisting of many steps in the solution of the inverse kinetic problem, means, in general, that some combinations of the unknown rate constants can be determined but not the constants themselves.

Let the initial system of ordinary differential equations be

$$\vec{y} = \vec{f}(\vec{k}, \vec{y}), \qquad \vec{y}(0) = \vec{y}_0 \tag{3.218}$$

where $\vec{y}$ is the n-dimensional solution vector, $\vec{k}$ is the m-dimensional vector of the parameters determining the right-hand parts of System (3.218), and y_0 is the initial value of $\vec{y}$.

This system, in a certain time, can be reduced to a simpler system of ordinary differential equations and a system of algebraic equations:

$$\vec{y}_1 = \vec{f}(\vec{k}, \vec{y}_1, \vec{y}_2) \tag{3.219}$$

$$\vec{f}_2(\vec{k}, \vec{y}_1, \vec{y}_2) = 0$$

where $\vec{y}_2$ is the vector of the fast variables for which the quasi-steady state approximation can be used and $\vec{y}_1$ is the vector of the slow variables. Then, in the case where the system

$$\vec{f}_2(\vec{k}, \vec{y}_1, \vec{y}_2) = 0 \qquad (3.220)$$

is soluble with respect to $\vec{y}_2$, we can substitute $\vec{y}_2(\vec{y}_1)$ into the first equation of Set (3.219) to obtain a system of equations including combinations of the components of the rate-constant vector $\vec{k}$. They can be determined unambiguously as a result of the solution of the inverse kinetic problem.

3.10.5 Classical trajectory method

Studies on the kinetics of chemical reactions in gases often require calculation of the cross-sections and probabilities for physicochemical processes involving heavy particles (atoms, molecules, or ions). The cross-sections can be determined with the use of both statistical and dynamical approaches. The statistical methods (for example transition-state theory or the Rice–Ramsperger–Kassel–Marcus (RRKM) theory) usually produce analytical expressions of the values concerned, while simulation of the dynamics of interaction of particles almost always requires the use of numerical methods. However, the category of processes and systems which can be studied with the dynamical approach is much wider. In some cases the applicability of one or another statistical method can be examined only by means of dynamical calculations.

One more advantage of dynamical calculations is that these are the only way to check whether a given potential energy surface (PES) adequately fits the elementary process under consideration.

Dynamical calculations are most frequently used in the form of the so-called 'classical trajectory method' [87, 205–207] in which the motion of heavy particles involved in the process to be simulated can be described with the classical equations of motion. Using the classical trajectory method, it is possible to study almost all processes in which heavy particles take part, except, perhaps for, processes at low temperatures when quantum mechanical tunnelling may become appreciable.

Major obstacles to the application of the classical trajectory method lie in the construction of potential energy surfaces adequately fitting the simulated elementary process.

3.10.6 Monte Carlo method

In tackling problems on high-velocity gas dynamics and chemical kinetics and some other problems it is necessary to know the fundamental features of different relaxation processes. These processes are the establishment of the Maxwell distribution, rotational and vibrational relaxation, dissociation, ionization, etc. It is essential here that the descriptions of the elementary event are known and the corresponding statistical problems are stated and solved [208].

Analytical solution of such problems is complicated by factors which can be separated into two groups. The first group of difficulties is related to the mathematical statement of the problems of physical and chemical kinetics. The question raised is whether classical mathematics is appropriate for the description of the physical phenomena that we are interested in. The second group is connected with the methods of solution of rate equations. All analytical methods are, in one way or another, related to the expansion of the desired quantities into series in terms of small parameters. In an appreciable number of cases of great theoretical and practical interest there is no possibility of distinguishing such parameters. However, the validity of perturbation theory itself seems to be a more critical issue. The procedures of expansion into series do not necessarily envisage consideration of higher-order terms, which can severely misrepresent the real physical picture.

The rapid advance of computers gave rise to a number of novel numerical methods, one of them being the Monte Carlo method (or the statistical trial technique) [209, 210]. Let us consider briefly the basic aspects of the method and its applicability to the problems of physical and chemical kinetics.

The Monte Carlo method is essentially a mathematical experiment. In some cases it consists of the construction of an artificial random process in such a way that the average of a random variable would correspond to the solution of a system of integrodifferential equations. It can also consist of the reduction of an original probabilistic physical process to a model permitting practical realization with a computer [210]. The most important advantage of the Monte Carlo method is that it gives the possibility of constructing models which help to get round serious, often insurmountable, obstacles encountered in the solutions of some problems with analytical methods. By using the Monte Carlo method one can succeed even in cases when there is no possibility of formulating relevant equations.

Let us turn now to the aspects of implementation of the Monte Carlo method in the solution of problems in chemical kinetics. A system is subdivided into a 'medium' and an ensemble of 'trial particles', the medium being described phenomenologically through such parameters as the concentration of the individual components, temperature, etc. Only the interaction of the trial particles with the medium is taken into account here. So far as the problems of kinetics are concerned, it may be concluded that systems consisting of a small addition of molecules of the gas under study to molecules of an inert gas acting as a heat bath can be investigated with this method. The concentration ratio of the heat bath to the additive has to be of such a value that collisions other than of the added molecules with the particles of the heat bath can be disregarded. Naturally, there are some cases when these simplifications can be, and must be, accepted. The construction of a nonlinearized model is a fundamental issue. This possibility exists in principle, and it is realized via implementation of the concept of 'periodic boundary conditions'.

3.10.7 RRKM method in unimolecular reaction rate theory

In many physicochemical processes occurring under nonequilibrium conditions, the Maxwell–Boltzmann distribution is disturbed and hence the calculation of the average characteristics of chemical reactions, namely the rate coefficients, presents a problem as to the determination of nonequilibrium distribution functions. For example, the calculations of nonequilibrium energy distribution functions of electrons and the vibrational level distribution functions of molecules are of great interest in studies on chemical transformations in low-temperature discharge plasmas [208, 211].

We shall briefly describe one of the most popular statistical theories of unimolecular reaction, namely the RRKM approximation. A detailed account of this theory is given in books devoted to unimolecular reactions [211, 212]. Using the formulae of the statistical RRKM theory, it is possible to calculate both the reaction rate coefficient (rate constant) for a unimolecular reaction and the resulting quasi-stationary function of the energy distribution in terms of different degrees of freedom.

The principal assumption in statistical RRKM theory is that the energy is almost instantaneously redistributed among all the quantal states of the same energy. This allows the rate of dissociation of the molecule from a given quantum state to be characterized solely by the energy of this state. However, not all of the molecular energy can be freely redistributed over the available degrees of freedom insofar as not all of these are equivalent; thus the concept of active and inactive degrees of freedom was introduced into the theory. Those degrees of freedom (for example rotational) whose quantal states remain unchanged during a reaction are generally referred to as inactive or adiabatic. Nonadiabatic vibrational degrees of freedom are considered as active, so there is no statistical exchange of energy between these and the inactive degrees of freedom.

The second assumption in RRKM theory is the hypothesis concerning 'strong' collisions. This means that the deactivation of an active molecule takes place on every collision, thus a large quantity of energy is transferred in collision, that is, sufficient for deactivation. The strong collision hypothesis makes it possible to omit details of the dynamic behaviour of a system from consideration.

3.10.8 Laplace transform

The form of the fundamental equations of chemical kinetics prompts the wide use of Laplace transformations [120, 121]. Thus, the rate coefficient of a bimolecular chemical reaction is the Laplace transform of the function $E\sigma(e)$ (σ is the reaction cross-section): it depends only on the relative translational energy of the molecules and on their orientation. The mathematical properties of the

Laplace transform are very convenient for practical use. The original functions have a physical meaning in kinetic problems satisfying all limitations, and can be mapped unambiguously, moreover, the transform is linear and additive. Most cross-sections significant for chemical kinetics have a threshold nature which necessarily leads to emergence of the Arrhenius term $\exp(-E_{th}/kT)$ in the equation for the rate coefficient [88, 120, 119].

3.10.9 Methods of the qualitative theory of differential equations (multiplicity of stationary states and the stability problem)

The methods of the qualitative theory of differential equations are widely used in various branches of science and technology, including the mathematical modelling of chemical reactors and processes.

The issue of the qualitative investigation of differential equations was first stated by Poincaré in 1885 in relation to the problems of celestial mechanics. The concept of the trajectory is of fundamental importance in these methods, especially on their application in nonlinear vibration theory by Andronov. The trajectory (in the two-dimensional case) is the curve

$$x = \varphi(t - t_0, x_0, y_0) = x(t)$$
$$y = \varphi(t - t_0, x_0, y_0) = y(t) \tag{3.221}$$

where the pair functions $x(t)$ and $y(t)$ are the solution of the system

$$\dot{x} = P(x, y), \qquad \dot{y} = Q(x, y) \tag{3.222}$$

extended to the maximum possible range of values.

Generalization to the multidimensional case is accompanied by a series of complications, but the principal idea of the approach remains the same. Mathematical models represented by differential equations are designed to give the possibility of determining the state of a system at any moment, provided that its state is known at a certain initial time. Models of this sort[†] are dynamical systems [213].

The states of a dynamical system in which the system does not change with time, or its properties are repeated periodically, are called steady or stationary (see section 3.9). Chemical processes (and chemical reactors) can have one or several steady states corresponding to the same values of their parameters. From the physical point of view, the existence of several stationary states for a dynamical system is caused by its nonlinearity. By varying the values of the parameters for a system of differential equations, both the number and the

---

† A mathematical model of some phenomenon claims to provide a true representation of certain features of the phenomenon provided that these features do not disappear on slight variation of the differential equations. Mathematical models satisfying this requirement are called 'crude', and the corresponding systems 'crude systems'. A more specific definition is as follows. Dynamic systems are termed crude if they conserve the qualitative character of the arrangement of phase trajectories for a sufficiently small change in the parameters involved in the differential equations.

stability of the equilibrium states of the system generally change [214]. The complete solution of the problem of the stability of a chemical process consists of subdividing the parameter space of its mathematical model into regions differing in the number and type of stability of equilibrium states [213, 215, 216–219].

3.11 KINETICS OF CHEMICAL REACTIONS INVOLVING VIBRATIONALLY EXCITED MOLECULES

A problem in the description of reactions involving vibrationally excited molecules (VEM) arises in the simulation of kinetic processes in systems at vibration–translational equilibrium (plasma chemical and laser chemical systems) [95]. The most consistent approach is the application of the general principles of nonequilibrium kinetics [88] and solution of the balance equations for the population of every vibrational level of the molecules participating in the chemical reaction. Such an approach enables one to simulate a series of reactions of diatomic VEMs with atoms in laser systems [137], with the excitation of a number of the first vibrational levels in the reactant molecules.

However, direct implementation of the general principles of nonequilibrium chemical kinetics to systems of chemically reacting polyatomic VEMs involving multi-stage mechanisms runs into substantial obstacles [96]. Firstly, the rate coefficients for a large number of processes which have to be taken into account in the solution of balance equations are unknown, and, secondly, there arise some mathematical problems related to the very large dimensionality of the system obtained. Hence, it is necessary to develop simplified approaches in the description of reactions of VEMs in complex multistage processes. One of these approaches was suggested by Zhivotov et al. [220]. It involves the following operations: (i) calculation of the level's rate coefficients for reactions of VEMs, (ii) self-consistent analytical calculation of the vibrational energy distribution function for the reacting molecules, and (iii) mutual solution of the equations for the concentrations of the components of the system together with the balance equations for vibrational and translational energies in the system.

(i) In describing the effect of vibrational excitation of reactants on the rate coefficient of a chemical reaction (see section 3.8) it is convenient to use an empirical factor α of the effectiveness of the vibrational energy which characterizes that part of the vibrational energy involved in overcoming the activation barrier of reaction. The microscopic rate coefficient (that averaged by the Maxwell–Boltzmann distribution for translational and rotational energies, but not averaged by the distribution for vibrational energy) [96] of a chemical reaction in this case is:

$$k_v = k_0 \exp\{\theta(E_a - \alpha E_v)(E_v - E_a)\} \tag{3.223}$$

where E_v is the vibrational energy of the VEM, k_0 is the pre-exponential factor, and θ is the Heaviside function.

A value for the coefficient α in a given reaction can be calculated by using the classical trajectory method [96] or can be determined with empirical rules suggested *ibid.* [96] or with a simple semi-empirical equation

$$\alpha \approx \frac{E_a^+}{E_a^+ + E_a^-} \tag{3.224}$$

where E_a^+ and E_a^- are the activation energies for the forward and reverse reactions respectively.

(ii) To obtain the macroscopic rate coefficient for a nonequilibrium reaction involving VEMs, it is necessary to integrate Eq. (3.223), making use of the vibrational energy distribution of the reactant molecules. In the simplest case of a Boltzmann distribution with a temperature T_v different from the translational temperature T, one finds

$$k = k_0 \exp(-F_a/\alpha T_v) \quad \text{at} \quad \alpha T_v > T. \tag{3.225}$$

Hartman [197] derived analytical equations for macroscopic rate coefficients of reactions of VEMs for boundary cases when a chemical reaction barely affects the character of the vibrational energy distribution of the molecules, or *vice versa* it determines its appearance.

(iii) Using the dependencies $k(T_v, T)$, a system of rate equations supplemented by the balance equations of vibrational and translational energy in a chemically reacting system is solved, values for $k(T_v, T)$, T_v, and T being determined self-consistently. The numerical solution, as usual, encounters some mathematical complications related to the stiffness of the system. However, there are special numerical methods for the adequate solution of stiff systems [96].

Examples of implementation of the above approach in simulation of the kinetics of complex plasma chemical processes under conditions of vibration–translational nonequilibrium can be found in [220].

3.12 KINETICS OF RADIATION-INDUCED PROCESSES AT THE CHEMICAL STAGE

As with conventional chemical processes, a quantitative description of the kinetics of a radiation-induced process with analytical equations can be carried out only for the relatively simple cases discussed in preceding sections of this chapter. In principle, these are those cases where the mechanisms includes one or two dominant reactions or where most of the reactions are suppressed by appropriate scavengers of transient species, which greatly simplifies the radiolysis scheme. So, most data on reaction rate constants of transient species have been obtained in model experiments. Nevertheless, a complete set of kinetic characteristics of a reaction cannot be provided even in model experiments insofar as some types of intermediate species cannot be detected directly or even indirectly.

Meanwhile, there seems to be an urgent need for the kinetic description of a radiation process as a whole, since neither optimization of the process nor any prediction of change in behaviour of the system under irradiation by varying the process conditions is possible without solution of the kinetic problem. The kinetic problem for a rather complex system as met in radiolysis can be solved, in principle, by simulation methods.

Radiation chemistry employs mathematical simulation methods for many different purposes. Some of the problems connected with the physicochemical (inhomogeneous) stage will be given some consideration in section 3.13. Here we shall concentrate on the homogeneous stage of radiolysis which begins in most liquids 10^{-8} to 10^{-7} s after the primary act; the basic features of radiation-induced chemical processes occurring at this stage can now be established experimentally for almost any liquid or gaseous system (see section 5.1).

The chemical stage of radiolysis generally involves a considerable number (a few dozen) of transient species and stable products for a process occurring in the bulk, each species being able to take part in many reactions. The kinetics of the process at this stage is described with a set of N ordinary first-order differential equations (where N is the total number of types of transient and final products of radiolysis of a given system). Such a set of equations can be solved only by numerical methods. The systems of ordinary differential equations under conditions of radiolysis are called 'stiff' since the concentrations of different species taking part in the process may differ by several orders of magnitude (see below) because of the difference in their reaction rate constants.

For practical modelling of the homogeneous stage in a non-stirred (no mass-transfer) system, the description of the chosen set of experimental data requires the following parameters to be defined: (i) the concentration of each component of the system at zero time, (ii) the composition of the transient and final products, (iii) the sequence of corresponding chemical reactions (the mechanism), (iv) the radiation yields for the transient and final products of the first generation (the first generation products means those species coming into existence at the moment chosen as zero time in the calculation), (v) the rate constants for reactions involved in the scheme, (vi) the dose rate as a function of time, and (vii) the range of integration over time.

Let us consider the above parameters in more detail.

(i) The concentration of every component of a system, especially if the system contains highly reactive solutes of impurities, must either be measured experimentally or calculated for the initial state of the system.

(ii), (iii) The choice of scheme for a process is that part of the simulation which cannot be formalized. It must be established by the researcher on the basis of experimental and theoretical investigation of every stage of radiolysis. This part includes the determination of all intermediates in the radiolytic transformations and identification of the reactions of these species. The adoption of the mechanism follows elaborate experiments on the system, since existing theoretical approaches are hardly able to design the complete scheme of reactions for a compound unless it has been properly studied. This experimental investigation includes testing the system under model conditions created with the use of different solutes, of freezing, of matrix isolation, of pulse radiolysis, of kinetic treatment of separate reactions or competing pairs of reactions, and of other methods of determination of the composition. The reaction scheme must include all possible reactions of identified and hypothetical transient and final products.

(iv) So long as the simulation is carried out only for the chemical stage of radiolysis, it should be noted that all in-spur reactions in a liquid have been completed by this time (in the gas phase there are no such processes) and the primary transient species have produced some secondary or tertiary intermediates and some final products. Naturally, a special series of experimental tests is needed to identify the first-generation products (for a given type of simulation procedure) and to determine their radiation yields. A necessary but not sufficient condition for reliability in the determination of yields at this point is the fulfilment of (a) atom $vs.$ atom material balance equation for species of the first generation and (b) the ion-charge balance.

(v) The rate constant must be defined for each of the reactions involved in the scheme. Usually, only few of the rate constants needed can be found in the literature. The rest of them must be either measured experimentally or estimated on the basis of some theoretical assumptions or by analogy. Some of the rate constants may be allowed to undergo variation during calculation.

(vi) Computer programs currently available have been worked out only for the case of the dose rate being uniform throughout the system. Meanwhile, it can be time-dependent for interrupted or pulsed irradiations or constant for continuous irradiation. Where the dose rate changes, its variation has to be established.

(vii) The range of integration over time also has to be defined since a radiolytic process develops in time. So long as we carry out calculations starting from 10^{-8} s, the step size should be of the same order. But if such a size is used throughout the entire period of the process considered, it will take up much machine time. Therefore, the step size of the integration is increased as the time increases, for example a step size of 10^{-8} s is used for a time interval of 10^{-8} to 10^{-7} s, one of 10^{-7} s for the next interval of 10^{-7} to 10^{-6} s, and so on.

A satisfactory description (within the range of experimental error) of a process with a mathematical model indicates, like all other kinetic methods, only the $authenticity$ of the reaction scheme and the parameters of the model, but cannot $prove$ the suggested mechanism. Rather good evidence for the correctness of the model can be the agreement between the calculated and experimental concentrations of transient species, or the fitting of the calculated and experimental dose dependences for the final products obtained on varying the external parameters (pulse duration, reactant concentration, dose rate, etc.). If one of the internal parameters has a tentative value, the coincidence of this value with that determined experimentally under conditions other than defined during construction of the model will also constitute a proof.

Of course, the simulation of an entire process, even only at the chemical stage of radiolysis, requires much information, the more complex the original chemical system and the more extensive its radiation-chemical conversion, the greater the amount of information needed. However, even with an incomplete knowledge of the internal parameters of the model, it is nonetheless reasonable to attempt a mathematical simulation in addition to an experimental study, at least for particular cases, since it can replace some part of the experimental operations, thereby saving time.

Mathematical simulations of the entire chemical stage of radiolysis have been carried out only for a small number of liquid and gaseous systems [221–227]. Consider as an example the mathematical simulation of the liquid-phase radiolysis of methanol in the absence of oxygen at room temperature [221, 222]. Much attention has been paid to methanol radiolysis, so the principal radiolytic transformations in this alcohol are believed to be well-established.

Let us consider first the simulation of the continuous γ-radiolysis of methanol up to high does, *ca.* 6 MGy (that is, 10% conversion for the average yield of methanol degradation of 5.6 in this dose range). The process involves 15 transient and 17 stable radiolytic products (Table 3.1). The first-generation products are represented by 7 transient and 5 stable products of conversion of methanol (the zero-point of calculation was 10^{-8} s). Some of these products can also be formed in secondary processes, but the yield listed in Table 3.2 refer to the first generation only. The charge balance and atomic balance are met, which can easily be checked with the table (the atomic balance is accurate to within 5%, that is, within the usual limits of experimental error in radiation chemistry). Consequently, the set of first-generation products may be considered to have been determined sufficiently reliably for use in further calculations.

A list of all reactions included in the scheme of methanol radiolysis is given in Table 3.2. There are two reactions, many of them competing with each other; also shown in the same table are the rate constants for the corresponding reactions. So far as some of the reaction rate constants could not be determined experimentally, their estimated values were used. Recombination rate constants for carbon-centred radicals were evaluated by analogy with the reactions of radicals of very similar structure known from the literature. Rate constants for Reactions 17, 19, 26, and 29 were taken as adjustable parameters of the model because no similar reactions with known rate constants were available. These adjustable parameters were varied until the calculated curves fitted the experimental ones within the range of experimental error. Since most analyses were carried out by GLC (accuracy 10%), the overall accuracy in the fitting was determined by a 10% change in the concentration of the product under

Table 3.1. Final and transient products of methanol radiolysis [222]

Transient species	Generation[†]	Stable products	Generation
e_{aq}^-	I (3.0)	C_2H_5OH	I (0.12)
$CH_3OH_2^+$	I (3.0)	CO	I (0.11)
$CH_3\dot{O}$	I (2.5)	$(CH_3)_2O$	I (0.25)
$\dot{O}H$	I (0.6)	H_2	I, II (1.3)
:CHOH	I (1.4)	CH_4	I, II (0.2)
H	I, II, (0.7)	CH_3OH	II
$\dot{C}H_2OH$	I, II, III (0.7)	H_2O	II, III
$\dot{C}H_3$	II	$(CH_2OH)_2$	II, III
$\dot{C}H_2O^-$	II	$HOCH_2CHOHCH_2OH$	II
CH_3O^-	II	$HOCH_2CHOHCH_3$	II
$HOCH_2\dot{C}HOH$	II	CH_2O	II
$CH_3\dot{C}HOH$	II	$CH_2(OH)OCH_3$	II
$HOCH_2CH_2\dot{O}$	III	CH_3CHO	III
$\dot{C}H_2CHO$	III	CH_3COOCH_3	III
$\dot{C}H(OH)OCH_3$	III	$HCOOCH_3$	III
		$(CH_2OHHCHOH)_2$	III
		$CH_2(OCH_3)_2$	III

† Values in parentheses refer to the yields of primary species.

consideration. It should be noted that only two of the reactions with adjustable rate constants, namely Reactions 17 and 26, exerted any appreciable effect on the character of the calculated curves. So far as the other adjustable constants are concerned, the reaction scheme turned out to be weakly-sensitive toward them.

The set of final radiolysis products also needs to be verified as regards elemental material balance. Among the stable products listed in Table 3.1, the concentrations of all substances except for water have been experimentally measured. The dose dependence for water was constructed from data on the material balance. The composition of the ultimate radiolytic products for moderate doses (to 0.1 MGy) may be regarded as sufficiently well-established, that is, all of the possible final products have been identified and their concentrations have been measured. This cannot be said, however, for dose ranges above 0.1 MGy. So, it might have been expected that the mathematical model would satisfactorily describe the dose dependences for the stable second-generation products and give a qualitative description of the dependence for the third-generation products. This was indeed the case.

Table 3.2. Chemical reaction scheme for mathematical simulation of methanol radiolysis [222]

No.	Reaction	$k/M^{-1}s^{-1}$ or s^{-1}
1	$e_s^- + CH_3OH \rightarrow CH_3O^- + H$	8.5×10^3
2	$e_s^- + CH_3OH_2^+ \rightarrow CH_3OH + H$	7×10^{10}
3	$e_s^- + CH_2O \rightarrow \dot{C}H_2O^-$	5.6×10^9
4	$e_s^- + CH_3OH \rightarrow \dot{C}H_3 + OH^-$	1.3×10^3
5	$CH_3OH_2^+ + CH_3O^- \rightarrow 2CH_3OH$	6×10^{10}
6	$CH_3OH_2^+ + \dot{C}H_2O^- \rightarrow \dot{C}H_2OH + CH_3OH$	$6 \times 10^{10\dagger}$
7	$CH_3OH_2^+ + OH^- \rightarrow H_2O + CH_3OH$	$6 \times 10^{10\dagger}$
8	$H + CH_3OH \rightarrow H_2 + \dot{C}H_2OH$	1.6×10^{10}
9	$H + CH_2OHCH_2OH \rightarrow CH_2OH\dot{C}HOH + H_2$	8×10^6
10	$\dot{O}H + CH_2OHCH_2OH \rightarrow CH_2OH\dot{C}HOH + H_2O$	1.4×10^9
11	$\dot{O}H + CH_3OH \rightarrow CH_2OH + H_2O$	8.4×10^8
12	$CH_3\dot{O} + CH_3OH \rightarrow CH_2OH + CH_3OH$	2.6×10^5
13	$CH_3\dot{O} + CH_2OHCH_2OH \rightarrow CH_2OH\dot{C}HOH + CH_3OH$	6.2×10^5
14	$\dot{C}H_2OH + \dot{C}H_2OH \rightarrow CH_2OHCH_2OH$	1.4×10^9
15	$\dot{C}H_2OH + :CHOH \rightarrow CH_2OH\dot{C}HOH$	$1 \times 10^{9\ddagger}$
16	$\dot{C}H_2OH + CH_2OH\dot{C}HOH \rightarrow CH_2OHCHOHCH_2OH$	$1 \times 10^{9\ddagger}$
17	$\dot{C}H_2OH + CH_2OHCH_2OH \rightarrow CH_2OH\dot{C}HOH + CH_3OH$	$10^\S$
18	$\dot{C}H_2OH + C_2H_5OH \rightarrow CH_3\dot{C}HOH + CH_3OH$	$10^\S$
19	$\dot{C}H_2OH + CH_3\dot{C}HOH \rightarrow CH_2OHCHOHCH_3$	$1 \times 10^{9\ddagger}$
20	$:CHOH \rightarrow CH_2O$	$1 \times 10^{6\dagger}$
21	$CH_2O + CH_3OH \rightarrow CH_2(OH)OCH_3$	10
22	$CH_2(OH)OCH_3 \rightarrow CH_2O + CH_3OH$	0.2
23	$CH_2O + \dot{C}H_2OH \rightarrow CH_2OHCH_2\dot{O}$	$1 \times 10^{4\dagger}$
24	$CH_2OHCH_2\dot{O} + CH_3OH \rightarrow CH_2OHCH_2OH + \dot{C}H_2OH$	$1 \times 10^{4\dagger}$
25	$\dot{C}H_3 + CH_3OH \rightarrow CH_4 + \dot{C}H_2OH$	2×10^2
26	$CH_2OH\dot{C}HOH \rightarrow \dot{C}H_2CHO + H_2O$	$1 \times 10^{2\S}$
27	$\dot{C}H_2CHO + CH_3OH \rightarrow CH_3CHO + \dot{C}H_2OH$	$2 \times 10^{2\S}$
28	$\dot{C}H_2CHO + \dot{C}H_2OH \rightarrow CH_3COOCH_3$	$1 \times 10^{9\ddagger}$
29	$CH_2(OH)OCH_3 + \dot{C}H_2OH \rightarrow \dot{C}HOHOCH_3 + CH_3OH$	10
30	$\dot{C}HOHOCH_3 + \dot{C}H_2OH \rightarrow HCOOCH_3 + CH_3OH$	$1 \times 10^{9\ddagger}$
31	$2CH_2OH\dot{C}HOH \rightarrow CH_2OHCHOHCHOHCH_2OH$	$1 \times 10^{9\ddagger}$
32	$CH_2(OH)OCH_3 + CH_3OH \rightarrow CH_2(OCH_3)_2 + H_2O$	2.5×10^{-10}

† The value of reaction rate constant was estimated by similarity.
‡ The average value for reactions of R + R type.
§ The adjustable rate constant to be varied to get the best fitting of calculated and experimental data

Fig. 3.14 shows the dose dependences for ethylene glycol, hemiacetal, and the sum of dimethyl ether and ethanol as secondary products. The curves are seen to fit the experimental data with 10%. Dose-dependences for other second-generation products and for one third-generation product, methyl formate, are also described with the required accuracy. There is a rather poor agreement between the concentrations of other third-generation products and the calculated values (with a 50% error) although the appearance of the dose-dependence curves is reproduced qualitatively fairly well. Obviously, it is necessary to extend the variation of the rate constants as well as to look, as mentioned above, for other possible final products in this dose range. In general, it may be stated that the mathematical model will describe satisfactorily the dose-dependences for penultimate generation of stable products.

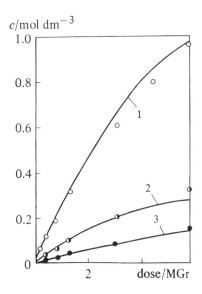

Fig. 3.14. Calculated dose dependences for ethylene glycol (1), hemiacetal (2), and dimethyl ether + ethanol (3) (circles refer to corresponding experimental data) [221].

The calculated dose-dependences for the transient products in the continuous γ-radiolysis of methanol are shown in Figs 3.15 and 3.16. As seen from the figures, the concentrations of transient species indeed reach their stationary values during radiolysis (see section 3.7), and these steady-state values are too small to be observed experimentally with current instrumental techniques under ordinary conditions.

Thus, the mechanism that we have just considered makes it possible to represent the dose-dependences for the first- and second-generation final products quantitatively and for the third-generation products qualitatively, that is, the

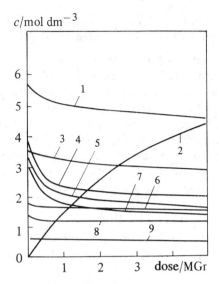

Fig. 3.15. Calculated dose dependences for some transient products from the liquid phase γ-radiolysis of methanol at room temperature and a dose rate of 5.6 Gy s^{-1} [222]. 1—CH$_3$OH$_2^+$ (10^9); 2—CH$_2$OHĊHOH (10^9); 3—ĊH$_2$OH (10^8)); 4—e_s^-(10^9); 5—H(10^{14}); 6—CH$_3$Ȯ (10^{13}); 7—CH$_3$O$^-$ (10^9); 8—ȮH (10^{17}); 9—:CHOH (10^{12}) (Numbers in parentheses refer to the order of magnitude of the concentration).

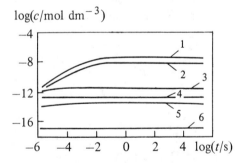

Fig. 3.16. Calculated time-dependences for the concentration of transient products in the liquid phase radiolysis of methanol at room temperature and a dose rate of 5.6 Gy s^{-1} [222]. 1—ĊH$_2$OH; 2 — CH$_3$OH$_2^+$; 3 — e_s^-; 4 — CH$_3$Ȯ; 5 — H·; and 6 — ȮH.

suggested scheme appears to be reliable. However, to estimate the degree of reliability, it is necessary to extend the experimental conditions, thus providing an additional verification. Good fitting of the calculated and experimental curves when varying external parameters, confirms the authenticity of the mathematical model considered, at least for those transient and stable products which have been verified. Computer programs for the calculation of the primal and inverse problems can be found elsewhere [227–230].

3.13 DIFFUSION-CONTROLLED RECOMBINATION KINETICS DURING EARLY EVENTS IN CONDENSED-PHASE RADIOLYSIS

As has been shown in section 1.5, the primary products of interaction of ionizing radiation with matter in a condensed state are distributed nonuniformly throughout the occupied volume. Among these primary species, the principal products (for a condensed phase) are ions and electrons; the following stages include also the solvated electron, radicals emerging owing to ion-molecule reactions, and excited species resulting from charge recombination. The diffusion of heavy particles (radicals, ions, and excited molecules) becomes noticeable after a time of 10^{-11} s. Because the mean distance between transient species in tracks is about nm, bimolecular reactions of the species occur within a period exceeding 10^{-10} s, as shown in Fig. 3.17, and are completed in a time of 10^{-8} s if the viscosity of the medium is *ca.* 1 mPa s. However, the mobilities (see Table 2.18) and, consequently, the diffusion coefficients of quasi-free electrons and holes are appreciably higher than those of ions, radicals, and other heavy particles, so that both reactions of quasi-free electrons with holes and even their capture by solutes can occur in shorter periods for which there is, as yet, no appropriate method of observation (contemporary pulse techniques make it possible to follow processes with a duration somewhat shorter than a picosecond). In terms of this, a study on the kinetics of early events in the effect of ionizing radiation, including photoionization, must cover periods from 10^{-14} to 10^{-8} s, and all reactions occurring within this period must be taken into account in the scheme of processes. The large number of reactions occurring and the nonuniformity in

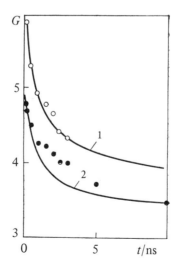

Fig. 3.17. Time-dependence for the yield of hydroxyl radicals (1) and hydrated electrons (2) [232] (curves are calculated as in [231] for ps pulse radiolysis by fast electrons).

the distribution of transient species by volume, greatly complicate the problem of the kinetic description of the early stages.

Since the processes of charge recombination and radical recombination play an important role in the early stages of radiolysis, the problem of creating a kinetic model of the early stages has become a problem of searching for a diffusion-controlled recombination model. We shall consider a model of this sort, implemented in a liquid-phase radiolysis by way of illustration. Models for other HEC processes have their own specific features.

The principles of the diffusion-controlled recombination model of radiolysis were set out by Kuppermann [233] and Byakov [234]. A solution of the kinetic model is found for one track-pattern of a chosen type (for example a single spur or a short track) and is then expanded to all tracks of this type in a unit volume. The concentration $c(r, t)$ of the i-th particle is a function of the distribution of the chosen transient particle at time t and distance r from the spur centre (if spherically symmetrical) or from the track axis (if of cylindrical symmetry). A nonlinear partial differential Eq. (3.226) is written for each type of transient species to show the change in its concentration with time. The equation includes the diffusion term for the chosen i-th species created by radiolysis and the terms describing the emergence of the i-th species in the first- and second-order reactions and its decay via a first-order process, via combination with each other and with other transient species, and by reaction with solutes:

$$\partial c_i / \partial t = D_i \Delta c_i + \sum_j k_j c_j + \sum_{j,l} k_{jl} c_j c_l - k_i c_i - \frac{n_0 - 1}{n_0} \times k_{ii} c_i^2 -$$

$$-\sum_j k_{ij} c_i c_j - \sum_m k_{im} c_i c_m, \tag{3.226}$$

where n_0 is the initial number of particles of the given type in a chosen unit track, D_i is the diffusion coefficient of the transient species, Δ is the Laplacian, k_i and k_j are the rate constants of the first-order reactions k_{jl}, k_{ii}, k_{ij}, and k_{im} are the rate constants for second-order reactions where the subscripts i, j, l, and m refer to the i-th and j-th (different from the i-th) transient species, substances, and scavengers present in solution, respectively.

Taking account of all the transient species and solutes present in the irradiated liquid, we obtain a system of equations of Type (3.226). Such a system can be solved only numerically with the use of powerful computers. To do this one must define (i) the distribution functions for the transient species at zero time, (ii) the initial number of transient particles in a unit track, (iii) the diffusion coefficients of the transient species, and (iv) the reaction rate constants for the species.

The reaction rate constants for recombination and scavenging of most of the transient species, like their diffusion coefficients, can be determined experimentally if we assume that the species to be at thermal equilibrium in the system under study. A rather less certain situation is when the distribution functions $c_i(r, 0)$ and the values of n_0 cannot be measured or determined directly in some way (see section 2.6).

Attempts have been undertaken repeatedly [233]. to calculate the time-dependence of the concentrations of transient species and radiation yields of final products for the radiolysis of water, using simplified sets of equations of Type (3.226) involving one, two, or more kinds of transient species. The effect of different parameters of the system and of radiation on the yield of transient and stable products has also been calculated. All these calculations give only a qualitative picture (see for example Fig. 3.17), and any quantitative agreement with experimental data has not yet been obtained. The reason for this lies in the inadequacy of existing representations of the spatial distribution of transient particles, the uncertainty in some parameters of the model, and the lack of information on some reactions in the early stages of radiolysis.

One of the most important techniques of investigation of the early events of radiolysis is the determination of the dependence of the radiation yields of molecular and radical products on solute concentration (the molecular products M here are meant as products resulting from the spur recombination of transient species and the radical products R as intermediate species escaping spur recombination, their yields being denoted G_M and G_R, respectively). For water radiolysis, it has been shown [235, 236] that the yields of molecular hydrogen and hydrogen peroxide G_M decrease in proportion to the concentration of solute scavenging their precursors raised to the power 1/3 provided this concentration is not too high (Fig. 3.18)

$$G_M = G_M^0 - G_M^0 q c_{Ac}^{1/3}$$
(3.227)

The same equation was obtained from the approximate solution of the two-radical diffusion-controlled recombination model for low solute concentrations [234], the empirical parameter q being expressed via parameters of the system

$$q = \left[\frac{1}{(1+b)^2} \frac{k_{Ac}}{k_c} \frac{V_0}{n_0} \right]^{1/3}$$
(3.228)

where b is defined from $b/(1 + b) = 2N_m/n_0$, N_m is the number of product molecules resulting from the recombination of transient species inside a unit track, V_0 is the effective volume of the track, k_c is the combination rate constant for the transient species, k_{Ac} is the rate constant for scavenging the species by the solute (scavenger) and c_{Ac} is the solute concentration.

The validity of Eqs (3.227) and (3.228) is confirmed by the ratio $q/G_{H_2O_2}$ fitting the rate constant ratio for $\dot{O}H$ scavenging by a solute to $\dot{O}H$ combination raised to the power 1/3 as shown in Fig. 3.19.

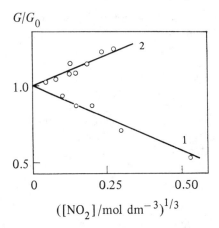

Fig. 3.18. Concentration-dependence for the yield of molecular hydrogen (1) from sodium nitrate solutions [236] and for the observed yield of reducing radical species (2) from potassium nitrate solutions [235] (γ-irradiation).

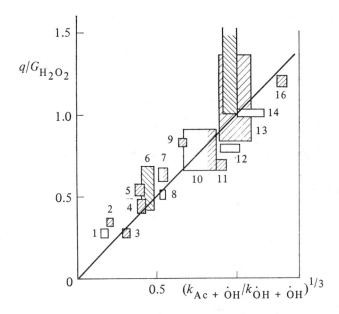

Fig. 3.19. Comparison of values of $q/G_{H_2O_2}$ and $(k_{Ac+\dot{O}H}/k_{\dot{O}H+\dot{O}H})^{1/3}$ for different $\dot{O}H$ - scavengers in acidic (dashed rectangulars) and neutral (blank rectangulars) aqueous solutions for γ-irradiation [237]. 1—H_2O_2; 2—Cl^- (pH = 2); 3—Cr^{3+}; 4—CH_3OH; 5—VO^{2+}; 6—Cl^- (pH = 1.3); 7—C_2H_5OH; 8— $(CH_2OH)_2$; 9—C_3H_7OH; 10—$CH_2CHCONH_2$; 11— Cl^- (pH = 0.4); 12—NO_2^-; 13—Br^-; 14—I^-; 15—Tl^+; and 16—hydroquinone.

Simultaneously with the reduction in the yield of recombination products, the measured yield G_R of transient particles escaping spur recombination increases [234]:

$$G_R = G_R^0 + G_R^0 b \left[\frac{1}{(1+b)^2} \frac{k_{Ac}}{k_c} \frac{V_0}{N_0} \right]^{1/3}.$$ (3.229)

This equation fits the experimental data, as shown in Fig. 3.18.

The approximate solution of Set (3.226) for two types of radical gives the following equation for high concentrations of solute:

$$G_R c_{Ac} = G_{(-A)} c_{Ac} - \frac{1}{2} G_{(-A)} \frac{k_c n_0}{k_{Ac} V_0} = \alpha - d c_{Ac}$$ (3.230)

where $G_{(-A)}$ is the total yield of transient species of a given type.

This equation is fulfilled fairly well for both polar and nonpolar liquid systems.

For geminate-ion recombination in the presence of solute, a theoretical consideration of the diffusion-controlled recombination model, taking account of the Coulombic interaction of charges, gives the equation [238]:

$$G_e = G_{fi} + G_e^0 \times \frac{\bar{r}_0 \sqrt{(k_{Ac} c_{Ac}/D)}}{1 + \bar{r}_0 \sqrt{(k_{Ac} c_{Ac}/D)}}$$ (3.231)

where G_e is the yield of separated ion pairs for a given solute concentration, G_e^0 is the total yield of ion pairs, G_{fi} is the yield of free ions, and $\bar{r}_0$ is the mean separation distance of the particles in a pair.

Eq. (3.231) coincides with an empirical equation found for the yield of quasi-free electrons trapped in nonpolar systems containing a scavenger [239]:

$$G_e = G_{fi} + G_e^0 \times \frac{\sqrt{(\delta c_{Ac})}}{1 + \sqrt{(\delta c_{Ac})}}$$ (3.232)

where δ is an empirical factor.

Comparing Eqs (3.231) and (3.232), we find $\delta = \bar{r}_0^2 k_{Ac}/D$. For the treatment of experimental data it is convenient to rewrite Eq. (3.232) as

$$\frac{1}{G_e - G_{fi}} = \frac{1}{G_e^0} + \frac{1}{G_e} \times \frac{1}{\sqrt{(\delta c_{Ac})}}$$ (3.233)

Fig. 3.20 shows that Eq. (3.233) is true not only for nonpolar but also for weakly polar systems. In highly polar liquids such as water, the equation is not followed.

So, in spite of the fact that the diffusion-controlled recombination model of radiolysis still fails to provide an accurate description of experimental data in its

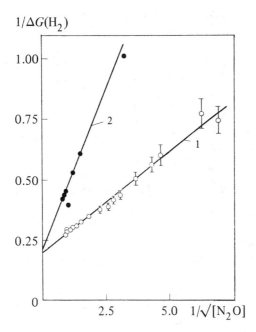

Fig. 3.20. Expression of experimental data on the radiolysis of cyclohexane (1) [240] and methanol (2) [241] by means of Eq. (3.233) (N_2O as the solute).

complete statement, in the form of the set of equations of Type (3.226), the relationships obtained by approximate solutions over a defined solute concentration range appear most useful in studies of the processes of radiolysis.

4

Fundamentals of photochemistry

Photochemistry deals with chemical processes occurring under the action of visible light (400–800 nm), near ultraviolet (100–400 nm), and near infrared (800–1500 nm) radiation. Man has encountered some photochemical processes from ancient times, including the photobleaching of dyes and the photodegradation of various materials. By means of vision, which is the photochemical process of transformation of the image on the retina to nerve impulses, one is able to receive information about one's immediate surroundings.

The first attempts of investigation and theoretical interpretation of the chemical action of light were made at the end of the 18th century when Senebi carried out a systematic study on the influence of light on different processes and assumed that the exposure time required to produce a certain chemical effect is inversely proportional to the light intensity. Empirical efforts undertaken to discover new light-induced reactions of inorganic and organic substances were continued in the 19th century, parallel to physicochemical studies on the laws governing photochemical reactions. A discovery of fundamental importance was made by Grotthus who showed that a chemical change can be caused only by the light absorbed, and, drawing an analogy between the actions of light and electricity, rejected the hypothesis of the thermal mechanism of the action of light.

A new stage in the development of photochemistry was opened with the advance of quantum theory. Thus ideas on the nature of the interaction of photons with atoms and molecules, on the nature of excited states, and on the mechanisms of primary photophysical and photochemical processes, appeared. Advances in the allied branches of science, quantum chemistry, spectroscopy, luminescence, and chemical kinetics as well as the development of novel instrumental techniques, enabled photochemistry to change from an empirical science to a subject with a sound theoretical foundation.

Important applications of photochemistry are related to photography, photolithography, and other processes of recording and processing information (images), with industrial and laboratory syntheses, the modification and stabilization of polymer materials, with quantum electronics, microelectronics, and the transformation of light to chemical energy.

The central point of modern photochemistry is the issue of the structure and properties of excited electronic states of organic and inorganic molecules and of the laws determining photochemical and photophysical processes occurring in these states [242–247]. With general concepts developed on this topic, photochemical reactions of different classes of organic and inorganic compounds become understandable, and the search for new reactions becomes informed. An excited electronic state is not a 'hot' modification of the ground electronic state. It differs from the ground state in electronic structure, geometry, and chemical properties [248–253]. Consequently, the excitation of molecules may cause not only quantitative but qualitative changes in chemical behaviour.

An important feature of photochemical reactions is the occurrence of competing physical processes of electronic energy degradation so fast that their rate constant exceeds 10^{10} s^{-1} [254–256]. Thus, the reactivity of excited molecules is determined not only by their reaction rate constant but by the ratio of the reaction rate to the rates of competing processes of energy degradation. In addition to the processes of energy degradation inherent to isolated excited species [257], excited molecules, when interacting with some reagents, turn out to have other, and frequently more effective, pathways of degradation of energy through the transition state. These pathways are provided by the emergence of additional degrees of freedom in the system, and the process itself is called induced deactivation [258–260].

The majority of photochemical reactions occur from thermalized lower excited states, either singlet or triplet, which also emit fluorescence or phosphorescence, respectively. This phenomenon, known as Kasha's rule, is due to the extremely fast rates of thermalization and internal conversion, from higher excited states to lower states of the same multiplicity, considerably exceeding the rates of most chemical reactions and of emission. However, there are some chemical reactions which occur from unrelaxed (Franck–Condon) excited states, which have been formed directly by photon absorption. These are, especially, certain dissociation and isomerization processes. Here the chemical reaction competes not with emission but with relaxation to a fluorescent or phosphorescent state, and the quantum yield of the reaction is independent of the lifetime of the luminescent state. If the intensity of light is high enough to populate higher excited states, then two-quantum and multi-quantum photochemical reactions are observed [2, 261].

There are two different types of primary reaction of excited molecules [242, 262–264]. If the electronic excitation is conserved in the elementary act, the reaction occurs on the potential surface of the excited electronic state, and the primary product is formed in its excited state, then this type of reaction is called adiabatic. The other, diabatic, type (sometimes incorrectly named nonadiabatic) refers to reactions where the primary elementary act is accompanied by the loss of electronic excitation, the system transfers from the potential surface of the excited electronic state to that of another, (most frequently the ground electronic state), and the primary product is formed in its ground state. In some cases adiabatic reactions can also result in a ground-state product if the potential surfaces of the ground and excited states are degenerate in the region corresponding to the product (as takes place on dissociation to atoms or radicals).

Most of the concepts of molecular reactivity in thermal reactions used in chemical kinetics are based on transition-state theory which is applicable strictly only to adiabatic reactions. Diabatic reactions can be considered from the viewpoint of the theory of radiationless processes which is not so well-developed, especially for polyatomic molecules.

Different approaches currently used in the estimation of the reactivity of excited molecules use both the 'static' representation of electron density in molecules (orbital classification, electron density distribution, free valence indices, atomic charges, etc.), and the 'dynamic' treatment of changes in some parameter (orbital symmetry, multiplicity, Gibbs free energy, localization and delocalization energy, etc.) in the course of transformation. The dynamic approaches seem more promising since they take into account the specificity of a particular process and enable the chemical transformation and the competing process of degradation of electronic energy to be considered together in some cases. From this point of view, reactions of excited molecules can be subdivided into two classes: allowed reactions (in terms of multiplicity, orbital symmetry, etc.) and reactions *forbidden* by these rules. For allowed reactions, a rather good correlation between the activation energy and the Gibbs free energy of the initial step is observed, which makes possible the quantitative prediction of values for the reaction rate. If the orbital symmetry is disturbed, then a significant potential barrier appears on the reaction path, the height of which is not related explicitly to the energy characteristics of the reaction. The rate of such a reaction can change appreciably even with slight variation in the structure and symmetry of the reactants. The same picture is observed for reactions accompanied by a change in multiplicity. The essential role here is played by factors affecting spin interactions, in particular, by the presence of heavy atoms or paramagnetic centres in either the reacting molecules or the medium, and even by the presence of an external magnetic field.

Let us consider the mechanisms and quantitative relationships describing the most important types of primary reaction of excited molecules, using examples of the most widely-studied compounds.

4.1 PHOTOINDUCED ELECTRON TRANSFER

Donor–acceptor electronic interaction plays a specific role in photochemistry. The donor or acceptor properties of a molecule are radically affected by excitation. The ionization potential decreases by the value of excitation energy, and the electron affinity increases by this quantity. Exchange interactions between molecules take place over distances up to 1 nm which exceed appreciably their kinetic radii. For all of these reasons any kind of chemical interaction begins as electronic. The processes of photoinduced electron transfer are the key steps in many photochemical reactions of organic and inorganic compounds in photo-sensitized reactions, in biological photosynthesis, and, especially, in the proce-dures for conversion of light into chemical energy which are currently under intensive study [3, 265].

Electron transfer after excitation by light can proceed via three different pathways [242, 250, 262–268] (Fig. 4.1): (i) direct photoionization of a mol-ecule producing a free or solvated electron, followed by its capture by another molecule of the medium, (ii) reaction of the excited molecule D* with an electron-acceptor A, and (iii) immediate transfer of the electron in the act of photon absorption on excitation with light of a wavelength corresponding to the charge-transfer band of donor–acceptor (charge-transfer) complexes.

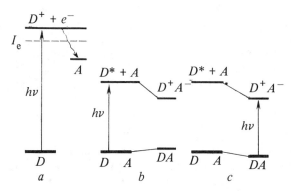

Fig. 4.1. Photoinduced electron transfer processes. a—direct photoionization; b—electron transfer in the excited state; c—photoexcitation of charge-transfer complex.

In the first case the photon energy must exceed the ionization potential I_e of the molecule. In the condensed phase this energy is by 1 to 2 eV less than it would be in the gas phase because of polarization of the medium by the charges formed. The ionization potential of a molecule nearly always considerably ex-ceeds the energies of the lower singlet and triplet states, so direct photoionization is energetically unfavourable and needs light of too short a wavelength.

The absorption of a photon with a wavelength corresponding to the charge-transfer band of complexes formed in the *ground* state by electron-donor and electron-acceptor molecules (this band is not necessarily that of longest wave-length) immediately results in transfer of an electron from the donor component of the complex to the acceptor (Fig. 4.1c). This type of electron transfer is simply the excitation of the compound molecule DA, where the binding is realized by means of a small transfer of charge (usually less than 0.1–0.2 of the electronic charge) to the charge-transfer state D^+A^- (where the degree of charge transfer is 0.8–0.9). If the ground state of the CT complex is highly polar (for example in radical-ion complexes), then back-electron transfer takes place and a weakly polar excited state is formed. The photon energy required for excitation in the CT band depends on the ionization potential of the donor, the electron affinity of the acceptor, the charges on the donor and acceptor molecules, and the solvent polarity [269–272].

There are the following relationships (in eV) for different donor–acceptor pairs

$$AD \qquad hv = I_e(D) - A_e(A) - 3.2 \pm 0.2 \text{ (in gas phase)} \qquad (4.1)$$

$$hv = I_e(D) - A_e(A) - 3.0 \pm 0.2 \text{ (in solution)} \qquad (4.2)$$

$$AD^- \text{ or } A^+D \quad hv = I_e(D) - A_e(A) - 1.2 \pm 0.1 \qquad (4.3)$$

$$A^+D^- \qquad hv = I_e(D) - A_e(A) - 5.4 \pm 0.4. \qquad (4.4)$$

The energy needed for photoinduced electron-transfer in a complex made up of neutral molecules is less than the ionization potential of the electron donor by several electronvolts. If the CT complexes include ions, it should be borne in mind that the parameters used in the above expressions for hv are the values in the gas phase for the ionization potential and electron affinity. The ionization energy and the energy of electron attachment for charged particles change on transition to the condensed phase by the value of the solvation energy of the ion, which equals a few eV. Therefore, the energy of the electron transfer in the complexes is less than the energy of photoionization of the donor also in these cases. The excited states of CT complexes can be stable (to give luminescence [273, 274]), or they can dissociate to radical ions or neutral radicals, depending on the charges on the donor and acceptor, or can react with other components of the system, that is, to participate in a photoinduced reaction.

Excited singlet and triplet states of molecules can interact with various donors to form exciplexes in weakly polar media (Figs 4.1b and 4.2), radical ions in polar media, or photoreduction or photo-oxidation products, otherwise their energy is degraded. For photoinduced electron-transfer to occur, the excitation energy of the donor and acceptor must exceed the difference in their redox potentials:

$$E^* \geq E^0 (D/D^+) - E^0 (A^-/A) \qquad (4.5)$$

which in terms of the ionization potential and electron affinity takes the form

$$E^* \geq I_e(D) - A_e(A) - 4.3 \text{ eV}. \qquad (4.6)$$

In liquid solutions, the process of photoinduced electron transfer is a sequence of the diffusion together with the electron transfer itself:

$$A^* + D \rightleftharpoons (A^*D) \rightleftharpoons A^-D^+ \rightarrow A^- + D^{+3}. \qquad (4.7)$$

The isoenergetic transfer of an electron in the collision complex $(A.D)^*$ results in the radical–ion pair A^-D^+ in a nonequilibrium configuration of the nuclei and solvent shell (that is, in the Franck–Condon state). Vibrational relaxation and relaxation of the solvent shell leads to formation of an exciplex, radical ions, or other products, depending on the conditions.

The elementary act of electron transfer is currently viewed as the isoenergetic sub-barrier transition (tunnelling) of the electron from donor to acceptor [275]. Under these conditions the reaction coordinate used is not the position of the electron but the nuclear coordinates of the reacting molecules and solvation environment. The rate of electron transfer is determined both by the activation energy (that is, the energy of the change in nuclear configuration necessary for equalizing the energy levels of the electron in the initial and final states) and by the probability of electron tunnelling.

In frozen solutions, where diffusion does not interfere with the kinetics of elementary process, experimental evidence has been adduced for the tunnelling mechanism of photoinduced electron-transfer [3, 194, 276]. The kinetics of decay of excited molecules in the presence of electron acceptors in frozen glassy solutions is nonexponential because of the variations in distance between the molecules of donor and acceptor which are randomly distributed in solution. The rate constant for electron tunnelling decreases exponentially with increasing distance r between the donor and the acceptor:

$$k = v\exp(-2r/a) \qquad (4.8)$$

where v is the frequency factor and a is the parameter characterizing the diminution of the electron wave function.

The kinetics of fluorescence decay $I(t)$ under these conditions is often described with an equation corresponding to the uniform distribution of molecules in solution:

$$I(t) = I_0 \exp(-k_0 t - A[\ln vt]^3) \qquad (4.9)$$

where k_0 is the rate constant of fluorescence decay for free molecules of the fluorophore, $A = a^3 10^{20}[Q]$ and $[Q]$ is the (molar) quencher concentration.

This equation may be linearized approximately as

$$\ln I(t_0) - \ln I(t) - k_0(t - t_0) = 3 \times 10^{20}\, \pi a^3 [\ln(vt)]^2 [Q]\ln(t/t_0) \qquad (4.10)$$

where t is any moment in time.

Eq. (4.10) is useful for determination of the parameters a and v. A typical value for a is $ca.$ 0.1 nm and for v $ca.$ 10^{14} s^{-1}. Within the excited singlet state lifetime (10–100 ns), the tunnelling of an electron can take place over 0.7 to 0.8 nm, that is, the donor and the acceptor may be separated by one or two solvent molecules. Within the lifetime of triplet states in the solid state (10^{-3}–10^2 s), electron tunnelling over distances of 1.3–1.8 nm can take place. Thus, photoinduced electron-transfer in solids does not require direct contact between the donor and acceptor molecules.

Rehm & Weller [277] analysed the relation between the rate constants for the reaction of excited molecules with electron donor and acceptors and their redox potentials. The dependence on the Gibbs free energy of electron transfer found for the quenching rate constants of some excited aromatic molecules in acetonitrile is shown in Fig. 4.2. Similar dependences for the quenching of excited singlet and triplet states by electron donors and electron acceptors for a wide variety of molecules in different solvents were found later by others. The dependence of $\log k_q$ on ΔG^o shown follows from the kinetic scheme of diffusion-controlled reactions and indicates the existence of a correlation between the activation energy of the electron-transfer stage and ΔG^o for the reaction. Thus, the correlation methods used so successfully in the chemistry of thermal reactions beginning from Brönsted have also found implementation in photochemistry. They have made it possible to predict reaction rate constants for excited molecules from the thermodynamic parameters of the reactants in the ground electronic states and from their spectral properties.

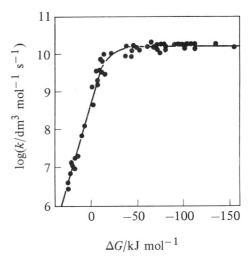

Fig. 4.2. Dependence of rate constant for fluorescence quenching on the Gibbs free energy of electron transfer [465].

The Gibbs free energy of electron transfer is given by the difference in redox potentials of the excited molecule and quencher as follows:

$$\Delta G^o = E^o(D^*/D^+) - E^o(A^-/A) - C$$
$$= E^o(D/D^+) - E^*(D) - E^o(A^-/A) - C \qquad (4.11)$$

or

$$\Delta G^{\circ} = E^{\circ}(D/D^+) - E^{\circ}(A^-/A^*) - C$$
$$= E^{\circ}(D/D^+) - E^{\circ}(A^-/A) - E^*(A) - C \qquad (4.12)$$

depending on which role, either donor or acceptor, is played by the excited molecule (C is the difference between the energy of the Coulombic interaction of the ions in the pair A^-D^+ and the solvation energy). If the rate constants and redox potentials are measured in the same polar solvent, acetonitrile, the C-value is nearly constant, that is, 0.15 ± 0.10 eV [278, 279]. From the kinetic Scheme (4.7) it follows that the apparent rate constant for quenching is given by

$$k_q = \frac{k_D}{1 + k_{-D}/k_R + k_{-D}k_{-R}/k_R k'}, \qquad (4.13)$$

where k_D and k_{-D} are the diffusion rate constants, k_R and k_{-R} are the rate constants for electron transfer and back-electron transfer, respectively, in the collision complex, and k' is the decay rate constant of the radical–ion pair A^-D^+ in dissociation or recombination.

Expressing the rate constants in terms of the Gibbs free energy of activation, we obtain

$$k_q = \frac{k_D}{1 + \dfrac{k_{-D}}{k_R^{\circ}}\left[\exp(\Delta G_R^{\ddagger}/RT) + \exp(\Delta G_R^{\circ}/RT)\right]} \qquad (4.14)$$

where $\Delta G_R^{\ddagger}$ and ΔG_R° are the Gibbs free energies of activation and of electron transfer, respectively, and k_R° is the frequency factor which is determined by the rate of dielectric relaxation of the solvent for reactions accompanied by charge transfer.

Rehm & Weller [277, 279] treated experimental data, using the following empirical relationship to determine $\Delta G_R^{\ddagger}$

$$\Delta G_R^{\ddagger} = \Delta G_R^{\circ}/2 + \sqrt{[(\Delta G_0^{\ddagger})^2 + (\Delta G_R^{\circ}/2)^2]} \qquad (4.15)$$

where the parameter $\Delta G_R^{\ddagger}$, corresponding to the Gibbs free energy of activation for the isoenergetic reaction, turned out to be about 10 kJ mol^{-1}, assuming that $k_R^{\circ} = 10^{11}$ s^{-1}.

A simpler equation for k_q can be obtained on taking into account that the reactions of excited molecules tends to be very fast and not very endoergic, in which case the last term in the denominator of Eq. (4.13) may be neglected. Then

$$k_R \approx \frac{k_{-D}}{k_D/k_R - 1}$$

and the Gibbs free energy of activation can be calculated directly from experimental data on quenching rate constants.

$$\Delta G_R^{\neq} = 2.303\, RT[\log k_R^0 - \log(k_{-D}/k_D) + \log(1/k_q - 1/k_D)] \qquad (4.16)$$

Fig. 4.3 shows the dependence of $\Delta G_R^{\neq}$ on ΔG_R^0 for the quenching of excited molecules by electron donors and acceptors. It is seen that this dependence can be described with a simpler linear equation

$$\Delta G_R^{\neq} = \Delta G_0^{\neq} + \alpha \Delta G_R^0 \qquad (4.17)$$

even better than with the Weller equation (Eq. (4.15)). Eq. (4.17) is more convenient for the treatment of experimental data since one does not require the parameter C of Eq. (4.11) for ΔG_R^0 (variation of this parameter leads simply to a shift in the straight line along the abscissa); thus it is possible to analyse the dependences of quenching rate constants by varying only one reaction component while disregarding the properties of the other component. Of course, the linear equation has no physical meaning for $\Delta G_R^0 < -\Delta G_0^{\neq}/\alpha$ and for $\Delta G_R^0 > \Delta G^{\neq}/\alpha$, but these limitations are insignificant for short-lived excited states when reactions with $k_q < 10^6\ \mathrm{M^{-1}\ s^{-1}}$ are not amenable to experimental study.

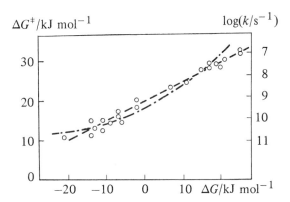

Fig. 4.3. Dependence of the Gibbs free energy of activation on the Gibbs free energy of electron transfer.

Analysis of many published data has shown that the value for the parameter α is approximately equal to 0.5. However, the deviation of $\Delta G^{\neq}$ from the general relationship may be as large as ± 5 kJ mol^{-1} for some molecules, and α ranges from 0.2 to 0.8 and can even be negative for a few particular series of compounds.

4.2 PHOTODISSOCIATION AND PHOTOSUBSTITUTION REACTIONS

Photodissociations are classified in terms of the nature of the bond rupture and by the mechanism of the process. Depending on the scheme of bond scission, one differentiates between homolytic and heterolytic reactions. The primary products of photodissociation can be radicals, biradicals, radical ions, and molecules. Interaction of the primary products of photodissociation with reagents present in a system leads to substitution products.

There are several ideas concerning the mechanism of photodissociation [280]. On absorption of a photon, a molecule can either dissociate immediately or form a relatively stable excited state which is able to luminesce. The nature of the excited states of simple molecules can be elucidated by analysis of the structure of their absorption and emission spectra.

Direct photodissociation takes place in those cases where the excited state (singlet or triplet) is dissociative. Such a process is adiabatic and can result in formation of one of the products in its excited state.

Dissociation products corresponding to a given electronic transition can be predicted for simple substrate molecules on the basis of correlation rules, of the convergence limits of band spectra, and of relevant thermochemical data.

Photosubstitution reactions are complex, being at least two-stage processes. They occur via photodissociation of a substrate and subsequent addition of a reactant, or abstraction of an atom from the reactant, or via addition of the reactant to the excited molecule and subsequent elimination of a substituted group. Examples receiving much current attention are the photosubstitutions occurring via a radical–ion chain mechanism where electron transfer is the primary step of the chain initiation [281–283].

The kinetic features of photodissociation and photosubstitution will be considered further, using the best-known types of reaction.

4.2.1 Homolytic reactions

A typical example of a homolytic photodissociation is the dissociation of hydrogen iodide (Fig. 4.4); there are three excited states, two of them producing an iodine atom in the ground state ($^3P_{3/2}$) and one giving an excited iodine atom ($^3P_{1/2}$). Any excess excitation energy over the dissociation energy of 3.05 eV is transformed mainly to the translational energy of the H atom. The relative quantum yield of excited iodine atoms is wavelength-dependent and has a value of 0.07–0.5 at 273.7 nm according to different sources [32].

Another example of direct photodissociation is given by the photolysis of nitrosyl chloride [32]. On visible-light excitation ($\lambda \sim 400$–630 nm) in the region of continuous absorption, the NOCl molecule enters a repulsive state which dissociates immediately, producing NO in the ground ($^2\Pi$) electronic state:

$$NOCl + h\nu \rightarrow NO\left(^2\Pi\right) + Cl \qquad (4.18)$$

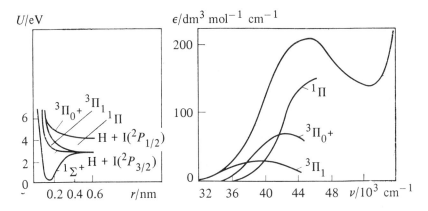

Fig. 4.4. Potential energy diagrams for a molecule of hydrogen iodide in different electronic states (a) and decomposition of the hydrogen iodide absorption spectrum into components from different transitions (b).

The chlorine atom thus formed reacts with another NO molecule

$$Cl + NOCl \rightarrow NO + Cl_2 \tag{4.19}$$

so that the quantum yield of photolysis is equal to 2. The excess excitation energy over the N–Cl bond energy (1.61 eV) is realized in the form of translational energy of the fragments and partially in the vibrational excitation of NO. On excitation in the vacuum UV region (*ca.* 150 nm), the dissociation of a higher excited state of NOCl produces NO in the $^2\Sigma^+$ excited state. The quantum yield of fluorescence is about 4%.

The energy of the Franck–Condon transition from the ground to the dissociative state generally exceeds 1.5 or 2 times the energy of a ruptured bond. The singlet–triplet splitting is quite large for dissociative states, and the energy of a dissociative triplet state is less than the energy of the corresponding singlet state by 0.5–1.0 eV. Based on this fact, an effective technique of promoting photodissociation by means of triplet sensitizers has been suggested [284].

Direct photodissociation in the gas phase occurs within a period of one vibration (*ca.* 0.1 ps), and its quantum yield is close to unity. The process of photodissociation in the condensed phase is more complicated. Firstly, the primary dissociation products are formed, surrounded by a solvent cage which prevents their separation and facilitates recombination to the initial ground state molecule [285]. Secondly, the process of dissociation of the excited molecule can occur through interaction with the solvent, especially when charged species are formed. Hence the quantum yield of direct photodissociation may be considerably less than unity in the condensed state.

Predissociation is observed when the bonding term of the excited state interacts with some dissociative term [286], that is, predissociation may be

represeneted as the intramolecular transfer of electronic excitation energy leading to formation of a dissociative electronically excited state. The rate constant
for such a transition depends on the matrix element β of the interaction of the
electronic and overlap integrals of the vibrational wave functions χ for the initial
(i) and dissociative (d) states [280]:

$$k = \frac{2\pi\beta^2}{h} \langle \chi_i | \chi_d \rangle^2. \tag{4.20}$$

The closer the point of intersection of the terms to the minimum potential energy
of the bonding state, the higher the rate of the transition. If the bonding and
dissociative states are of different multiplicity, the spin selection rule leads to a
very strong decrease in the rate constant. The dissociative state may have the
nature of a zwitterion or intramolecular charge transfer, in particular, for ionic
photodissociation reactions. For the possibility of qualitatively estimating the
interaction of different terms, the principle of orbital symmetry conservation is
used in the construction of correlation diagrams [238, 243, 287].

Predissociation is apparent in the absorption spectra of simple molecules in
the vanishing of the vibrational structure in a certain wavelength region. The
fluorescence quantum yield decreases drastically in this region of the excitation
spectra.

The classical example of predissociation is given by the photolysis of SO_2,
which exhibits an absorption beginning from 390 nm. On excitation with light of
$\lambda > 220$ nm, a broad fluorescence band is observed with λ_{max} $ca.$ 360 nm and a
quantum yield close to unity. On excitation at shorter wavelengths, the quantum
yield of fluorescence decreases dramatically and dissociation products emerge:

$$SO_2 + h\nu \rightarrow SO + O,$$
$$SO_2 + O \rightarrow SO_3. \tag{4.21}$$

At low pressures, a reduction of the lifetime of the excited state from 40 ns
(220 nm excitation) to 8 ns (215 nm) is observed simultaneously.

Dissociation in some gas-phase reactions is assumed to occur through a 'hot'
ground (bound) state. There are two possible mechanisms for this type of dissociation [280]. In both cases it is necessary for the excitation energy to exceed the
bond energy appreciably. Upon internal conversion, the energy of electronic
excitation is transformed to vibrational energy localized mainly on those bonds
which differ significantly in their equilibrium distance in the ground and excited
states. Because of the large excess of vibrational energy, the dissociation of such
bonds becomes possible. If the amount of vibrational energy is insufficient for
dissociation of a given bond to occur, then its redistribution over other bonds of
the molecules takes place and the dissociation of weaker bonds can result (quasi-
equilibrium dissociation) [288].

So far as complex molecules are concerned, there are no theoretical approaches yet available to make quantitative predictions of the rate constant and of the quantum yield of photodissociation.

4.2.2 Heterolytic reactions

The simplest scission reactions of a chemical bond with formation of a new bond are proton transfer and hydrogen-atom abstraction. Proton transfer is a classical example of heterolytic reactions related to the acid–base properties of molecules, while hydrogen-atom abstraction exemplifies homolytic reactions.

Proton transfer reactions occurring both from the ground and excited states fall into two classes, depending on the nature of the atoms participating in the transfer. Proton transfer between atoms capable of hydrogen bonding, that is, possessing lone electron pairs, occurs with a very small activation energy (< 20 kJ mol^{-1}) and, correspondingly, with a very large rate constant ($\geq 10^{10}$ s^{-1}). If there is no possibility for hydrogen bonding, for example when at least one of the reactants is a C—H acid or a C-base, then the activation energy increases strongly (> 50 kJ mol^{-1}) and the rate constant falls by many orders of magnitude. In this section we consider photoinduced proton-transfer reactions via a hydrogen bond between heteroatoms. Many of these reactions occur adiabatically to result in electronically excited products [29, 30, 289]. Perhaps the presence of hydrogen bonding makes the perturbation of the potential surface of the excited state in proton transfer too small to lead to diabatic transitions.

Proton transfer in liquid solutions either results in the formation of free ions or ion-pairs, or finishes at the stage of formation of a hydrogen-bonded complex, depending on the solvating power of the solvent. Thus, excited hydrogen-bonded complexes are formed in the interaction of excited 2-naphthol molecules with diethylamine in hexane and in other alkanes of low polarity and polarizability [290]. The formation of ion pairs is observed in toluene, and the separation of ions in the highly polar dimethylformamide. These changes are clearly displayed in the emission spectra and fluorescence decay time (Fig. 4.5). Low temperatures also facilitate the production of more polar forms [291]. Solvated ions are usually formed in proton solvents.

In liquid solutions, proton transfer from a molecule of an acid AH^m to a molecule of a base B^n (m and n are charges) may be considered as a sequence of the formation and decay of a hydrogen-bonded complex[†] and the kinetic stage of the proton transfer itself [292, 293]:

$$\text{AH}^m + \text{B}^n \rightleftharpoons \text{AH}^m.\text{B}^n \rightleftharpoons \text{A}^{m-1}\text{HB}^{n=1} \rightleftharpoons \text{A}^{m-1} + \text{HB}^{n=1} \quad (4.22)$$

If the solvent acts as one of the reactants, then the first stage of this scheme is eliminated since all substrate molecules in the ground state are already H-bonded

---

[†] In protic solvents, the hydrogen-bonded complex may also include one or two solvent molecules, thus the proton transfer occurs via a chain of hydrogen bonds over the solvent molecules.

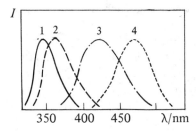

Fig. 4.5. Fluorescence spectra for 2-naphthol (1), hydrogen-bond complex (2), ion pair (3), and naphtholate ion (4).

to the solvent. The apparent rate constant of the interaction of AH with B is expressed through the rate constants for diffusion of the reactants k_D and products k'_D and for the forward- and backward-proton transfer k_R and k_{-R}, respectively, in the complex

$$k = \frac{k_D}{1 + k_{-D}/k_R + k_{-D}k_{-R}/k_R k'_D}. \tag{4.23}$$

The diffusion rate constants (of formation and dissociation of the hydrogen-bonded complex, k_D and k_{-D}) for species with charges of m and n, are equal to

$$k_D = \frac{4\pi N' D mne^2}{r\varepsilon kT}[1 - \exp(-mne^2/r\varepsilon kT)] \tag{4.24}$$

and

$$k_{-D} = \frac{3Dmne^2}{r^3\varepsilon kT}[1 - \exp(-mne^2/r\varepsilon kT)] \tag{4.25}$$

respectively, where r is the distance between the reactants in the reaction complex, D is the sum of the diffusion coefficients for the reactants, ε is the permitivity of the solvent, e is the electronic charge, and N' is the number of molecules in a millimole.

For $m = 0$ and/or $n = 0$ these equations transform to

$$k_D = 4\pi N' r D \tag{4.26}$$

and

$$k_{-D} = 3D/r^2 = 3k_D/4\pi N' r^3. \tag{4.27}$$

The analysis of the experimental data for aqueous solutions shows that most photoinduced transfers are characterized regardless of the charge on the reactants, by the same dependence of the rate constant on the Gibbs free energy of the proton transfer ΔG_R^0 (which is distinguished from ΔG^0 for the overall reaction by the difference in the Gibbs free energies for the dissociation of the initial and final H-bonded complexes) [293].

The relation between the Gibbs free energy of activation calculated by taking $k_R^0 = 10^{11}$ s^{-1} and the Gibbs free energy of reaction (Fig. 4.6) is equally well-described by the empirical Rehm–Weller equation (4.15) and by the theoretically based Marcus equation [303]

$$\Delta G^{\neq} = \Delta G_0^{\neq}(1 + \Delta G_R^0 / \alpha \Delta G_0^{\neq})^2 . \tag{4.28}$$

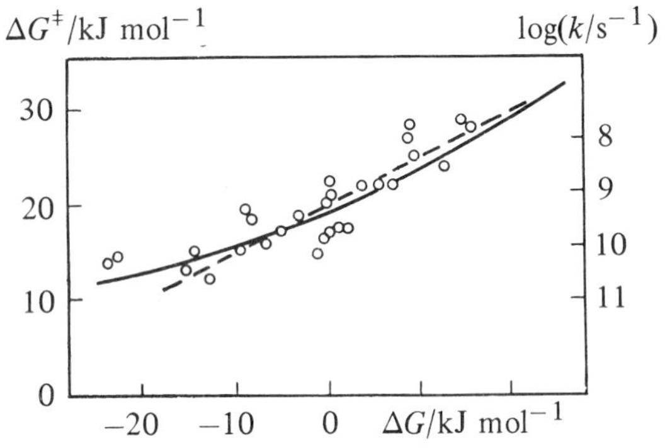

Fig. 4.6. Dependence of the Gibbs free energy of activation on the Gibbs free energy of proton transfer in a complex with hydrogen bonding. (Solid line refers to Eq. (4.28), dashed line to Eq. (4.29).)

The linear approximation

$$\Delta G_R^{\neq} = \Delta G_0^{\neq} + \alpha \Delta G_R^0 \tag{4.29}$$

also provides good agreement.

The value for α is *ca* 0.5, and k_R^0 is considerably less than kT/h and agrees with the characteristic time of the dielectric relaxation of water of 8.3 ps (at 298 K). Thus, proton transfer over a hydrogen bond is limited not by the motion of the proton itself but by the relaxation processes in the solvation shell related to the translocation of charge during reaction, as follows from the theoretical concepts by Marcus [294] and by Levich, Dogonadze *et al.* [295].

The role played by solvent relaxation in proton transfer was confirmed by studies on photoinduced proton transfer in the solid state. Since diffusion in the solid state is impossible, for photoinduced proton transfer to occur the reactants must form a pre-existing hydrogen-bonded complex in their ground states. Proton transfer takes place on excitation of complex of this type only when the Gibbs free energy decrease in the reaction (assuming equilibrium solvation) is 5–20 kJ mol^{-1}, depending on the type of reactant [296]. It is know that this excess energy needed for solid-state reactions is related to the solvation energy of the dipole formed on proton transfer [296]. Rate constants for proton transfer are greater than 10^{10}–10^{11} s^{-1} even at 77 K, which indicates that the activation barrier is low, that is, ≯ few kJ mol^{-1}.

The possibility of proton transfer occurring via a tunnelling mechanism has been under discussion for some time. However, no large isotope effect typical for the tunnelling process has been found in reactions of this type, despite numerous studies.

Not long ago, the adiabatic reaction of the protonation of a carbon atom in an aromatic ring was discovered. Diabatic protonation has long been known. In the photolysis of aromatic hydrocarbons in aqueous solution, isoptic exchange of the hydrogen atoms in the aromatic ring was observed [297]. The latter reaction was assumed to occur via addition of a proton or deuteron to the aromatic system to form a σ-complex, with subsequent elimination of the substituted atom in the form of a cation. Detailed investigations showed that the photosubstitution of aromatic hydrocarbons did not produce an excited σ-complexes, that is, the reaction proceeds diabatically. Meanwhile, an emission band assigned to the excited σ-complex was observed in a study on aromatic amines (3-aminopyrene and 2-naphthylamine) in acidic solutions. Thus, the adiabatic protonation of carbon atoms appears to be possible in the presence of sufficiently strong electron-donating substituents in the aromatic system.

One of the best-known and most-widely studied examples of the heterolytic photodissociation of a carbon–carbon bond in organic molecules is the photodissociation of arylmethane derivatives [299].

If R_3CX-type compounds containing one, two, or three aryl groups (X = an anionic group e.g. CN, OR, OCOR', etc.) are irradiated in polar media, heterolytic dissociation occurs, resulting in cations R_3C^+ and anions X^-. The photolysis of triphenylmethane leuco-dyes produces stable ions of the dyes. Less-stable cations have also been detected and studied by means of flash photolysis.

A large number of photodissociation reactions of arylmethane derivatives were quantitatively studied under different conditions, the quantum yields measured, and the dissociation rate constants for singlet and triplet states determined [299]. Heterolytic dissociation is observed only in rather polar solvents ($\varepsilon > 15$). The dependence of the dissociation rate constant on the medium polarity shows that the dipole moments of the transition state are from 3×10^{-29}–8×10^{-19} C m for different compounds. Thus a considerable separation of charge (corresponding approximately to the formation of a contact ion-pair) and rearrangement of the solvation shell take place in the transition state. The activation energies are small and do not exceed 20 kJ mol^{-1}.

The presence of electron-donating substituents in the aryl groups facilitates the photodissociation and strongly increases the rate constants. However, the position of the substituents in the ring has no appreciable effect. For instance, the rate constants and quantum yields for the photodissociation of *ortho- meta-*, and *para-* methoxyphenyldiphenylacetonitrile are nearly equal, and range from

$4 \times 10^7 - 7 \times 10^7 \, s^{-1}$ and from 0.2–0.3 (in ethanol) respectively. Triphenylacetonitrile does not photodissociate under these conditions, but dimethylaminophenyldiphenylacetonitrile does so, with a quantum yield of unity. An increase in the electron affinity of the anionic group X enhances the quantum yield of photodissociation, the nitriles, methoxy derivatives, and chlorides reacting from their singlet excited states, but acetates dissociating from the triplet state. In contrast to the photoinduced transfers of a proton or electron, there is no simple relation between the rate constants and the energetics of the primary reaction for the photodissociation of arylmethane derivatives.

The predissociation mechanism seems to be most probable for a reaction consisting of a transition from a locally-excited state of the aryl group to a state of charge transfer from the aryl to the anionic group (a zwitterion) with subsequent homolytic dissociation of the C–X bond. This mechanism is consistent with the substituent effect and of the nature of group X on the quantum yield and rate constant of photodissociation.

Most arylmethane derivatives give ground-state cations upon photodissociation, that is, the process occurs diabatically. As mentioned above, adiabatic heterolytic photodissociations of C–C and C–O bonds have also been observed for the leuco-derivatives of rhodamine, pyronine, and acridine [300–302]. It should be noted that the reaction of the leuco-lactone of rhodamine C turned out to proceed from the unrelaxed Franck–Condon excited singlet state.

4.3 RADICAL PHOTOREACTIONS OF EXCITED MOLECULES

Certain types of excited electronic state (for example the n, π^* states of carbonyl and heterocyclic compounds) are known to have unpaired electrons located in spatially-separated molecular or atomic orbitals. The electronic structure and properties of such excited states are similar to those of radicals [303–306]. Thus, typical for ketones in the excited n, π^* state is the reaction of hydrogen atom abstraction from other molecules. It is interesting to compare the rate constants for the abstraction of hydrogen by a ketone in its triplet state and by *tert*-butoxyl radicals (Table 4.1).

A close similarity is seen for these reactions. One may also note the typical dependence for radical reactions of the rate constant (or activation energy) on the energy of the ruptured bond (Fig. 4.7) [307] which is analogous to the Polanyi rule. Radical pairs are formed in the abstraction of a hydrogen atom by triplet-state ketones [308].

The radical reactions following hydrogen atom abstraction finally result in reduction of the original compounds, for example ketones to alcohols or pinacols:

$$Ar_2CO \xrightarrow[RH]{hv} Ar_2\dot{C}OH \begin{cases} \longrightarrow Ar_2C(OH)C(OH)Ar_2 \\ \longrightarrow Ar_2CHOH \end{cases}$$

Table 4.1. Reaction rate constants for abstraction of hydrogen by triplet state ketones and by *tert*-butoxyl radicals [307, 409]

Ketone or radical	Hydrogen donor	$k/M^{-1} s^{-1}$
$(CH_3)_2CO$ (n, π^*)	$(CH_3)_2CHOH$	1×10^6
$PhCOCH_3$ (n, π^*)	$(CH_3)_2CHOH$	2×10^6
Ph_2CO (n, π^*)	$(CH_3)_2CHOH$	1×10^6
	$(C_4H_9)_3SnH$	5×10^7
PhC_6H_4COOPh (n, π^*)	$(CH_3)_2CHOH$	1×10^3
$(CH_3)_3C\dot{O}$	$(CH_3)_2CHOH$	2×10^6

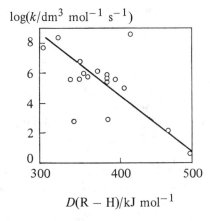

$$\log(k/dm^3 \ mol^{-1} \ s^{-1})$$

$$D(R - H)/kJ \ mol^{-1}$$

Fig. 4.7. Reaction rate constant for abstraction of a hydrogen atom by triplet molecules of benzophenone as a function of the C–H bond energy of the hydrogen donor.

In molecules with hydrogen atoms located at the γ-position, the intramolecular abstraction of hydrogen is observed, for example

4.4 CONCERTED PHOTOREACTIONS

Concerted photoreactions, in which the formation and cleavage of bonds occur simultaneously (synchronously) so that the reactants transform into the products in a single stage through a cyclic transition state, obey the Woodward–Hoffman rules of conservation of orbital symmetry. With this rule it is possible to estimate qualitatively the energy barrier to various stereochemical changes, based on the symmetry correlation of the electronic wave functions for the initial and final states [287]. There are also formally different but are related to approaches to the theoretical treatment of concerted reactions [53, 310–314] which can be applied to both symmetrical and unsymmetrical systems. According to the rule of conservation of orbital symmetry, dimerization (cycloaddition) of two π-bonded molecules to form a four-membered ring (as well as the back-reaction of the concerted opening of the ring) is forbidden (that is, it has a high energy barrier) in the ground electronic state, and is permitted (that is, has a low energy barrier) in the lowest electronically excited state. *Vice versa*, the addition of a double bond to a conjugated diene system, leading to formation of a six-membered ring (the Diels–Alder reaction) is permitted in the ground state and forbidden in the excited state. Similarly, other reactions involving the concerted rearrangements of bonds (sigmatropic rearrangements) proceeding via a four-membered cyclic transition state are forbidden in the ground electronic state and allowed in the excited state, while reactions occurring through a six-membered cyclic transition state are allowed in the ground states but forbidden in the excited state.

This consideration of concerted reactions from the standpoint of the principle of conservation of orbital symmetry assumes that these reactions occur adiabatically. However, the energetics of many such reactions which proceed with a high quantum yield (for example the photocyclization of butadiene to cyclobutene [315]) is such that the adiabatic reaction is highly endoergic. Consequently, reactions of this type have to occur diabatically with a passage to the ground-state potential surface. Strictly speaking, the principle of orbital symmetry conservation cannot be applied to diabatic reactions. The correctness of the predictions based on this principle is explained by the fact that the close mutual approach of atoms necessary for formation of new bonds, and for the transition to the ground electronic state already takes place on the potential surface of the excited state and does not require much activation energy (Fig. 4.8).

4.5 PHOTOISOMERIZATION AND PHOTOREARRANGEMENTS

Many complex molecules in their excited states undergo changes in their geometrical configuration or experience a redistribution of the bonds between their atoms, resulting in isomerization or rearrangement of the molecules. So long as

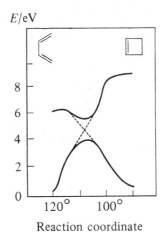

Fig. 4.8. Potential-energy diagram for cyclization of butadiene.

these processes are intramolecular, they often predominate over other photo-reactions or photophysical processes. The character of such isomerizations or rearrangements is determined by the properties of the potential surfaces of the excited state. Peaks on the potential-energy surface of an excited state do not generally coincide with those of the ground electronic state, which brings about a change in the geometry or even rearrangement of the chemical bonds of a molecule. In addition, the potential energy surfaces of excited electronic states usually exhibit a type of 'funnel' leading (via a diabatic transition) to the maxima of the ground-state potential energy surfaces, whence the molecule can get to different surface minima corresponding to different isomers.

A large number of photoisomerization and photorearrangement reactions are known. We shall consider here only the most important, which illustrate all the typical features of these processes. Reactions of this kind are represented by three major classes: (i) stereoisomerization, referring to a change in the geom-etry of a molecule without a major change in its molecular framework, (ii) valence isomerization, referring to the reorganization of the bonds in some lim-ited group of atoms (usually, the synchronous shift of bonds in a chain), and (iii) rearrangement itself, which results in a more significant change of molecular structure (usually, rearrangements occur not as a single event but as a stepwise process, including the intermediate formation of transient radical and/or ionic species).

A typical example of photochemical stereoisomerization is the *cis–trans*-isomerization of unsaturated compounds [316]. The energy of the ground elec-tronic state of alkenes adopts a minimum value if all four of the substituents at the double bond are in the same plane, so that the p-orbitals are coplanar to provide maximum overlap. In the excited state of an alkene, the energy will be minimal for the orthogonal configuration of *p*-orbitals when their overlap is minimal (Fig. 4.9). Thus the molecule in its excited state is twisted from the planar to the orthogonal configuration. After returning to the ground electronic state, the molecule arrives at the planar configuration once more, and it can assume either the *cis*- or *trans*- configuration with almost equal probability.

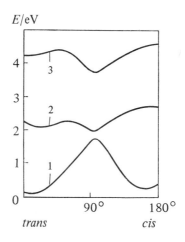

Fig. 4.9. Energies of ground state (1), excited triplet state (2), and excited singlet state (3) of stilbene as functions of the angle of twist around the central C–C bond.

In the presence of aromatic substituents conjugate to the double bond, the potential energy surface of the excited state can exhibit a local minimum even for a near-planar configuration. For instance, *trans*-stilbene and many of its derivatives emit fluoresence, that is, they have a sufficiently stable excited singlet state with a lifetime of several ns, corresponding to the *trans*-configuration. The *trans–cis* photoisomerization of these compounds takes place after their transition to the triplet state. The *cis*-isomer has no minimum for the near-planar configuration, it does not exhibit fluorescence, and isomerizes from the excited singlet state. The *cis*- to *trans*- ratio of the quantum yield of isomers is strongly dependent on the relative position of the minimum on the potential energy surface of the excited state with respect to the maximum on that of the ground state.

Absorption spectra for *cis*- and *trans*-isomers are, in general, not identical and can even be quite different for some compounds, so it is possible to control photoisomerization, directing it back and forth, simply by varying the excitation wavelength, unlike thermal isomerization which always proceeds in only one direction toward the thermodynamically more stable isomer (usually, the *trans*-iosmer).

Making use of the phenomenon of *triplet–triplet* energy transfer and of the existence of a large energy gap between the excited singlet and triplet states of alkenes, it is possible to carry out sensitized stereoisomerization by means of typical triplet sensitizers such as benzophenone.

Valence isomerization [317] is an attribute of many unsaturated compounds, and it usually proceeds by a synchronous mechanism, most frequently via a four-membered cyclic transition state. One of the relevant examples, the photo-cyclization of butadiene to cyclobutene, was mentioned in the previous section. In addition to cyclobutene, bicyclobutane is formed from another butadiene conformer:

Many similar examples are known where photoisomerization results in four-membered ring formation or the shifting of double bonds. Efforts are being made to use the highly endothermic isomerization reaction of norbornadiene to quadricyclane:

for the purposes of solar energy storage. The main problem is to find an efficient sensitizer absorbing light, since norbornadiene itself absorbs only UV radiation of $\lambda < 250$ nm which makes up less than 1% of near-terrestrial solar radiation.

The photoisomerization of aromatic hydrocarbons yields very constrained structures; this reaction is a significant pathway for the degradation of electronic excitation energy [318]:

Similar unstable photoisomers appear to be intermediates in the photoisomerization of polysubstituted benzenes, resulting in migration of the substituents together with the corresponding ring carbon atoms:

Phototautomerism is observed for many functional derivatives of aromatic compounds such as salicylaldehyde derivatives, *o*-nitrocompounds, etc. [319, 320]:

Cyclic unsaturated ketones are characterized by a specific type of rearrangement leading to formation of bicyclic systems and to the migration of substituents originally located at the γ-position. These rearrangements are believed to occur via a transient biradical or zwitterion [321].

Phenol esters form *o*-acylphenols on UV irradiation as a result of the photochemical Fries rearrangement [322]:

4.6 PHOTOREACTIONS OF COORDINATION COMPOUNDS

The most typical reactions of coordination compounds are ligand substitution (exchange), isomerization, and oxidation–reduction processes [247, 323].

On excitation by light of wavelength corresponding to the *d–d* transitions of transition-metal ions, the substitution of one ligand by another, present in the solution, takes place [323, 324]. Substitution of the ligand by solvent (solvation or hydration) and the substitution of water in aqua-complexes by an anion (annation) are also possible. Substitution is assumed to occur either by a S_N1 mechanism, that is via initial elimination of one ligand, or by a S_N2 mechanism, that is, through the synchronous addition of a new ligand with elimination of the original one.

The possibilities of photochemical synthesis has induced great interest in the photochemistry of metal carbonyls. When the metal–ligand charge-transfer band is excited, the elimination of one CO molecule takes place, followed by addition of some other organic molecule or fragment. The transient products of photolysis of mixed metal carbonyl complexes can act as effective catalysts for the stereospecific reactions of addition and polymerization.

The photoisomerization of coordination compounds can result in formation of stereoisomers and optical isomers or in a change in the ligand atom coordinated to the metal. Stereoisomerization most frequently occurs via the intermediate solvation of a complex or by an intermolecular mechanism. Intramolecular isomerization through a tetrahedral transition state is known for some square-planar complexes.

Intramolecular oxidation–reduction reactions in the coordination compounds of transition metals occur on excitation by light with λ corresponding both to ligand–metal or metal–ligand charge-transfer bands and to d–d transitions, but the former usually have an intensity higher by several orders of magnitude. A well-known example of such a redox process is the photoreaction of *tris*(oxalato)ferrate(III) used in chemical actinometers:

$$2Fe(C_2O_4)_3^{3-} + h\nu \rightarrow 2FeC_2O_4 + 3C_2O_4^{2-} + 2CO_2.$$

Intermolecular redox reactions are characteristics of those metal ions which have a sufficiently long lifetime in the excited state, for example, of uranyl ions or ruthenium(II) complexes:

$$(UO_2^{2+})^* + R_3N + 4H^+ \rightarrow U^V + 2H_2O + R_3N^+ \cdot.$$

4.7 LASER-INDUCED REACTIONS

The invention of lasers immediately raised the question of the possibility of the stimulation of chemical reactions by means of laser radiation. Laser radiation differs from optical radiation obtained from other sources (heat, gas discharge, etc.), especially as regards coherence, monochromaticity, and high power. The possibilities of using the unique monochromaticity for the realization of selective (resonance) photoprocesses and of using a high enough intensity to carry out multi-photon processes was found to be especially fruitful, so that they have become stimuli for extensive studies in these fields [261, 325]. Considerable progress has been achieved in the selective excitation of different atoms and molecules, including those which differ from each other very slightly, for example in their isotopic composition [261]. As a result, the great interest in the latter process is related to efforts to work out substantially new, more productive, methods of isotope separation.

The selective excitation of isotopic atoms and molecules with monochromatic light from conventional sources was described as early as 1922 [326–329].

Using the reactions of excited mercury atoms with oxygen, it became possible to carry out the photochemical separation of mercury isotopes [328, 329]. Laser excitation is capable of the selective ionization of atoms, thus providing the possibility of employing the reactions of the ions, or their deviation by electric or magnetic fields, for their separation.

The selective, stepwise, ionization of atoms is carried out by resonance excitation of one of the levels by the first photon and the subsequent ionization of the excited atom by the second photon (or by several photons) to the ionization continuum (Fig. 4.10a). The small Doppler width of atomic levels (about 10^{-1} cm^{-1}) permits not only different isotopes but different nuclear isomers to be selectively excited. To enhance photoionization, various versions of the stepwise ionization procedure have been suggested, making use of autoionization states and the ionization of Rydberg states by IR-radiation or by a strong electric field [330–332]. A good example of selective stepwise photoionization is the ionization of either ^{235}U or ^{234}U atoms at natural isotopic abundance by simultaneous irradiation of the atomic beam with the 591.5 nm light of a tunable dye laser and the 210–310 nm UV radiation of a mercury lamp [333].

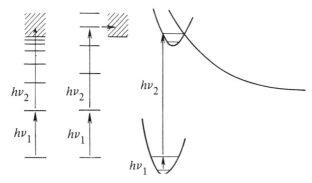

Fig. 4.10. Energy diagrams for the selective stepwise ionization of atoms (a) and selective two-stage photodissociation of molecules (b).

The selective excitation of molecules is achieved with preliminary resonance excitation of rather narrow vibrational levels by an IR-laser, with subsequent excitation to electronically excited states by UV-radiation (Fig. 4.10b). The first step provides the desired selectivity, while the second gives the electronic excitation needed for a substantial change in the reactivity of the molecules and for initiation of their transformations. Under the action of intense IR-radiation, the vibrational sublevels of the ground electronic state are populated and the electronic absorption spectra are red-shifted where the UV excitation is carried out. Notable examples are given by the selective photoionizations of ammonia ^{15}NH$_3$ and of trifluoroiodomethane composed of ^{12}C and ^{13}C isotopes [334] observed under the simultaneous action of carbon dioxide IR-laser radiation and

UV radiation from either an electric spark discharge or a nitrogen or excimer laser:

$$NH_3 + h\nu_1 \rightarrow NH_3^- \xrightarrow{h\nu_2} NH_3^* \rightarrow \dot{N}H_2 + H$$

$$CF_3I + h\nu_1 \rightarrow CF_3I^- \xrightarrow{h\nu_2} CF_3I^* \rightarrow \dot{C}F_3 + I$$

(the symbol ~ denotes vibrational excitation).

Again, the stepwise *cis-trans* isomerization of 1,2-dichloroethylene under the simultaneous action of IR-radiation from a CO_2 laser and UV-radiation from a KrF excimer laser has been described [335].

Following the stepwise multi-photon laser excitation to higher excited electronic states, the normal processes of ionization, dissociation, and fragmentation of molecules are observed which have no essential difference from the stepwise two-quantum reactions which occur in solid matrices under the action of radiation from conventional sources [335].

The stepwise excitation of higher vibrational levels of the ground electronic state is carried out by means of pulsed IR laser radiation with a power density of 10^6–10^9 W cm^{-2} acting on polyatomic molecules in the vapour phase at low pressure in the absence of collisional deactivation (Fig. 4.11). On increasing the energy density of the laser pulse above a threshold, luminescence and chemical transformations are observed. The quantum efficiency rises sharply (in proportion to the n-th power of the energy density where $n \geq 3$ in the initial region). The value of the energy density for the quantum efficiency as it becomes less than the experimental limit of detection of products is called the threshold. For the stepwise multiphoton absorption of IR laser radiation by polyatomic molecules, there is no significant decrease in the absorption coefficient with increasing excitation energy, unlike the two-level system or the anharmonic oscillator characterized by the saturation effect. Such behaviour of polyatomic molecules indicates the gradual excitation of higher and higher vibrational levels. Owing to the interaction of the different vibration modes in a molecule, a quasi-continuum of higher vibrational levels of the molecule is formed, and the absorption does not decrease, because of any change in the resonance frequency of the

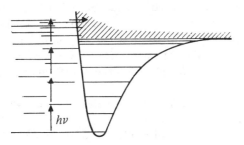

Fig. 4.11. Diagram for stepwise multiphoton excitation of polyatomic molecules to the vibrational quasi-continuum and to dissociation.

transitions between the higher levels, whereas such a decrease is typical of simple molecules. The anharmonicity can also be compensated at the expense of rotational energy. Gradually, the molecule gains an energy exceeding its dissociation threshold. This energy is essentially distributed over many degrees of freedom of the molecules since many of the composite vibrations participate in the quasi-continuum. For dissociation to come about it is necessary that this energy is localized at one bond. Under these conditions, the dissociation rate constant must exceed the reciprocal of the laser pulse duration (that is, it must be greater than $10^7 \, s^{-1}$).

The processes of dissociation are successfully described by RRKM theory. Dissociation of one of the bonds predominates over molecular fragmentation, although the latter may need a few times less energy (perhaps because molecular reactions, unlike dissociation/recombination, have a much higher potential barrier). Most often, the scission of the weakest bond in a molecule takes place regardless of the group whose vibrations were excited.

After absorption of a pulse of laser radiation, the function of the energy distribution of the molecules differs significantly from that typical of a heated gas. Two kinds of ensemble of molecules are formed: 'cold' molecules differing little from the original ones, and 'hot' molecules appearing mainly in their higher vibrational levels. It is this second ensemble which undergoes chemical transformation.

Suggestions have been made, and efforts repeatedly undertaken, to make use of the characteristic narrow bands assigned to different types of bond in the vibrational spectra of molecules for selective multiphoton excitation of some particular bond and for its selective rupture or activation. However, this attractive idea cannot be realized in practice because of the strong coupling of the higher vibrational levels forming the quasi-continuum, which leads to the very fast redistribution of energy between the different vibrational modes and prevents selective excitation of a selected bond. The action of pulsed laser radiation of moderate intensity in the fundamental absorption bands of the visible and UV regions of molecules in the vapour phase and in solution approximates to the normal one-photon absorption, but a high concentration of short-lived excited species can be created in this way, so that the interaction of these species with each other and with components of the reaction medium can be observed. Lasers of ns, ps, and fs pulse duration are of great importance to the investigation of mechanisms of photochemical and photophysical processes, the direct detection of excited species and of other intermediates, and studies on the kinetics of their reactions.

The application of laser radiation has not only promoted elaboration of the technique of selective excitation but has provided a great advance in our ideas about intramolecular and intermolecular processes of relaxation and exchange of energy [336].

5

Fundamentals of radiation chemistry

The action of radiation of a substance, material, or chemical system has the name 'irradiation' and the chemical effects caused by irradiation are called 'radiolytic' or 'radiation-chemical' effects. The entire process is called 'radiolysis'. The multiplicity of chemical effects in radiolysed matter has brought about the development of a new field of science—radiation chemistry.

This branch of chemistry arose on the eve of the last century together with the discovery of radioactivity (in fact, radioactivity itself was revealed because of the photographic, hence radiation-chemical effect of α-, β-, and γ-rays). The first stage in the development of radiation chemistry appeared to be a period of pure curiosity, when researchers simply wondered what would happen if some substance underwent the action of α-particles. During this period it was shown that a substance may suffer destruction, polymerization, and redox transformation under irradiation. The first monograph on radiation chemistry [337] was published as early as 1928. The invention of nuclear reactors raised serious practical challenges to be solved by radiation chemistry. From this point radiation chemistry advanced rapidly, so that it has now become a developed branch of high-energy chemistry.

Radiation chemistry tackles the same typical problems as the other branches of HEC considered in this book, that is, (i) investigation of the features of production and transformation of highly reactive transient species, (ii) the determination of the resistance of substances, materials, and chemical systems to the action of ionizing radiation (radiation stability), (iii) the induction of chemical processes consuming the energy of ionizing radiation (radiation synthesis, radiation processing of materials), and (iv) the development of equipment and technologies for carrying out radiation-induced processes.

Since the effects observed on irradiation are related to the energy absorbed, that is, to the dose, then experimental data are usually represented as dose-dependences of some quantitative characteristics of the effect, for example product concentration, viscosity of the system, etc. This type of dependence is also called a dose curve, accumulation curve, or degradation curve, including the cases when the dependence is actually linear. The representation of experimental data in the form of dose curves is convenient both for the qualitative and quantitative description of a radiation-induced chemical process. The quantitative characteristic is the radiation yield G, or the number of species (molecules, ions, radicals, etc.) undergoing change (produced, destroyed, or transformed) per 100 eV of energy[†] absorbed by the substrate. Thus, the radiation yield, unlike the mass or percentage yield used in synthetic chemistry, is an energetic quantity.

The radiation-induced effect for every system depends on many factors such as the concentration of the components in the system, temperature, or mass transfer, which can vary during irradiation, so that the yield in general is considered as a function of dose $G = f(D)$. Thus the initial yield G_0 and the current (apparent) yield G need to be distinguished. The initial yield is the one extrapolated to zero dose, while the current yield is that at a given dose, which is determined either by graphical differentiation of the dose curve or by taking the derivative of an analytical function describing the dose dependence. Finding the value for the current yield is not a simple problem for nonlinear dose curves. Graphical integration introduces a considerable error into the G-value to be determined, owing to the presence of random errors even in the initial pseudo-linear dose range. Determination of G by numerical differentiation of an analytical function describing the experimental data seems more correct. In those cases where the analytical function can be obtained from a model of the process, its parameters are determined by the least squares method. Since there is usually no model available for the process under study, one is obliged to look for an approximation function in the form of an n-th order polynomial or to apply the spline method [338].

The initial yield for the transient products of radiolysis (electrons, ions, and excited species) of matter has a value of *ca.* 10. The highly reactive transients interact with each other, or decompose, or react with the principal substrate or additives, so that three main possible situations are found:

(i) The transient products react mainly with each other to regenerate the original molecule. Then the apparent radiation yield for transformation of the substrate will be low compared with that of the transient species. Substances, materials, and systems of this type are usually regarded as radiolytically stable.

(ii) When interacting with each other or/and decaying, the transient species produce stable substances different from the original substrate. In this case, the value for the radiation yield of transformation of the irradiated substance will be of the order of that for the transient species. The same is true for the transient products reacting with additives.

(iii) The transient species are capable of chain propagation in the

---

† $G = 1$ is equivalent to 1.036×10^{-7} mol J^{-1}.

transformation of the irradiated substance or solute. Under these conditions, the yields of final products will significantly exceed the values for the transients.

A tremendous bulk of data on values for the radiation yields of transformations of various substances, materials, and chemical systems is available. These G-values range from 10^{-6}–10^{8} for different systems (Table 5.1).

Table 5.1. Radiation yields of final products from different types of substance

Substrate	G
Metals and ceramics	10^{-6}–10^{-4}
Glasses	10^{-4}–0.1
Polymers with conjugated bonds	10^{-4}–10^{-2}
Alkali metal halides, silicates, and sulphates	10^{-4}–1
Alkali metal nitrates, chlorates, and perchlorates	0.1–5
Saturated polymers	0.1–10
Aromatic hydrocarbons and aromatic heterocycles	0.1–1.0
Aliphatic hydrocarbons and aliphatic heterocycles	0.1–10
Oxygen- and sulphur-containing organic aliphatic compounds	3–10
Water	0.02–1.2
Aqueous solutions of inorganic and organic compounds	1–15
Monomers	10^{2}–10^{6}
Aliphatic compounds in presence of oxygen	5–10^{3}
Chlorine and sulphur dioxide solutions in alkanes	10^{3}–10^{6}
Chlorine solution in benzene	10^{6}–10^{8}

Irradiations can be continuous (that is, a constant flow of radiation), variable (interrupted radiation, when a period of irradiation is followed by a pause), or pulsed, when a single brief exposure carries a high dose).

The radiation field G for continuous irradiation is directly related to the rate v of the overall chemical process induced by the action of the radiation on a system. By definition, $G = dn/dD$ (where n is the number of molecules). The dose D can be written as $D = \dot{D}dt$ (where $\dot{D}$ is the dose rate and t the time). Then $G = dn/\dot{D}dt$ and $dn = G\dot{D}dt$. The rate of the overall chemical process is $v = dn/dt$ and $dn = vdt$. Thus we get

$$v = G\dot{D}\frac{1}{100N},$$

where $100\,N_A$ is the proportionality constant introduced to get v into units of $dm^3\ mol^{-1}\ s^{-1}$.

5.1 THREE STAGES IN RADIATION-INDUCED CHEMICAL PROCESSES

The entire process of radiolysis, from the initial impact of ionizing radiation on matter to completion of the process of formation of stable products, is normally considered as three consecutive steps.

(i) The first stage of radiolysis is the interaction of ionizing radiation with matter and formation of the first-generation transient products. This is the physical stage of radiolysis.

(ii) The second stage includes processes occurring within the period of the spur lifetime (see section 1.6). This is the physicochemical or inhomogeneous stage.

(iii) The third stage involves processes occurring in the bulk of the system after erosion of the tracks by diffusion. This is the chemical, or homogeneous stage.

The physical stage of radiolysis is quite clearly defined with respect to time: the event constituting the interaction of ionizing radiation with matter is rather fast, that is, shorter than 3×10^{-16} s (the time required for a 10 eV electron to travel 0.3 nm, corresponding to the diameter of most molecules). The next two steps are distinguished with respect to time rather less definitely and, sometimes, not even distinctly.

In the gas phase, the primary radiolytic products become distributed homogeneously within less than a picosecond, owing to the high values of the diffusion coefficient, so that the inhomogeneous stage is not developed. In liquids, both stages are observed. The inhomogeneous stage is believed to be over after 10^{-8}–10^{-7} s in a liquid with a viscosity of about 1 mPa s. As the viscosity increases, the diffusion coefficient decreases and the time for erosion of the various types of track increase. In solids, especially at low temperatures, the diffusion coefficients are very small and the process of radiolysis may be delayed at its second stage without reaching the homogeneous stage, except on warming (a process called 'annealing').

The time taken for the homogeneous stage, involving mainly radical recombination reactions, may vary rather widely since the rate of recombination depends on the total concentration of radicals and, hence, upon the dose rate applied. Thus, in an aqueous solution of organic matter, the G-value for radicals is 6 at a dose rate of 1 Gy s^{-1}, and the recombination rate constant is equal to 10^9 M^{-1} s^{-1}. Then, using the steady-state approximation (see section 3.7), we get for the lifetime of organic radicals

$$\tau = \left(\frac{2 \times 100 N}{kG\dot{D}} \right)^{1/2} \approx 0.07 \text{ s}.$$

When increasing the dose rate by two orders of magnitude, the lifetime becomes smaller by one order, and if the recombination rate constant is appreciably less than the diffusion rate constant, then the lifetime can be prolonged indefinitely.

Consideration of radiolytic processes as a function of time is justified not only for pulsed irradiation, but for continuous radiolyses since this consideration refers to a unit track pattern (the tracks do not overlap at a dose rate of below 10^6 Gy s^{-1}). When the homogeneous distribution is attained, naturally, the various types of track erode and overlap.

One can distinguish the following features of radiation chemistry as differing from other branches of HEC related to other means of supplying energy to the molecules of a substance:

(i)   Both the ionization and excitation of the molecules of the substance are primary events,
(ii)   the presence of an inhomogeneous stage in condensed media,
(iii)   multiple types of chemically active species (ions, radical ions, excited ions, excited molecules, etc.),
(iv)   a variety of reactive species of each particular type,
(v)   the lack of selectivity of radiation-induced chemical processes,
(vi)   the possibility of initiating a chemical process in a substance in any physical state.

Listed below are the time scales of events occurring in the first and second stages of radiolysis and some other quantitative characteristics of relevant processes in polar and nonpolar compounds at room temperature (the time is given in seconds):

Time scales of events in first and second stages of radiolysis

	Water	Methanol	Hexane
Impact of ionizing radiation on matter: ionization and excitation of molecules	$< 3 \times 10^{-16}$	$< 3 \times 10^{-16}$	$< 3 \times 10^{-16}$
Period of intramolecular vibration	0.91×10^{-14}	0.91×10^{-14}	1.12×10^{-14}
Decay of superexcited states of molecules	$< 10^{-12}$	$< 10^{-12}$	$< 10^{-12}$
Thermalization of non-ionizing electrons	$ca.\ 10^{-13}$	$ca.\ 10^{-13}$	$ca.\ 10^{-13}$
Solvation of electron	$\leq 3 \times 10^{-13}$	10×10^{-12}	$< 5 \times 10^{12\dagger}$

Time scales of events in first and second stages of radiolysis

	Water	Methanol	Hexane
Reactions of parent ions (fragmentation, ion–molecule reactions, isomerization)	$< 10^{-12}$	$< 10^{-12}$	$< 10^{-12}$
Diffusion of molecules to 0.3 nm (the start of radical recombination in tracks)	6×10^{-12}	1×10^{-11}	7×10^{-12}
Half-life of electrons in tracks	5×10^{-9}	$7 \times 10^{-8\ddagger}$	$1 \times 10^{-11\dagger}$
Diffusion of molecules over 3 nm	6×10^{-10}	1×10^{-9}	7×10^{-10}
End of track processes (homogenization of spatial distribution of radicals)	*ca.* 10^{-8}	$10^{-7\ddagger}$	–
Lifetime of excited singlet molecules	$< 10^{-11}$	$< 10^{-11}$	6×10^{-10}
Quantum yield of fluorescence	$< 10^{-10}$	$< 10^{-10}$	7×10^{-4}
Total initial yield of ion pairs	6.7 ± 0.5	≥ 4.4	≥ 4
Electron yield at $t = 30$ ps	4.7 ± 0.2	2.5 to 3.2	0.08 to 0.18
Apparent yield after completion of track reactions	2.8	1.9	0.08 to 0.18

† Data for cyclohexane.
‡ $T = 185$ K.

It is seen that many of the numerical quantities are similar. This fact allows us to consider the radiolyses of liquids of differing chemical natures firstly in general and then taking into account the effect of a parameter such as polarity insofar as it strongly affects the ionic component of radiolysis.

Since the velocity of ionizing particles is always high (according to the definition of the term 'ionization', it is higher than the velocity of an electron in an atomic orbital, that is, about 10^8 cm s^{-1} for valence electrons), and the size of molecules is small (< 1 nm), then the interaction time is negligible, that is, considerably less than the period of intramolecular vibrations. According to the Franck–Condon principle, no chemical process is possible at this stage because chemical processes are associated with the rearrangement of atoms in molecules; thus the only processes which can occur are those involving rearrangement of the electronic subsystem of atoms and molecules, that is, physical processes, which is why the first stage of radiolysis has acquired this name.

The processes of rearrangement of the electronic subsystem of molecules are those of electronic excitation and ionization. These two processes produce from a molecule M the following types of chemically-active species: ions M^+ (or holes h^+ in the condensed phase), excited ions M^{+*}, doubly charged ions M^{2+} (the probability of double ionization increases with the number of atoms in a molecule), electrons possessing some kinetic energy, singlet and triplet excited states M^* (higher excited singlet and triplet states can also be generated), and superexcited states M^*_{su} possessing an energy above the first ionization potential of the molecules M, and plasmons, which can emerge in the condensed state as a cooperative superexcited state. The processes occurring in the physical stage of radiolysis may be denoted as follows:

$$M \overset{\wedge\wedge\rightarrow}{} M^+, h^+, M^{+*}, M^{2+}, e^-, M^*, M^*_{su} \qquad (5.1)$$

where the symbol $\overset{\wedge\wedge\rightarrow}{}$ refers to the action of ionizing radiation.

The total number of species generated and their yield depend on the chemical composition of the system and on its physical state, as well as on the type and energy of the ionizing radiation.

Since there is no available technique for direct observation of the physical stage, information on the processes represented by Eq. (5.1) has been inferred from data obtained by different methods for times exceeding 10^{-13} s which is the best time resolution currently attainable.

The first of the processes which should be discussed is the behaviour of the free electron and the hole (see also [339]). The electron has a definite kinetic energy at its initial moment of generation and is thermalized (see section 1.2.2) by travelling some distance from its 'parent' hole. The hole also has a high mobility, though less than that of the electron (see Table 2.14), so they move together as a bound pair of charges. Owing to their difference in mobility, the distance between the hole and its electron increases with time up to the moment when the electron is thermalized. Under these conditions most electrons remain with the limits of the Onsager radius (see section 2.6) and, if no supplementary events occur, each of these electrons must return to its own hole to recombine. This is the situation in nonpolar media. In polar systems, the solvation of the electrons in traps (structure defects) of the system is possible. The diffusion coefficient of the solvated electron is by several orders of magnitude smaller than that of the quasi-free electron; thus the velocity of its motion decreases drastically. The velocity of movement of the hole decreases simultaneously so that the hole is transformed to the radical cation $M^{+\cdot}$ (perhaps, a slowly-moving hole can also be transformed into the radical cation at structural imperfections of a liquid system). The charge recombination is considered to be a very fast process so long as it involves quasi-particles, that is, the hole and the quasi-free electron. The yield of solvated electrons is less than the initial yield of ion pairs although the period of solvation is quite short.

When the diffusion of species with a mass close to that of the parent molecule becomes noticeable (after *ca.* 10^{-10} s), in addition to the reactions of the solvated electrons and the transformation of the hole to a radical cation, other processes take place in the irradiated system such as the fragmentation and isomerization of radical cations whenever possible, the degradation of superexcited states, and ion–molecule reactions of the radical cations producing free radicals and stable ions (see sections 2.3 and 2.4). So far as the transformation of primary excited molecules is concerned, it is known that higher excited singlet and triplet states undergo rapid transition to lower states. Charge-recombination also produces lower excited singlet and triplet states. For the polar compounds exemplified earlier in this section, the excited states decompose to give radicals and molecules (see section 2.2) and are deactivated quite rapidly. In the case of nonpolar substances the lifetime of charged species is much longer. Thus after only a few ps have passed, we have a collection of second-generation transients differing considerably from that of the first generation as given by Eq. (5.1), namely, M^+, e_s^-, e_{qu}^-, ^SM, ^TM, MH^+, fragment ions, radicals, and stable final products.

After passing the picosecond time domain, diffusion becomes appreciable and the period of bimolecular reactions in the track (where the concentration of species is *ca.* 10^{-2} M [340]) begins. The main reactions occurring during this period are those of radicals and radical ions (between each other) and various cross-reactions. While the spatial distribution of transient species in track patterns (spurs, blobs, and short tracks—see section 1.5) is non-uniform, we have the period when the effect of the type of radiation is displayed, which is known as the LET effect (see section 1.2.1). This is most clearly seen if the radicals participating in recombination reactions have essentially different properties, as is the case for water and methanol radiolysis.

The time scale of the events given above is typical for neat substances which are either weakly or non-reactive as regards their radiolysis products (the low reactivity means the absence of reactions with transient species during the inhomogeneous stage of radiolysis). If a compound can react with the short-lived products (thus, carbonyl compounds or halohydrocarbons efficiently capture even quasi-free electrons, while alcohols and alkylamines react rapidly both with the holes and radical cations by proton transfer, etc.) then the corresponding reaction substitutes a transient species from the previous generation by one from the next generation for the time of in-track reactions. Particles formed under these conditions have different rate constants and thus the role of spur recombination in the total recombination processes changes.

In every single-track pattern, not all the radicals recombine to form final combination/disproportionation products, but some escape into the bulk of the system. The yields of solvated electrons in water and methanol illustrate this feature. Before the beginning of radical combination in water, the G-value for e_s^- is 4.7 (already less than the initial ion-pair yield), and after completion of the track processes it is 2.8. A similar picture is observed in methanol. The products arising at the end of the physicochemical stage are classed as molecular and radical products. The molecular products are those final products which are formed in tracks during the second stage of radiolysis, their yield being denoted G_M. The radical products refer to those radicals escaping into the bulk, their yield being denoted as G_R. The next-generation radical products can also emerge in the bulk of the system, when their yield is specified as $G(R)$. Ultimately, all radicals in the bulk will produce final combination products with a yield denoted $G(M)$.

The yields of radical and molecular products are usually determined by means of the scavenging technique (see section 3.11). However, the solute (acceptor) Ac and its concentration c_{Ac} have to be selected in order to react with radicals only in the bulk, that is, so that the characteristic time of radical scavenging $\tau = 1/kc_{Ac}$ (where k is the rate constant for the reaction of the given radical with the chosen solute) will be longer that the time τ_t needed for track processes to finish. If the scavenging time τ_{Ac} is less than τ_t, then the solute will change the G_R/G_M ratio for different types of track: G_M will be lower and G_R higher than these in the absence of scavenger in accordance with the relationships described in section 3.13.

Whereas the processes of the second stage can hardly be simulated mathematically, the third, homogeneous stage of radiolysis can be successfully described with the use of a mathematical model, provided that the mechanistic scheme, the yield of transient species, and their reaction rate constants are known (see sections 3.10 and 3.12).

5.2 GAS-PHASE RADIOLYSIS

Gas-phase studies have dealt with the radiolyses both of true gases and the vapours of water, alcohols, and some other substances. These processes have been investigated in relation both to fundamental problems, (the study of the elementary events of radiolysis) and applied topics (manufacture of certain chemical products, for example nitrogen fixation [341], production of hydrogen from water [342], removal of sulphur dioxide from air [343], etc.). The sets of products of the radiolytic conversion of some inorganic gases and gaseous mixtures, with their radiation yields, are given in Table 5.2. We shall not consider particular mechanisms of radiation-induced chemical processes in the gas phase,

Table 5.2. Products of radiation-induced chemical processes in gases [341], 344–347] at room temperature and ambient pressure. Radiation yield of degradation (for pure substances) and product yields (for mixtures)

O_2	O_3	6 to 14
$O_2 + O_3$ (0.2–2%)	O_2	Chain process
CO	$CO_2, O_2, C_n, C_nO_{n-1}$	9 to 10
CO_2	CO, O_2	0.01 (3.6)[†]
		0.01 (8 + 1) ($\geq 120°C$)
H_2O	H_2, O_2	8.5 (400°C)
H_2S	H_2, S_n, H_2S_n	7 to 11.5
NH_3	N_2, H_2, N_2H_4	2.7 to 4.7
HBr	H_2, Br_2	9.8
HCl	H_2, Cl_2	8.0
N_2O	N_2, NO	11 to 13
NO	N_2, NO_2, N_2O, N_2O_3	9 to 15 .
NO_2	N_2, O_2, NO, N_2O	12 (on lowering pressure to 2 kPa $G = 53$)
$H_2 + O_2$ (2:1)	H_2O	3 (chain process above 200°C)
$N_2 + H_2$ (1:3)	NH_3	0.6 to 0.9
$CO + O_2$	CO_2	Chain process
$H_2 + Cl_2$	HCl	Chain process
$SO_2 + O_2$ (1:100)	SO_3	8 to 10 ($\geq 200°C$)
$N_2 + O_2$ (4:1)	N_2O, NO_2	2.2

† G-values given in parentheses are those determined in the presence of scavengers

which are described adequately in relevant reviews and monographs [341, 344–350]. It should be noted only that gas-phase processes involve the participation of a large number of transient species: ions, atoms, and excited molecules. For example, even such a relatively simple system as oxygen involves at least eight species: the excited molecule $O_2\left(^2\Sigma_i^-\right)$, excited oxygen atoms, $O(^3P)$ and $O(^1D)$, the ions O^+, O_2^+, and O_2^-, the molecule O_3, and the excited molecule O_3^*. Fortunately, the reaction kinetics of transient species in the gas phase have been sufficiently well-studied that the processes of gas-phase radiolysis in a number of cases can be described with adequate mathematical models [344]; this is not too surprising since the inhomogeneous stage of radiolysis plays no role.

5.3 RADIOLYSES OF WATER AND OTHER INORGANIC LIQUIDS

The radiolysis of water has been studied more than that of any other compound [351]. This is associated with its important industrial use, with the fact that it is considered to be a model substance for polar compounds, and that it is the major component of biological systems, of which the radiolytic behaviour is of great interest in cancer therapy. The ultimate products of water radiolysis are hydrogen, hydrogen peroxide, and oxygen. Hydrogen peroxide is assimilated to a relatively small concentration and then decomposes into water and oxygen, so that with large doses water gives only hydrogen and oxygen in their stoichiometric ratio of 2:1.

The sequence of elementary events of water radiolysis is shown in Fig. 5.1. The primary act results in molecular ions $H_2O^{+}\cdot$, electrons, and excited molecules. The electrons, after thermalization, undergo either hydration (to become e^-_{aq}) possibly via the intermediate step of formation of a weakly-bound electron e^-_{IR} (see section 2.5), or recombine with the parent ions (in fact, they can also recombine with the holes h^+, but there is no definite information on the topic):

$$H_2O^{+}\cdot + e^- \to H_2O^*. \tag{5.2}$$

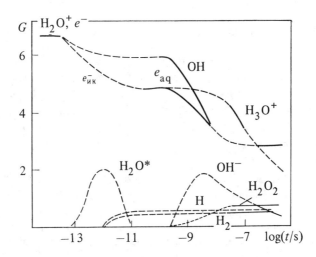

Fig. 5.1. Time-dependence of yields of transient and final products from water radiolysis by radiation with LET 0.2 eV nm^{-1} [351].

The nature of the excited states formed has not been specified, but, judging from the experimental data available, they take part in the reactions:

$$H_2O^* + H_2O \to H_2 + 2\dot{O}H \tag{5.3}$$

$$H_2O^* + H_2O \to 2H + 2\dot{O}H \tag{5.4}$$

$$H_2O^* + H_2O \to deactivation. \tag{5.5}$$

The yield of Reaction (5.2) is equal to 2.0 ± 0.2. Since the picosecond yield of hydrated electrons is known to be 4.7 ± 0.2, the total ionization yield is 6.7 ± 0.5 ion pairs per 100 eV, which is considerably higher than the G-value of 3.4 for the ionization of water vapour. This difference agrees with the change in the first ionization potential on transition to the liquid phase (see section 1.6).

The excited states of water, both primary and those produced via Reaction (5.2), seem to have a short lifetime, and Reactions (5.3)–(5.5) finish after 10^{-11} s. Thus, the final product of radiolysis, hydrogen, emerges before the spur recombination reactions begin (see sects. 1.5 and 3.13). Simultaneously, the ion–molecule reaction takes place:

$$H_2O^{+\cdot} + H_2O \rightarrow H_3O^+ + \dot{O}H. \tag{5.6}$$

The spur recombination reactions start in tenths of a nanosecond as shown by the direct technique of picosecond radiolysis [232]. The principal reactions are as follows:

$$e_{aq}^- + \dot{O}H \rightarrow OH^- \tag{5.7}$$

$$\dot{O}H + \dot{O}H \rightarrow H_2O_2 \tag{5.8}$$

Other possible combination reactions of the species $\dot{H}$, $\dot{O}H$, and e_{aq}^- contribute little to the spur processes. At the end of the track reactions ($ca.\ 10^{-8}$ s), we have the following collection of products for γ-radiolysis: H_2 and H_2O_2 as molecular products, H, $\dot{O}H$, and e_{aq}^- as radical products, and OH^- and H_3O^+ as ionic products with radiation yields as given below, where $G(-H_2O)$ is the yield of decomposition of water:

pH	$G_{e_{aq}}$	G_H	G_{OH}	G_{HO_2}	G_{H_2}	$G_{H_2O_2}$	$G(-H_2O_2)$
0.4	0	3.6	3.0	0.02	0.45	0.80	4.50
7.0	2.8	0.5	2.8	–	0.45	0.70	4.20

Further, at the homogeneous stage of radiolysis, the above products enter various mutual reactions:

$$H_3O^+ + OH^- \rightarrow 2H_2O \tag{5.9}$$

$$e_{aq}^- + H_2O_2 \rightarrow \dot{O}H + OH^- \tag{5.10}$$

$$e_{aq}^- + e_{aq}^- \rightarrow H_2 + 2OH^- \tag{5.11}$$

$$e_{aq}^- + H \rightarrow H_2 + OH^- \tag{5.12}$$

$$\dot{H} + H_2O_2 \rightarrow \dot{O}H + H_2O \tag{5.13}$$

$$\dot{H} + \dot{H} \rightarrow H_2 \tag{5.14}$$

$$\dot{H} + \dot{O}H \rightarrow H_2O \tag{5.15}$$

$$\dot{O}H + H_2O_2 \rightarrow H_2O + H\dot{O}_2 \tag{5.16}$$

$$H\dot{O}_2 \rightleftharpoons O_2^- \cdot + H^+ \tag{5.17}$$

$$H\dot{O}_2 + O_2^- \cdot H_2O \rightarrow H_2O_2 + OH^- \tag{5.18}$$

$$H\dot{O}_2 + H\dot{O}_2 \rightarrow H_2O_2 + O_2 \tag{5.19}$$

$$\dot{O}H + H_2 \rightarrow H_2O + \dot{H} \tag{5.20}$$

$$H\dot{O}_2 + \dot{H} \rightarrow H_2O_2 \tag{5.21}$$

$$H\dot{O}_2 + e_{aq}^- \rightarrow HO_2^- \tag{5.22}$$

$$HO_2^- + H_2O \rightarrow H_2O_2 + OH^- \tag{5.23}$$

$$H\dot{O}_2 + \dot{O}H \rightarrow H_2O_3 \rightarrow H_2O + O_2 \tag{5.24}$$

$$\dot{H} + O_2 \rightarrow H\dot{O}_2 \tag{5.25}$$

$$e_{aq}^- + O_2 \rightarrow O_2^- \cdot \tag{5.26}$$

Owing to the occurrence of Reactions (5.7)–(5.26), only three final products H_2, H_2O_2, and O_2, remain in the system at the homogeneous stage of radiolysis, their yields depending quite significantly on the radiolysis conditions. For instance, the G-values for H_2 and O_2 from γ-irradiated pure water are equal to 0.02 and 0.01 respectively, whereas in oxygen-saturated water, the yield of hydrogen under otherwise identical conditions is 0.42. When irradiated with α-particles, pure water produces H_2 with a yield of 1 which changes only slightly on saturating the system with O_2. If γ-irradiated water contains scavengers of the hydrated electron, of the H-atom, and of $\dot{O}H$ radical, then these solutes capture the radical products of radiolysis, thus preventing their recombination and destruction of the molecular products by the radical species. The radiation yield of reduction–oxidation or other relevant conversions of solutes corresponds to the observed G-value of water decomposition.

The yields of the radical and molecular products appear to be constant: they are identical for all dilute aqueous solutions, but depend on the solution pH (the acidity effect) and on the type of radiation (the LET effect). The latter effect is shown in Fig. 5.2 where LET-values for electrons and protons are taken as those given in Tables 1.6 and 1.8 respectively, and this for ^{210}Po α-particles is 90 eV nm^{-1}. The observed decrease in radiation yield of the radical products, and the increase in the molecular products, happens because the fraction of single pairs decreases, while that of short tracks increases, with increasing LET (see section 1.5) and the probability of recombination in the latter types of pattern is higher than in, for example, spurs.

The radiation-induced conversion of different molecules in dilute aqueous solution results from their reactions with the radical products of water radiolysis and with hydrogen peroxide, for example:

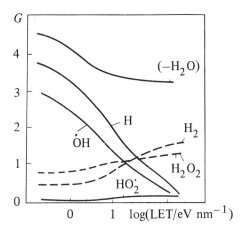

Fig. 5.2. LET-dependence for yields of radical and molecular products from water radiolysis at pH = 0.4 [352].

$$\dot{H} + Fe^{3+} \rightarrow H^+ + Fe^{2+},$$

$$\dot{H} + CH_3CH_2OH \rightarrow H_2 + CH_3\dot{C}HOH,$$

$$\dot{H} + C_6H_6 \rightarrow \dot{C}_6H_7,$$

$$\dot{O}H + Fe^{2+} \rightarrow OH^- + Fe^{3+},$$

$$\dot{O}H + NO_2^- \rightarrow OH^- + \dot{N}O_2,$$

$$\dot{O}H + CH_2{=}CH_2 \rightarrow HOCH_2\dot{C}H_2,$$

$$\dot{O}H + CH_3COCH_3 \rightarrow H_2O + CH_3CO\dot{C}H_2,$$

$$e^-_{aq} + RCl \rightarrow RCl^-{\cdot} \rightarrow \dot{R} + Cl^-,$$

$$e^-_{aq} + Tl^+ \rightarrow Tl^o,$$

$$e^-_{aq} + NO_3^- \rightarrow NO_3^{2-}{\cdot},$$

$$Fe^{2+} + H_2O_2 \rightarrow Fe^{3+} + \dot{O}H + OH^-,$$

$$Ce^{4+} + H_2O_2 \rightarrow Ce^{3+} + H^+ + H\dot{O}_2.$$

At low concentrations when a solute does not interfere with spur reactions (see section 3.13), it is convenient to use the yields of radical and molecular products of water radiolysis for determination of the G-values for some of the products of solute reactions. For instance, the radiation-induced chemical conversion of methanol in aqueous solution is described by the reactions:

$$\dot{H} + CH_3OH \rightarrow \dot{C}H_2OH + H_2,$$

$$\dot{O}H + CH_3OH \rightarrow \dot{C}H_2OH + H_2O,$$

$$e_{aq}^- + CH_3OH \rightarrow CH_3O^- + \dot{H},$$

$$H_2O_2 + e_{aq}^- \rightarrow \dot{O}H + OH^-,$$

$$\dot{C}H_2OH + \dot{C}H_2OH \rightarrow HOCH_2CH_2OH.$$

The radiation yield of ethylene glycol (EG) may be expressed via the G-values of the radiolytic products of water with the use of the steady-state method (see section 3.7):

$$G(EG) = 3.1 = \tfrac{1}{2}(G_H + G_{OH} + G_{e_{aq}^-}) = \tfrac{1}{2}(0.5 + 2.8 + 2.8) = 3.05.$$

The conversion of a solute, because of its interaction with the radiolytic products of any solvent, is known as the indirect action of radiation. In dilute solutions, the transformation of a solute occurs solely owing to the indirect action of radiation. In concentrated solutions, where the electron fraction ε (the electron fraction is the ratio of the number of electrons belonging to a given component of a system to the total number of electrons in the whole system) of the solute becomes significant, the direct action of radiation on the solute also becomes evident. Thus, there emerge products of the ionization and excitation of the solute which are *not* formed by indirect action. The primary products from each component of a solution or a mixture of substances caused by direct action are formed in proportion to their electron fraction. To date, the radiation-induced transformations caused by indirect action have been studied for very many organic and inorganic substances in aqueous solution (see, for example, [42, 69, 352]). The mechanism of direct action has been studied much less [353].

The radiolyses of other inorganic liquids may be treated in the same manner as water. It is possible to distinguish radical and molecular products, to describe on this basis the transformation of solutes under irradiation, and to differentiate effects occurring in spurs (tracks) and in the bulk of the system. But, in general, the radiolyses of other inorganic liquids have been investigated in much less detail than water. Stable and transient radiolytic products identified for some inorganic liquids are listed in Table 5.3.

5.4 RADIOLYSES OF ORGANIC COMPOUNDS

The action of ionizing radiation on organic compounds has been studied for so many substances that even merely listing them would take up the whole of this chapter. Therefore, we shall confine ourselves to a brief consideration of the most basic types of organic compound, namely hydrocarbons, and the behaviour

Table 5.3. Transient and final products in the radiolysis of inorganic liquids [347]

O_2 (−196°C)	O_3	—
CO (−196°C)	CO_2, C_nO_{n-1}	—
HCl (−78°C)	H_2, Cl_2	H, H_{hot}, Cl
HBr (−78°C)	H_2, Br_2	H, H_{hot}, Br
NH_3 (−33°C)	H_2, N_2, N_2H_4	e_s^-, H, NH_2, N_2H_3
$N_2 + O_2$ (−196°C)	O_3, N_xO_y	—
N_2O (−90°C)	N_2, O_2, NO_2	—
N_2O_4	O_2, N_2, N_2O	—
HNO_3	HNO_2, NO, NO_2, O_2	NO_3
KNO_3 (320°C)	KNO_2, O_2	—
SO_2 (−78°C)	S_n, SO_3, O_2	—
H_2S (−78°C)	H_2, S_n, H_2S_2	$H_2S_2^-$·
H_2SO_4	SO_2, H_2SO_5, $H_2S_2O_8$	HSO_4^-·
H_2O_2	H_2O, O_2	—
$NaCl$ (850°C)	Cl_2	e_s^-, Cl_2^-·

of other substances will be presented in terms of the effect of a substituent group on a hydrocarbon molecule.

Table 5.4 shows the G-values of the products from C_6 hydrocarbons of different structure. It is quite clear even at a glance that the set of products and their yields are dependent on the degree of branching, of the presence of a ring, a double bond, or aromaticity.

The following general conclusions can be drawn.

(i)   The products are diverse, involving both C–H and C–C bond scission.

(ii) H_2 is the product common to all aliphatic compounds, $G(H_2)$ decreases with increasing numbers of double bonds, for example from 5.0 for hexane to 1.4 and 0.036 for hexene and benzene, respectively.

(iii) The products derived from carbon-chain scission are observed for all alkanes, appearing as hydrocarbons with the number of carbon atoms both less and more than in the parent compound.

(iv)   All hydrocarbons form dimers, whereas unsaturated compounds also give 'polymers' (condensation products).

(v)   The G-values for the products of ring-opening in cycloalkanes are low. Thus, their contribution is only 4% for cyclohexane, but, in aromatic hydrocarbons, the yield of ring degradation is much lower.

(vi)   The total yield of conversion of hydrocarbons is similar for all alkanes, while it is lower by almost an order of magnitude for aromatics.

(vii) The set of products of degradation of the carbon chain is determined by its degree of branching.

(viii) The products of ring degradation include many unsaturated compounds; products of the isomerization of the carbon chain are also formed.

At present, the following scheme of radiation-induced transformations in alkanes is usually adopted [354]:

Localization of hole $\qquad\qquad$ $RH \xrightarrow{} h^+, e^-, RH^*$

$$h^+ + RH \rightarrow \dot{R}H^+$$

Charge-recombination processes $\quad$ $h^+ + e^- \rightarrow RH^*$

$$\dot{R}H^+ + e^- \rightarrow RH^*$$

$$r^+_{(1)} + e^- \rightarrow r_{(1)}(-H) + \dot{H}$$

Fragmentation of ions $\qquad\qquad$ $RH^+ \rightarrow R^+ + \dot{H}$

$$\rightarrow r^+_{(1)} + r^+_{(2)}$$

Ion–molecule reaction $\qquad$ $r^+_{(1)} + RH \rightarrow r_{(1)}H + R^+$

Degradation of primary and secondary excited states $\qquad$ $RH^* \left\{ \begin{array}{l} \rightarrow \dot{H} + R \\ \rightarrow \dot{H}^* + R \\ \rightarrow H_2 + R(-H) \\ \rightarrow r_{(1)}H + r_{(2)}(-H) \\ \rightarrow r_{(1)} + r_{(2)} \end{array} \right.$

Radical–molecule reactions $\qquad$ $\left. \begin{array}{l} \dot{H}(\dot{H}^*) + RH \rightarrow H_2 + \dot{R} \\ r + RH \rightarrow rH + \dot{R} \end{array} \right\}$ $\qquad$ (5.27)

$$\dot{R}_{(1)} + RH \rightarrow R_{(1)}H + \dot{R}_{(2)}, \dot{R}_{(3)}, \dot{R}_{(4)} \qquad (5.28)$$

Radical combination reactions $\qquad$ $\dot{R} + \dot{R} \left\{ \begin{array}{l} \rightarrow RR \\ \rightarrow R(-H) + RH \end{array} \right.$ $\qquad$ (5.29)

$$r_{(1)}(r_{(2)}) + r_{(1)}(r_{(2)}) \left\{ \begin{array}{l} \rightarrow r_{(1)}r_{(2)}; r_{(1)}r_{(1)}; r_{(2)}r_{(2)} \\ \rightarrow r_{(1)}(-H) + rH \\ \rightarrow r_{(2)}(-H) + rH \end{array} \right.$$

$$r_{(1)}(r_{(2)}) + \dot{R} \left\{ \begin{array}{l} \rightarrow r_{(1)}R; r_{(2)}R \\ \rightarrow r_{(1)}(-H) + RH \\ \rightarrow r_{(2)}(-H) + RH \\ \rightarrow r_{(1)}H + R(-H) \\ \rightarrow r_{(2)}H + R(-H) \end{array} \right. \qquad (5.30)$$

Table 5.4. G-values for radiolysis products of C_6 hydrocarbons [40, 76]

Product	Hexane	3-Methyl-pentane	2,3-Dimethyl-butane	Cyclo-hexane	Hexene-2	Benzene
Total radicals	4.8–5.9	2.5	2.9	4.8–5.5	—	0.7
Hydrogen	5.0	3.4	3.5	5.6–6.4	1.4	0.036
Methane	0.18	0.20	0.99	0.01	0.13	0.012
C_2–C_5 (saturated)	1.28	2.59	3.16	0.04	0.18	—
C_2–C_5 (unsaturated)	0.65	1.71	3.71	0.08	0.28	—
Acetylene	—	—	—	0.02	0.05	0.024
Isohexanes	—	0.78	—	0.28	—	—
Hexenes	2.06	0.82	—	0.38	—	—
Cyclohexene	—	—	—	3.3	—	—
Cyclohexa-diene	—	—	—	—	—	0.027
C_7–C_{11}	0.43	1.36	1.12	—	—	—
C_{12}	1.29	0.98	0.05	$1.8^a +$ 0.11^b	5.0	$0.07 -$ 0.12^c
'Polymer'	—	—	—	—	—	ca. 1
$G(-RH)^d$	6.6	7.4	5.1	7.8	10.2	1

(a) Bicyclohexyl.
(b) Hexylcyclohexane and 6-cyclohexylhexene-1.
(c) Biphenyl and two phenylcyclohexadiene isomers.
(d) As calculated from the carbon atom balance.

where RH is the parent hydrocarbon, h^+ is the hole, RH^* is the excited molecule (either singlet or triplet), $\dot{R}H^+$ is the radical cation, R^+ and r^+ are ions with the original and shorter carbon chains, respectively, $\dot{H}$ and $\dot{H}^*$ are hydrogen atoms in thermal equilibrium and in a 'hot' state, respectively, rH and r(–H) the alkanes and alkenes having shorter carbon chains than the parent alkane, $r_{(1)}r_{(2)}$, rH, etc. are hydrocarbon molecules, RR is the dimer, and R(–H) is the congruent alkene. This scheme describes more or less satisfactorily the principal features of alkane radiolysis.

The radiation yield of hydrocarbon degradation increases with increasing temperature of irradiation as shown in Fig. 5.3 for heptane radiolysis. The relative contribution of different radiolytic products also changes, thus the G-value for C_7–C_{12} hydrocarbons decreases, while that of the others increases in accordance with competing Reactions (5.27) and (5.30). It is interesting to note that the ratio of the yields of heptyl radicals R_1, R_2, R_3, and R_4 (the subscript denotes the sequential number of the carbon atom carrying the unpaired electron if counting from either end of the molecule), also varies with increasing temperature. The yield of the terminal radical $R_{(1)}$ decreases, but those of the R_2–R_4 increases (see Fig. 5.4). The radical yield ratio for radicals R_1–R_4 arising from C–H bond

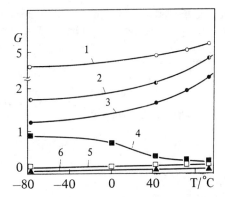

Fig. 5.3. Effect of temperature on the yields of products from heptane radiolysis: 1—hydrogen; 2—total C_1-C_6; 3—total dimers C_{14}; 4—total intermediate hydrocarbons C_7-C_{13}; 5—methane; and 6—total C_6 [355].

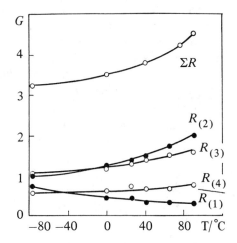

Fig. 5.4. Effect of temperature on radiation yield of formation of heptyl radicals $R_{(1)}$-$R_{(4)}$ and on total radical yield ΣR[355].

rupture should be, to a first approximation, 3:2:2:1, that is, corresponding to the number of relevant C–H bonds in the molecule. If irradiated at room temperature and at a normal dose rate of $ca.$ 10^{16} eV cm^{-3} s^{-1}, heptane gives a ratio of 0.6:2:2:1 owing to the rather effective conversion (Reaction (5.28)) of radicals $R_{(1)}$ into radicals $R_{(2)}$, $R_{(3)}$, and $R_{(4)}$. Increase of temperature disturbs the theoretical ratio even more. However, an increase in dose rate, resulting in reduction of the lifetime of radicals with respect to recombination (Reaction (5.29)) and, correspondingly, undermining the role of the abstraction Reaction (5.27), brings, on the contrary, this ratio closer to the theoretical, making it 2:2:2:1. A similar effect is produced by addition of radical scavengers.

The rate of radiolytic degradation increases with increasing temperature, and the process at some temperature adopts a chain character (radiation-thermal cracking [358]) for all hydrocarbons. The chain process is described by the following general equations:

$$\left.\begin{array}{l} RH \xrightarrow{\sim\!\!\!\wedge\!\!\!\sim} \dot{R} + \dot{H} \\[2mm] \dot{H} + RH \longrightarrow H_2 + \dot{R} \end{array}\right\} \text{(radiation-induced)} \qquad\qquad \begin{array}{l}(5.31)\\[4mm](5.32)\end{array}$$

chain initiation

$$RH \xrightarrow{\;t^o\;} \dot{R} + \dot{H} \qquad \text{(thermal)} \qquad\qquad (5.33)$$

chain propagation $\dot{R} + RH \longrightarrow Product + \dot{R}$ $\qquad\qquad\qquad$ (5.34)

chain termination $\dot{R} + \dot{R} \longrightarrow RR$ $\qquad\qquad\qquad\qquad\qquad$ (5.35)

For those organic and inorganic systems where a chain process develops on irradiation, four distinct temperature regions can be found (Fig. 5.5). The first is the radiation region, where radicals are generated only by irradiation (Reactions (5.31) and (5.32)), and Reaction (5.34) does not occur. There is no chain process in this region, and the activation energy is equal to only the few kJ mol^{-1} typical for the overall process of radiolysis. The second region is the radiation-thermal region: here Reaction (5.34) becomes possible and the process becomes a chain (5.34). The third, thermo-radiation, region is displayed when radical $\dot{R}$ is thermally unstable and can decompose, that is, we have a branching chain process

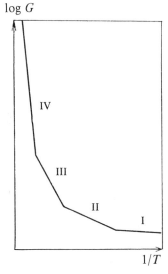

Fig. 5.5. Effect of temperature on yields of radiation-induced oxidation of hydrocarbons [357]. Temperature regions: I—radiation region; II—radiation-thermal region; III—thermo-radiation region; and IV—thermal region.

with an activation energy of *ca*. tens of kJ mol^{-1} characteristic of Reaction (E_a is now of the order of a few hundred kJ mol^{-1}). The fourth is the thermal region where the rate of initiation by Reaction (5.33) exceeds the rate of radiation initiation. The activation energy in this temperature range is determined purely by the thermal process. The features of the temperature and dose-rate effects (the rate of a chain process depends on the rate of radical generation which, in turn, is dose-rate dependent) are reviewed by Saraeva [357] for the radiation-induced oxidation of hydrocarbons.

The introduction of a functional group into a hydrocarbon results in the appearance of specific features in the radiolysis. These features, if compared with those of relevant hydrocarbons, may be represented in a general form as follows (Table 5.5 shows the final and transient products for some monosubstituted derivatives of hydrocarbons):

(i)   the products are more diversified,
(ii)   products are formed by the degradation of the functional group,
(iii)   elimination of the functional group can take place, when the transient products will include alkyl radicals,
(iv)   bifunctional derivatives can form due to the combination of particular radicals,
(v)   insofar as the functional group is either an electron donor or electron or hole acceptor, radical cations or radical anions are produced as transients,
(vi)   for compounds with a sufficiently long carbon chain, the radicals formed result not only abstraction of hydrogen from the position α to the functional group but also from the β-, γ-, δ-, etc. sites,
(vii)   the functional group may protect the hydrocarbon moiety [358].

The mechanisms of radiation-induced transformations of functionalized hydrocarbons have been studied much less than that of hydrocarbons themselves, except, perhaps, for methanol.

One further problem of interest is that of radiation stability [59]. The specific term comes from materials science. Here we shall consider radiation stability as the ability of a given substance to endure a certain irradiation dose without significant change in those properties pertinent to a specific technical function. The radiation stability (resistance) can also be described by a radiation yield G: the lower the G-value, the more resistant to radiation is the substance. Table 5.1 gives examples of both very stable and very unstable compounds. Most substances have a yield of primary radiolytic products of *ca*. 10, so that the degradation field is of the same order. If a compound is used for some purpose, then the problem of decreasing its yield of degradation will arise. This problem needs to be solved for each group of materials separately since the mechanisms of radiation-induced transformations are quite different and depend on the conditions of operation (temperature, physical state, etc.). The radiation stability of liquids can

Table 5.5. Transient and final products from radiolysis of derivatives of alkanes [347].

Substrate	Final products	Transients
Alcohols		
primary	H_2, H_2O, alkanes, diols, aldehydes	$R\dot{C}HOH$, $R\dot{O}$, e_s^-
secondary	H_2, H_2O, alkanes, ketones	$RR'\dot{C}OH$, $R\dot{O}$, e_s^-
tertiary	H_2, alkanes, ketones	$R_3C\dot{O}$
Aldehydes	H_2, alkanes, CO, polymer	$R\dot{C}O$, R, $R\dot{C}HOH$
Ketones	H_2, alkanes, aldehydes, ketones, alcohols, diketones, acids	R, $R\dot{C}HCOR'$ radical-cations, radical-anions
Acids	H_2, alkanes, CO, CO_2, aldehydes, esters, dicarboxy acids	Radical-anions, R, $R'\dot{C}HCOOR$
Ethers	H_2, alkanes, alcohols, aldehydes, diethers	$R'\dot{C}HOR$, radical cations, e_s^-
Esters	H_2, alkanes, ethers, aldehydes, diesters	
Amines		
monoalkyl	H_2, alkanes, diamines, NH_3	$R\dot{N}H$, $R'\dot{C}HNH_2$, $R\cdot$, e_s^-
dialkyl	H_2, alkanes, diamines	$R_2N\cdot$, $R'\dot{C}H\dot{N}HR'$, $R\cdot$, e_s^-
trialkyl	H_2, alkanes, diamines	$R_2N\cdot$, $R'\dot{C}HNR_2$, $R\cdot$
Amides	H_2, CO, alkanes, amines	e_s^-
Nitriles	H_2, alkanes, HCN, cyanogen, dinitriles	$R\dot{C}HCN$, radical anions
Thiols	H_2, alkanes, disulphides, sulphur	R, $R\dot{S}$, radical - cations and radical- anions
Thioethers	H_2, alkanes, disulphides	R, $R\dot{S}$, $R'\dot{C}HSR$, $R\dot{C}HCH_2SR'$
Disulphides	H_2, alkanes, monosulphides	R, $R\dot{S}$, radical - cations, and radical-anions
Monoalkyl-halides	H_2, alkanes, alkenes, dihalides	R
Polyalkyl-halides	H_2, alkylhalides	$H\dot{a}l$, $R\dot{C}HHal$, $R\dot{C}Hal_2$, etc.
Alkyl-sulphonyl chlorides	H_2, alkanes, HCl, SO_2	—

be affected only by changing the fate of the transient species. It is impossible to suppress completely the radiation-induced destruction of a substance, but in many cases it can be decreased by an order of magnitude or even more. In liquid hydrocarbons, for instance, the method of scavenging of excited species is used to transfer a reactive form of the parent excited molecule to the inactive form of an excited solute; again, electron- or hole-scavengers are introduced to restore the parent compound or to prevent the formation of the active form of its excited state. Under these conditions the solute is definitely destroyed, but the main purpose of enhancing the radiation stability of the base compound as a source of desirable material properties is fulfilled.

5.5 RADIATION-INDUCED POLYMERIZATION AND RADIATION EFFECTS IN POLYMERS

Polymerization is a chain process consisting of the addition of a reactive chemical species (radical or ion) to a monomer molecule, resulting in formation of a larger radical or ion capable of adding to another monomer molecule. Monomers are typically compounds with multiple bonds, such as $C=C$, $C\equiv C$, $C=O$, and $C\equiv N$, and some cyclic compounds like alkene oxides, lactones, and lactams. On irradiation, both ions and radicals are formed which function as effective initiators for polymerization, providing either ionic or radical process:

$$\mathrm{>C{=}C<} + M^- \longrightarrow M{-}\overset{|}{\underset{|}{C}}{-}\overset{|}{\underset{|}{C}}{}^-,$$

$$\mathrm{>C{=}C<} + M^+ \longrightarrow M{-}\overset{|}{\underset{|}{C}}{-}\overset{|}{\underset{|}{C}}{}^+,$$

$$\mathrm{>C{=}C<} + R\cdot \longrightarrow R{-}\overset{|}{\underset{|}{C}}{-}\overset{|}{\underset{|}{C}}\cdot,$$

$$\mathrm{>C\overset{O}{\overset{\wedge}{-}}C<} + R\cdot \longrightarrow RO{-}\overset{|}{\underset{|}{C}}{-}\overset{|}{\underset{|}{C}}\cdot.$$

Radiation-induced polymerization may be carried out under various conditions [359–361], for example in gaseous, liquid, and solid states, in a pure substance (bulk polymerization), in solution, emulsions, and suspensions, or even at liquid helium temperature [362]. The techniques of radiation-induced polymerization made it possible to manufacture a large number of different polymers and polymer-based materials.

Polymerization induced by radiation has some specific features differentiating it from normal polymerization [363].

(i)   The polymer obtained does not contain the products of degradation of initiators,

(ii)   so long as the generation of reactive intermediates under irradiation does not actually depend on temperature, the radiation-induced polymerization can be carried out at any temperature down to that of liquid helium,

(iii)   the effect of impurities such as transition-metal ions is considerably less,

(iv)   the process can be controlled by varying the dose rate. The possibility of terminating initiation at any chosen moment provides conditions for preparing a polymer of ultra-high RMM with a narrow distribution of RMMs.

As regards the effect of ionizing radiation, polymer molecules may be considered as appropriate hydrocarbons or their derivatives, taking into account the situation that all the events take place in the solid phase. Because of its large size, one polymer molecule can simultaneously generate several activated (ionized or excited-state) sites which, being separated by many carbon atoms, do not interact with each other and so may be considered as independent, active (ionic, radical, or electronically excited) centres. Owing to the slow transfer of active centres in a solid, their lifetimes in polymers are large compared with the liquid phase. Hence, radicals live indefinitely at room temperature in many polymers, so that it is necessary to take into consideration post-irradiation effects for irradiated polymeric materials.

The action of ionizing radiation on high polymers was found to cause the following changes:

(i)   Formation of intermolecular bonds, or cross-linking of molecules and the emergence of a network structure,

(ii)   formation of intramolecular links between different sites of the same polymeric molecule,

(iii)   bond scission in the main chain and loss of the side groups (polymer degradation),

(iv)   the emergence and disappearance of vinylene ($RCH=CHR$), vinyl ($RCH=CH_2$) and vinylidene ($RC(CH_3)=CH_2$) unsaturation and of conjugated double bonds,

(v)   isomerization and cyclization,

(vi)   gas evolution,

(vii)   radiation-induced oxidation,

(viii) a change in crystallinity,

(ix)   linking of the polymer molecules to additives.

Aspects of the radiation-induced transformations in polymers are considered in many monographs [364–366]. All of the above-listed processes are shown to occur simultaneously on irradiation of any polymeric material, but polymers can be divided into two classes according to which process predominates, that is, either the enhancement of structure (cross-linking) or structural degradation. Some examples from both classes are shown in Table 5.6. It is evident that those polymers undergoing degradation have quaternary carbon atoms in their structure. If each carbon atom is bound to at least one hydrogen atom, the polymers are referred to as cross-linking under irradiation. The G-values for processes in polymers are largely determined by the physical state of the polymer, that is, as either crystalline or amorphous, by the irradiation temperature, and by the presence of moisture, oxygen, and other impurities.

High polymers are widely used as construction materials operating in irradiation fields, so there is a problem as to the elimination or, at least, reduction of undesirable effects induced by radiation. This problem is solved on the basis of principles similar to those developed for hydrocarbons, that is, mainly with the use of scavengers of reactive transient species. These scavengers are called *antirads* if they suppress radiation-induced degradation, *antioxidants* if preventing oxidation, and *thermostabilizers* if helping to sustain operational temperatures. It is possible to impose both external protection by introducing additives and internal protection by introducing appropriate functional groups into the polymer molecules [366]. Some specific techniques are also used such as structural chemical thermostabilization, or creation of a well-formed transmolecular structure [364].

Table 5.6. Polymer structure in relation to the direction of radiation-induced processes [359]

Cross-linking	Degradation
Polyethylene ~CH_2~	Polyisobutylene ~CH_2–$C(CH_3)_3$~
Polypropylene ~CH_2–$CH(CH_3)$~	Poly-α-methylstyrene
Polystyrene ~$CH_2CH(Ph)$~	~CH_2–$C(CH_3)$ (Ph)~
Polyacrylates ~CH_2–$CH(COOR)$~	Polymethacrylates
	~CH_2–$C(CH_3)$ (COOR)~
Polyacrylamide ~CH_2–$CH(CONH_2)$~	Poly(methacryl amide)
	~CH_2–$CH(CH_3)$ $(CONH_2)$~
Poly(vinyl chloride) ~CH_2–$CH(Cl)$~	Poly(vinylidene chloride)
Poly(vinyl alcohol) ~CH_2–$CH(OH)$~	~CH_2–$C(Cl)_2$~

5.6 RADIATION-INDUCED CHEMICAL EFFECTS IN SOLIDS

Processes caused by ionizing radiations in solids have certain specific features related to the properties of the solid state, such as its regular structure (crystallinity and crystal lattice imperfections, both original and radiation-induced), the electronic interaction between the particles constituting a solid (from which an important role played by quasi-particles such as excitons, phonons, conduction electrons, and holes, emanates), the restrained diffusion of large particles, etc. Radiation-induced effects in solids are studied both by radiation chemists and radiation physicists. Radiation physics deals mainly with radiation effects on metals [367], semiconductors [368, 369], inorganic glasses [370, 371], and dielectrics based on ionic crystals such as alkali metal halides and oxides [372, 373]. As regards the mechanism of radiolytic processes, radiation physics covers radiation-induced structural defects, that is, stabilized electrons in the form of F-, M-, and X-centres or of colloidal metal, hole species in the form of different V centres, the role of excitons in the production of radiation defects (termed 'colour centres' as they exhibit a characteristic band in the visible spectrum), etc. In radiation chemistry much attention is paid to the formation of radical products in radiolysis, to their formation kinetics and their reactions [366, 377], and to their transformations under the combined action of light and ionizing radiation [378]. This dichotomy of the research area between radiation physics and chemistry has occurred historically, but it is now gradually fading as topics of study and experimental techniques have become almost identical. In this section we shall not try to cover all types of effects induced by radiation in solids, but shall consider only a few paradigm examples.

In a liquid, the radiolytic products are distributed homogeneously after the physicochemical stage is over. This is proved at least by the fact that the rate constants for reactions of the transient species at this stage have the same value as when they are generated by methods giving an unquestionably homogeneous distribution. In crystals, by contrast, the physicochemical stage for particles of a size close to the lattice parameters is not completed even at room temperature. However, the G-values for stabilized particles in many cases are low, for example less than unity for alkali metal halide crystals, sulphates, aluminates, alkali metal silicates, etc., instead of a value of few particles per 100 eV as expected at the first stage of radiolysis. This discrepancy gave birth to the idea that a geminate ion-pair, on recombining, produces an exciton (the Onsager radius in solid dielectrics is the same as in hydrocarbons, that is, about 30 nm, which agrees with this assumption) which is capable of migration through the matrix and of generating the stabilized electron and hole centres on structural imperfections. Even now it remains unclear which type of defect is responsible for decay of the excitons, that is, either point defects (with a uniform distribution) or dislocations (a nonuniform distribution of the stabilized species).

The radiation yields of trapped transient species in matrices containing ions capable of capturing the electron or hole (nitrates, perchlorates and chlorates of the alkali metals) are *ca.* 10, that is, they correspond to complete or nearly complete stabilization of the species appearing at the primary stage. The ionic components of radiolysis have to be taken into account in these matrices, though one cannot neglect the partial role also played by excitons [375].

On irradiation with electrons, induction of electric charge occurs in solids; an electric field is thus created in their bulk. Relevant data obtained by an acoustic probe technique for two dielectrics, lithium fluoride and poly(methyl methacrylate) (PMMA), are shown in Fig. 5.6. The induced field is seen to be nonuniform inside the sample and to depend on the irradiation dose (curves 1 and 2, curve 2 corresponding to the limit of electrification), on the type of dielectric (curves 1 and 4), on the ratio of range of the electron to the target thickness (curves 3 and 4), and on some other factors. The induced field is caused by the accumulation of charges due to the external flow of electrons, and its spatial distribution is also related to charge transfer under the action of the external field [382, 383]. Undoubtedly, the induced electric field affects the reaction kinetics of charged species, both in the bulk and on the surface of a solid dielectric.

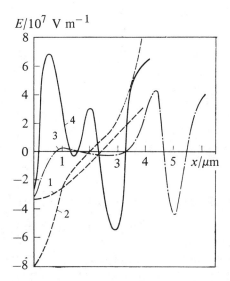

Fig. 5.6. Induced electric field after electron irradiation of lithium fluoride (1,2) and PMMA (3,4): 1—$D = 0.006$ C m^{-2}; 2—$D = 0.1$ C m^{-2}; 3—$D = 0.001$ C m^{-2}, sample thickness x is larger than the range of 1 MeV-electrons; 4— $D = 0.01$ C m^{-2}, sample thickness is less than the electron range [379].

A solid is not a space wholly filled with atoms, and even for close packing in an ideal crystal there is some free space consisting of cavities of complex form and of channels between them. This space is termed the 'free volume'. When a constituent ion or molecule of a solid undergoes degradation producing a radical

with a size close to the characteristic free volume, then these radicals have a
certain chance of escaping through the abovementioned channels from their site
of formation and thus avoid recombination with the original molecule or ion. In
accordance with this, the yield of radiolytic products must increase with increasing free volume in compounds of the same type. This has indeed been observed
for the final products from some salts of inorganic acids [384]. There is also an
example where the effect of free volume is apparent for radicals. As shown in
Fig. 5.7, the G-values for the total yield of free radicals for different acetates fit
a straight line with an intercept corresponding to a free volume numerically
equal to the size of the $\dot{C}H_3$ radical.

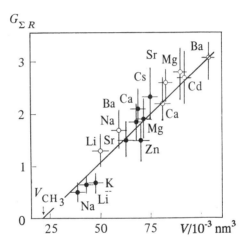

Fig. 5.7. Dependence of total yield of radicals from acetates on free volume of crystal
[383]. ●—anhydrous salt; ○—crystalline hydrate.

5.7 RADIATION CHEMISTRY OF HETEROGENEOUS SYSTEMS

Heterogeneous systems are composed of parts differing from each other in physical or chemical properties (phases) so that at the phase boundaries separating
neighbouring phases, one or more of the physicochemical properties (interatomic
distance, physical state, light transmittance, volume charge, concentrations of
impurity, etc.) undergoes an abrupt change. The basic types of heterogeneous
system are (i) polycrystals, epitaxies, and mechanical mixtures of solids, (ii) any
solid body of small dimensions in a gaseous atmosphere, (iii) solids where part
of the volume is occupied by a crystal phase and the other by an amorphous
phase, composite materials, (iv) solids contacting with liquids, and (v) disperse
systems (sols, gels, suspensions, aerosols, emulsions and foams).

The existence of an interface between the phases of components of a system
and the discontinuity of the physicochemical properties at the phase boundaries,

significantly complicates the scheme of radiation-induced chemical processes. It is necessary to consider the processes in each phase separately and the joint effect of the phases on the processes and properties. Gromov & Kotov distinguish five zones in the system 'solid + absorbed compound' [384]: the solid (I), a monolayer (II), a layer of thickness corresponding to the mean diffusion distance from the surface of the solid for an atom or ion (III), a layer of thickness corresponding to the range of Auger electrons (IV), and a zone affected by the electric field of the absorbent (V). To complete the picture, the zone where the migration of energy from the solid to the absorbate can occur, should be considered. If needed, several zones corresponding to the transfer of charge, atoms, or excitation from absorbed compounds are also included. It is also necessary to accommodate any change in physicochemical properties of both components with increasing distance from the phase boundaries ('edge effects'). Thus, the scheme of radiolytic processes in heterogeneous systems appears rather intricate. The current state of the topic is discussed in [383]. The advanced development of instrumental methods for studying surfaces and surface phenomena provides the hope that the radiolyses of heterogeneous systems will be studied in more detail and become clearer.

6

Fundamentals of plasma chemistry[†]

6.1 THE CONCEPT OF A PLASMA

A plasma is a partly or wholly ionized state of matter in which a system contains free positive (ions) and negative (electrons and, rarely, ions) particles so that their concentrations on average are practically equal. The presence of charged and excited species in a plasma and their interactions bring about some specific physical and chemical features causing the behaviour of a plasma to differ from that of an ordinary gas, thus giving a reason to consider it as a particular 'fourth' state of matter. The principal property of a plasma is its quasi-neutrality. This property is exhibited by starting from a certain volume and a certain time, depending on the scale of separation of the charged particles. The charge separation leads to emergence of a plasma oscillation whose period determines the time scale t_0, A space scale l, known as the Debye radius, is determined by the distance covered by a particle in thermal motion in a time t_0.

In addition to its quasi-neutrality, the plasma state is characterized by two more properties which are (i) the collective electrostatic interaction of charged species (for a sufficiently rarefied gas under normal conditions there is only the interaction of two particles) and (ii) the strong dependence of the parameters of the plasma on electric and magnetic fields.

Both gases and semiconducting solids can exist in the plasma state. In a solid plasma, the electrons and holes act as free, charged particles. We shall concentrate on gas plasmas. One may distinguish between a wholly-ionized (stripped) plasma and a part (weakly) ionized plasma, the latter often being called a 'low-temperature plasma'.

A low-temperature plasma is obtained under laboratory or industrial conditions with the use of specially designed reactors, or plasmotrons (a detailed description of these is given in Chapter 7).

[†] Chapter 6 was written in cooperation with Dr F. B. Vursel.

In the scientific literature, one can meet the terms 'equilibrium and nonequilibrium', 'high-temperature and low-temperature', and 'hot and cold' plasma. The latter two terms explicitly or implicitly involve the concept of temperature. Temperature in statistical physics is defined for a system whose probability of being in a state with energy E is proportional to $\exp(-\beta E)$, where $\beta = 1/kT$ and k the is Boltzmann constant. Only systems like this are said to have one and only one temperature.

If a plasma is in a state of complete thermodynamic equilibrium, it is described with one and only one temperature. However, a partial thermodynamic equilibrium is frequently observed in a plasma. This is caused by the fact that energy exchange between particles of equal mass occurs more efficiently than between particles of essentially different mass, for example between ions and electrons. In a molecular gas plasma, the exchange of energy between different degrees of freedom of molecules can also be hindered. Therefore, under certain conditions, a plasma may be characterized not by one but several temperatures: electronic, vibrational, rotational, and translational temperatures.

Thus in a plasma created by a shock wave in high electric or magnetic fields, as well as in a molecular gas plasma, where an important part is played by excitation of the internal degrees of freedom of its molecules, there cannot be even partial equilibrium, in which case none of the above mentioned temperatures is meaningful. Such a plasma must be characterized by energy distribution functions for the corresponding particles or by their mean energies.

It seems appropriate here to clarify the terms used in the literature. A nonequilibrium plasma is one characterized by nonequilibrium concentrations and/or different energies (temperatures) of its components, or a plasma where there is partial equilibrium.

However, even in an equilibrium plasma, local regions of nonequilibrium can exist, for example in sites where mixing of the plasma with the surroundings takes place, or near the walls of a channel where the plasma flows, or at the site of formation of a condensed phase. A plasma where the temperature of the heavy particles is $\leq 10^5$ K is considered as a low-temperature plasma, while one with $T > 10^5$ K is taken to be a high-temperature plasma. A hot plasma is the same as a high-temperature plasma. A cold plasma is a nonequilibrium plasma where the temperature of the heavy particles is less than 10^3 K and the mean energy of electrons is *ca.* several eV.[†] The composition of a plasma is characterized by its degree of ionization. At equilibrium the degree of ionization is in terms of temperature and pressure. Ionization occurs owing to encounters of highly energetic heavy particles with each other and with electrons or to ion–molecule reactions, photoionization, etc. In a low-temperature plasma containing neutral and charged species, in addition to the usual gas molecular collisions

† 1 eV $\approx 10^4$ K

occurring at short distances, long-range encounters occur which are caused by electromagnetic interaction between the charged particles.

What are the values for the parameters of plasmas currently attainable? Different types of plasma generator (see Chapter 7) can provide a plasma state for almost any gas at a pressure of 10^1 to 10^7 Pa. The gas temperature can be varied from near-absolute zero to *ca.* 10^4 K at a concentration of charged species of 10^7–10^{17} particle cm^{-3}, with a mean energy of 10^{-1}–10^1 eV. The fraction of species excited via their internal degrees of freedom can also be large, ranging from fractions of a per cent to a few tens per cent. The velocities of plasma jets can be varied over very wide limits, from values close to zero to several km s^{-1}, so these jets can have a large dynamic head. Plasma flows are also characterized by large enthalpies reaching 10^3 kJ mol^{-1} for diatomic gases [87]. The energy gap between different components (for example the energy of electrons with respect to that of molecules) in a nonequilibrium plasma may reach several orders of magnitude.

Processes can occur in a low-temperature plasma which are not found in conventional chemistry. These are nonequilibrium processes. They play an increasingly important role in the processing of plasmas and, in particular, are used for manufacturing solid materials with unusual (nonequilibrium) structure and unique properties (ultradisperse powders and films). Plasmochemical procedures are available for the surface treatment of materials, including the surface modification of metals, semiconductors, and dielectrics (silication, nitridation, aluminizing, etc., and ion implantation, plasmoelectrolytic processes, etc.). Etching techniques in electronics and the application of plasma chemistry in medicine are also based on the specific physicochemical properties of a reacting nonequilibrium plasma. Nonequilibrium concentrations of reactants and of reactive transients and products can be created in this type of plasma which provide, in particular, an extremely high selectivity of reaction; nonequilibrium energy distribution functions for different components and nonequilibrium populations of excited rotational, vibrational, and electronic levels are also possible.

In this chapter we shall deal mainly with low-temperature plasmas.

6.2 KINETIC FEATURES OF PLASMOCHEMICAL PROCESSES

The kinetics of both equilibrium and quasi-equilibrium plasmochemical processes are a particular case of nonequilibrium chemical kinetics which have been described in detail in Chapter 3. When tackling the problems of plasmochemical kinetics, it is appropriate to use the mathematical simulation methods discussed in sections 3.10 and in the monographs [94, 170] where not only the procedures for calculation but the relevant computer programs are given.

Let us emphasize some features typical of plasmochemical kinetics. The characteristic times of physical, physicochemical, and chemical processes in a low-temperature plasma are similar (or of the same order of magnitude).

Consequently, they affect each other and therefore, plasmochemical processes appear as a multichannel problem with the channels interacting in a different way at different times and with a different energy. To illustrate this we give data on the characteristic times of processes in the energy range corresponding to apparent temperatures of 3×10^3–1.5×10^4 K:

Mean free path time of molecules (coincides with Maxwellization time)	*ca.* 10^{-9} s
Period of vibrational relaxation of molecules	*ca.* 10^{-7} s
Relaxation period in dissociation of O_2	*ca.* 10^{-7} s
Mean lifetime before an effective collision with a threshold energy of *ca.* 2.5 eV.	*ca.* 10^{-6}–10^{-5} s

In most cases it is necessary to distinguish in plasmochemical kinetics translational, vibrational, rotational, and electronic temperatures (note that in many instances the concept of temperature is wholly meaningless in nonequilibrium stationary or relaxing systems). This causes us to abandon the use of the Arrhenius reaction rate constant (a function of temperature) in plasmochemical kinetics.

Finally, in a highly nonequilibrium plasma (for example in a glow discharge plasma where the mean electron energy is *ca.* 3–5 eV and the temperature of heavy particles is *ca.* 500–800 K) nonequilibrium phase transitions can be observed, especially gas–solid transitions.

Electrons and neutral particles or electrons and ions in a plasma exchange their energies. Energy exchange between electrons and ions occurs as a result of Coulombic encounters. These processes can lead to chemical reactions and hence investigation of the mechanisms and probability of such induced reactions is one of the major tasks of plasmochemical kinetics [88, 124].

6.3 MECHANISMS OF CHEMICAL REACTIONS IN PLASMAS

The presence of charged and excited species in a weakly ionized plasma and their reactions are principal features of the kinetics and mechanism of plasmochemical reactions. The formation and decay of these species occur in the processes of excitation, dissociation, ionization, deactivation, and recombination [199, 386].

6.3.1 Rotational excitation of molecules
Molecules can be excited to their rotational states on collision with electrons or heavy particles.

Since a fraction of the energy transferred on the collision of particles is proportional to the ratio of their masses, the induction of rotational transitions to the ratio of their masses, the induction of rotational transitions on electron impact is weakly effective owing to the small electron mass. The cross-section of this process for molecules composed of identical atoms (homonuclear molecules) in the energy range with a threshold value to 10^{-2}–10^{-1} eV does not exceed 10^{-17} cm^2. The cross-section for excitation or rotational transitions in heteronuclear molecules is larger by nearly an order of magnitude. The rotational transitions of molecules can take place upon resonance formation and dissociation of a negative molecular ion at electron energies of several eV. The cross-section reaches a value of 10^{-16} cm^2 in this case, but it transpires that vibrational excitation is more significant. Also, the width of the resonance peaks is small, and this channel barely makes a contribution to rotational excitation.

Let us consider when the role of electron impact in the population of rotational levels becomes essential. Calculations show that, for example, in nitrogen and oxygen this takes place at a large degree of ionization ($ca.$ 10^{-1}) provided that the electron and gas temperatures are equal. In hydrogen and carbon monoxide this occurs at a degree of ionization of 10^{-2}. Thus, in commonly used low-temperature plasmas with a relatively small degree of ionization ($ca.$ 10^{-3}), electron impact barely affects the population of the rotational levels.

The excitation of molecules to rotational levels in collisions with heavy particles is more effective. The number of collisions needed for establishment of equilibrium between rotational and vibrational degrees of freedom depends on the mass of the molecule and ranges from several hundred (for H_2 and O_2 molecules) to unity (for example, for CF_4).

6.3.2 Vibrational excitation of molecules

According to calculations, direct vibrational excitation of molecules by electron impact is also weakly effective (cross-section $ca.$ 10^{-17}–10^{-16} cm^2). However, appreciably larger cross-sections have been obtained experimentally for this process. The cross-sections exhibit resonance peaks related to the formation of an unstable negative ion. This is illustrated in Fig. 6.1 which shows cross-sections for electron-impact induced excitation to the first eight vibrational states of N_2 in the ground electronic state.

Vibrational excitation also takes place upon the collision of heavy particles. A typical dependence of the probability of vibrational quanta exchange (the v–v process) and that of energy exchange between vibrational and translational degrees of freedom (the v–T process) on the sequential number of the vibrational level is shown in Fig. 6.2 for collisions of unexcited nitrogen molecules.

The relatively low probability of vibrational excitation in collisions with heavy particles increases the role of electron impact in this process. For instance, excitation rates into the first vibrational level of N_2 by electron impact and by collision with heavy ions are very similar if $T_e = T_v$ at different degrees of ionization, χ_e.

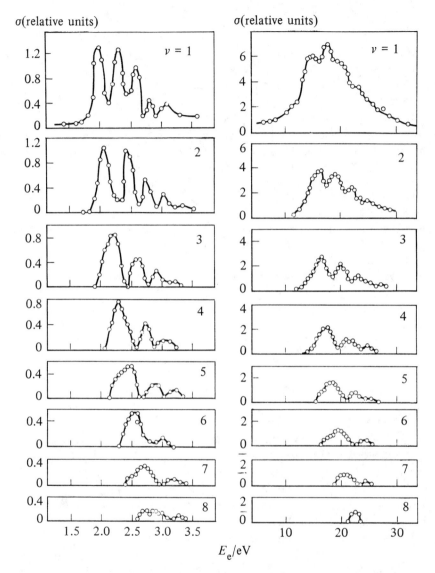

Fig. 6.1. Cross-sections for vibrational excitation to levels $v_1 = 1$–8 in collision of $N_2(X^1\Sigma_g^+)$ molecules with electrons (at $E_e = 2.3$ eV the sum of the cross sections is equal to 5×10^{-15} cm^2).

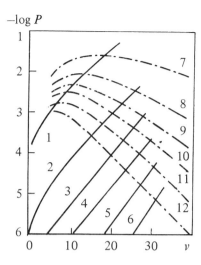

Fig. 6.2. Probabilities of $v-T$ transitions (1–6) and $v-v$ transitions (7–12) in encounters of nitrogen molecules depending on the sequential number of the level at different temperatures: 1, 7 – $T = 3340$ K; 2, 8 – 1300 K; 3, 9 – 1000 K; 4, 10 – 750 K; 5, 11 – 500 K; 6, 12 – 334 K.

$T/10^3$ K	χ_e	$T/10^3$ K	χ_e
1	1.4×10^{-5}	8	2.3×10^{-4}
3	2.0×10^{-5}	10	4.3×10^{-4}
5	7.0×10^{-5}	15	4.1×10^{-3}

In a low-pressure nonequilibrium plasma, the temperature of the electrons is usually considerably higher than the gas temperature. Under these conditions, consideration of the vibrational excitation of the N_2 molecule by electrons $(e-v)$ and via vibrational-translational $(v-T)$ and vibrational-vibrational $(v-v)$ transitions gives a rather complicated time-dependent picture of the population and depopulation of vibrational levels.

It emerges that the population distribution of vibrational levels cannot be described by one value of the vibrational temperature since it is non-Boltzmann. In addition, it is nonstationary. However, for a degree of dissociation χ_e greater than 10^{-4} the distribution becomes close to Boltzmann with $T_e \approx T_v$.

6.3.3 Electronic excitation of molecules
Various semi-empirical equations for calculating cross-sections of electron-impact induced excitation of electronic levels have been suggested in the literature. To calculate the probability of excitation to different electronic levels of a molecule, it is necessary to average the cross-sections by the electron energy distribution functions. Assuming the Maxwell function for excitation of the

electronic states of N_2, the probability of induction of a forbidden transition is 1.5–2 times higher than that of allowed transitions for large distances between the levels and to decrease with a decreasing difference in energy of the vibronic levels. Therefore, the role of forbidden transitions in the kinetics of population of low-lying levels may be considerable. Cross-sections for electronic excitation by the direct transfer of translational energy in collision with heavy particles possessing the energy of a low-temperature plasma are usually small. These cross-sections take a maximum value at energies of relative motion of particles exceeding 10^3–10^4 eV. The rate coefficients for deactivation of electronic levels of atoms and vibronic states of molecules depend on the type of colliding particle and range from 10^{-13}–10^{-9} cm^3 s^{-1} (for $T_{trans} \approx 300$ K). These coefficients do not usually depend strongly on temperature.

6.3.4 Dissociative attachment of electrons to molecules

The dissociation of molecules can also occur as a result of the formation of an unstable negative molecular ion by attachment of slow electrons to molecules in accordance with scheme

$$e^- + AB(i) \rightleftharpoons AB^-(m) \rightarrow A^-(l) + B(n)$$

where the letters in parentheses denote a set of quantum numbers describing the internal state of the molecule or atom. This process in a low-temperature plasma usually yields an ion in its ground state.

The maximum values for the cross-section of this process are *ca.* 10^{-14} cm^2 for H_2 at $E = 10$ eV and *ca.* 10^{-23} cm^2 for D_2 at $E = 3.75$ eV. The cross-sections and rate coefficients can be affected in different ways by temperature (T_e, T_v, and T_{gas}), depending on the form of the potential energy curves of the molecules and negative ions.

We have considered the different processes involving electrons, which result in the dissociation of molecules, and the cross-sections for these processes. Their role in low-temperature plasmas depends on the populations of levels and on the energy distribution function of the electrons. In gas discharge plasmas, the mean energies of electrons usually do not exceed 5–7 eV, so dissociation occurs mainly from electronically-excited states and for some molecules which easily form negative ions, for example, halogen-containing molecules, by dissociative electron-capture.

As the pressure is increased to its ambient value and above, the degree of ionization of the plasma increases and the average electron energy decreases. Under these conditions the role of stepwise excitation of vibrational and vibronic levels and of stepwise ionization becomes crucial.

6.3.5 Stepwise dissociation of molecules excited by electron impact
Dissociation from the ground state

Dissociation in this case occurs from higher vibrational levels on collision with different particles. Its rate depends on the ratio of the rate of dissociation from upper levels to the excitation of the molecule to its vibrational levels. It is obvious that under these conditions the dissociation rate may be determined by the rate of transition between low-lying levels. This rate will increase on excitation of the vibrational levels by electron impact, as confirmed by calculations on the vibrational relaxation of molecules, paying regard to $e-v$, $v-v$, and $v-T$ processes. It even transpires that under these conditions the rate of dissociation of nitrogen from the ground state can be greater than the rate of dissociation from an electronically-excited state in low-pressure gas discharges. This process was found to be very effective for diatomic homonuclear molecules.

Dissociation from electronically excited states
In calculations on this process two problems are usually solved successively: first, the relaxation of the ground-state vibrational levels is studied, and then the excitation of electronic states, disregarding the vibrational relaxation of the latter. The assumption of instantaneous dissociation after excitation of the relevant state is frequently made. On comparing the results of calculations of rate coefficients for stepwise and direct dissociations of molecular nitrogen shown in Fig. 6.3, it is seen that the stepwise dissociation is more effective. The experimental values for dissociation rate of nitrogen mixed with argon measured in an ambient pressure low-current-arc plasma are in good agreement with the calculated data.

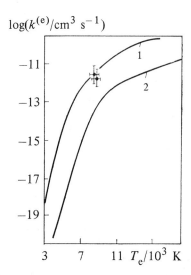

Fig. 6.3. Rate coefficients $k^{(e)}$ for electron-impact induced dissociation of nitrogen as a function of electron temperature. 1—Via stepwise excitation of electronic sates; 2— by direct excitation to electronic states from ground state (circles denote experimental data).

6.3.6 Recombination of uncharged heavy particles yielding molecules

Recombination is the reverse of dissociation. It can occur in the gas phase (homogeneous recombination). It has been established experimentally that homogeneous combination produces electronically-excited molecules. This type of combination plays an important part at moderate and high pressures. Heterogeneous combination is significant in low-temperature plasmas under reduced pressures. It occurs upon the collision of atoms from the gas phase with those adsorbed on the surface; its rate depends on the number of free absorption centres.

6.3.7 Dissociative recombination of molecular ions and electrons

The main process of neutralization of charged particles in a molecular gas-based plasmas is the dissociative recombination of the parent ions and electrons:

$$M_2^+ + e^- \rightarrow M(l) + M(n). \tag{6.1}$$

The energy evolved in recombination is partly consumed in excitation of the product particles, while the other part increases their kinetic energy. This process is also significant in atomic gas-based plasmas, since molecular ions and even complex aggregate ions (or clusters) are formed at relatively low temperatures. Clusters are also formed rather effectively in molecular gas plasmas:

$$A_n^+ + A_2 \rightarrow A_{n+1}^+ + A \tag{6.2}$$

The reaction rate coefficients for dissociative recombination are large (10^{-8}– 10^{-5} cm^3 s^{-1}), and they depend on the electron temperature and gas temperature. These processes also cause departure from ionization equilibrium in low-temperature plasmas. In plasma physics, the equilibrium concentrations of ions and excited species are calculated by using the Saha and Boltzmann equations with the electron temperature (provided that it can be determined with sufficient accuracy).

In a spatially uniform isothermal plasma, that is, when $T_e \approx T_{gas}$, a nonequilibrium distribution of atoms in terms of levels and nonequilibrium ionization is produced by plasma radiation. The latter decreases the population of the radiative levels, that is, it causes deviation from the Boltzmann population of the quantum levels of molecules.

In a spatially-nonuniform isothermal plasma, the nonequilibrium degree of ionization and the decrease in level population are caused by the diffusion of charged particles. Encounters with heavy particles in an isothermal plasma decrease the departure from equilibrium. Collisions with heavy particles in a nonisothermal plasma, on the other hand, enhance the departure from equilibrium at $T_e \approx T_{gas}$. However, if this channel is major then the degree of ionization and level population are determined by the Saha and Boltzmann equations.

Dissociative recombination and ion–molecule reactions strongly disturb the ionization equilibrium in a plasma, and these perturbations result in additional formation and decay of charged and excited species.

6.3.8 Decomposition of hydrocarbons in a nonequilibrium plasma
Thus, we have considered some mechanisms of plasmochemical reactions. Their role in each particular chemical process can be different, so that a special kinetic investigation is required anyway. Below, we try to show how a typical study of this sort is carried out [96].

The objective was the experimental determination of the mechanism and kinetics of decomposition of hydrocarbons in the positive column of a nonequilibrium dc glow-discharge plasma under reduced pressure. The behaviour of small additions of cyclohexane to the argon plasma was investigated. This hydrocarbon was chosen because of its symmetry. As a result, a relatively small number of decomposition products can be expected. Low concentrations of cyclohexane inhibit the effect of secondary processes and lead to stabilization of the plasma parameters.

A specially designed reactor was used for determination of the plasma parameters, the types of radical, the gas-phase products, and the polymers deposited on supports. The plasma parameters and the rate of polymer formation were measured by using electrostatic probes, and the gas temperature with thermocouples. The radicals were frozen out in traps and were detected by ESR. The gaseous products were analysed chromatographically. The polymeric products were identified by IR spectroscopy, electron microscopy, and ESR.

It was established that under experimental conditions, the gas temperature in the plasma without additives did not vary, and was equal to room temperature. However, the addition of 0.01% cyclohexane to argon resulted in a dramatic change in the plasma parameters (concentration of electrons, their mean energy, and the types of ion). This is related to the change in the mechanism of ionization. Analysis of the mechanism entailed involved calculation of all the energetically-possible processes of ionization, based on the experimentally found plasma parameters and literature data on the cross-sections and rate coefficients for relevant processes. The rates of processes obtained (unless their contribution was smaller than 1%) for one of the operational modes are listed in Table 6.1, which shows that the major contribution to ionization in neat argon is made by stepwise ionization from the electronically excited state of argon induced by electron impact (process 3) and in the collision of excited long-lived argon atoms (process 8).

The cyclohexane molecules effectively deactivate the excited argon atoms so that their concentration decreases dramatically. As a result, the rate of ionization of argon also decreases. The major channel of ionization in the mixture is process 9, or the ionization of cyclohexane by excited argon atoms (the so-called Penning ionization). It is this process which causes the observed changes in

Table 6.1. Rates of ionization processes in argon and argon–cyclohexane mixture at 133 Pa, 1.0 mA, and c_0 of 1.0% vol.

		$v/10^{13}$ s^{-1}	
No.	Process	Ar	Ar + c-C$_6$H$_{12}$
1	$Ar + e^- \rightarrow Ar^{+\cdot} + 2e^-$	2	0.01
2	$Ar + e^- \rightarrow Ar^* + e^-$	1500	510
3	$Ar^* + e^- \rightarrow Ar^{+\cdot} + 2e^-$	75	1.5
4	$Ar^* + Ar \rightarrow 2Ar + h\nu$	16	0.2
5	$Ar^* + 2Ar \rightarrow 3Ar + h\nu$	2	0.02
6	$Ar^* \xrightarrow{\text{wall}} Ar$	2	0.02
7	$Ar^* + e^- \rightarrow Ar + e^- + h\nu$	1200	26
8	$Ar^* + Ar^* - \begin{cases} \rightarrow Ar + Ar^{+\cdot} + e^- \\ \rightarrow Ar_2^+ \cdot + e^- \end{cases}$	3	0.0003
9	$Ar^* + c\text{-}C_6H_{12} \rightarrow c\text{-}C_6H_{12}^+ \cdot + Ar + e^-$	—	480
10	$c\text{-}C_6H_{12} + e^- \rightarrow 2e^- + c\text{-}C_6H_{12}^+ \cdot$		1
		9×10^{14}†	3×10^{15}†
		7×10^{11}‡	7×10^9‡

† Rate of ion neutralization at the walls, in cm^{-3} s^{-1}.
‡ Concentration of metastable argon atoms, per cm^3.

plasma parameters and in the ion composition. In a plasma of neat argon, the basic ions are argon ions, while in the plasma of the mixture they are the molecular ions of cyclohexane.

Analysis of the mechanism of cyclohexane decomposition was carried out in the same manner as in the study of ionization. The calculated values for the decomposition rates from experimental and literature data and the overall rate of decomposition observed in experiments are given below.

$$v/10^{15} \text{ cm}^3 \text{ s}^{-1}$$

Total decomposition rate	7.5×1.4
Electron impact-induced dissociation through excitation of optically allowed transitions	8
Electron impact-induced dissociation of vibrationally excited molecule	0.1
Dissociation through stepwise vibrational excitation	0.2

$$v/10^{15} \text{ cm}^3 \text{ s}^{-1}$$

Electron impact-induced dissociation ionization	0.001
Dissociative charge-transfer from argon ions to cyclohexane molecule	0.003
Dissociation on collision with fragment radicals	0.001
Dissociative wall recombination of molecular ions of cyclohexane	2

It is obvious that the principal channels here are the electron impact-induced dissociation of cyclohexane and the dissociative recombination of its molecular ions at the walls.

The experimental dependence of the decomposition rate of cyclohexane on its concentration is described by the equation

$$-\mathrm{d}c/\mathrm{d}t = k_{\mathrm{diss}}N_e c + k_i[\mathrm{Ar}^*]cZ \qquad (6.3)$$

where k_{diss} is the rate coefficient for electron impact-induced dissociation of cyclohexane, $k_i[\mathrm{Ar}^*]c$ is the rate of ionization, and Z is the probability of dissociative wall recombination of cyclohexane ions.

The decomposition of cyclohexane gives 12% mass of gaseous products and 88% mass of polymer. The radicals in the gaseous products are mainly $\dot{\mathrm{H}}$ and c–C_6H_{11}· which are formed by the abstraction of a hydrogen atom from cyclohexane molecule. The gas-phase products themselves are represented by hydrogen, ethylene, acetylene, cyclohexane, and divinyl. Ethylene, acetylene, and divinyl are formed as a result of degradation of the cyclohexane ring, and cyclohexene is produced by the elimination of H_2.

The condensation products were in the form of a yellow film insoluble in organic solvents and concentrated acids and alkalis. Chemically, it was composed of carbon and hydrogen. The presence of CH_3 and CH_2 end-groups, of a cyclohexane ring, and of a carboxyl group was detected in the film.

Perhaps the polymer is formed as follows. An excited cyclohexane molecule decomposes to a cyclohexyl radical and hydrogen atom in the gas phase. The radical then diffuses to the walls. Owing to the action of electrons, ions, and plasma radiation, the bonds of the cyclohexane ring are ruptured at the walls, and species with unpaired electrons are formed. Some of them recombine with neighbouring free radicals, which leads to polymerization, and the rest combines with cyclohexyl radicals from the gas phase.

6.3.9 Effect of electron-energy distribution on rate coefficients of plasmochemical reactions

We shall now consider the influence of functions of the energy distribution of particles on reaction rate coefficients for the case of plasmochemical reactions in electric discharges in the intermediate pressure range. Under these conditions the initiation of reactions is mainly by free electrons.

The electronic component in an electric discharge gas plasma acts as an agent transferring energy from an external source to the gas, and it simultaneously participates in the processes of redistribution of this energy between the different degrees of freedom of the particles. Under these conditions the nature of the electron energy-distribution function (EEDF) slightly affects the total flux of energy from the external source to the plasma. Important factors here are the concentration n_e and the mean energy $\bar{\varepsilon}$ of the electrons. So far as the distribution of the flux by the different degrees of freedom of heavy particles is concerned, it is largely dependent on the character of the EEDF, which can be essentially non-Maxwellian.

It is therefore necessary to know the real EEDFs (Fig. 6.4). Noting that plasmochemical reactions are multichannel processes, the thresholds of different channels differ considerably from each other. Changes in the high-energy component of the EEDF will result in corresponding changes in the rate of high-barrier reactions, leaving unchanged the rate of low-threshold reactions. The same will also take place on changing $\bar{\varepsilon}$, provided that the character of the EEDF is maintained.

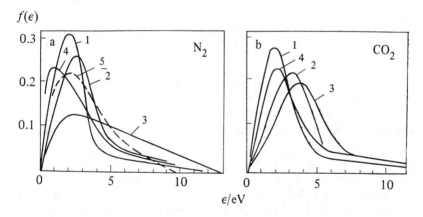

Fig. 6.4. Data on measurements of electron-energy distribution function $f(\varepsilon)$: *a.* 1— $\varepsilon = 2.9$ eV; 2—$N_2:O_2 = 8:2$ mixture, $\varepsilon = 3.7$ eV; 3—the same at 4:6, $\varepsilon = 5.3$ eV, pressure of 250 Pa, discharge current of 30 mA; 4 and 5—Druvenstein and Maxwell distribution curves, respectively, $\varepsilon = 2.9$. *b.* 1—$\tau = 0.23$ s, $\varepsilon = 3.2$ eV; 2—$\tau = 0.4$ s, $\varepsilon = 4.2$ eV;3—$\tau = 1.5$ s, $\varepsilon = 4.8$ eV; discharge current for 1–3 is of 30 mA; 4— $\tau = 0.23$ s,$\varepsilon = 4.4$ eV, discharge current of 75 mA.

It should be emphasized that for determination of k_i it is necessary to know the total EEDF whereas a theoretical or experimental study is usually concerned with its isotropic part. In most cases of quasi-uniform gas discharge plasmas, such data are sufficient since the anisotropic part of the EEDF is small.

The EEDF is determined by the chemical composition of the plasma gas, by the distribution of molecules in terms of internal degrees of freedom, by the degree of ionization $\alpha = n_e/N$ (where N is the concentration of heavy particles of the plasma), and by the intensity, spatial distribution, and time-dependence of the electric $\vec{E}$ and magnetic $\vec{B}$ fields (frequencies) as well as by the geometry of the discharge.

Inelastic collisions the electron distribution in the range of increasing cross-section of inelastic scattering of electrons on heavy particles.

Examples of EEDFs in the nitrogen dc discharge plasma calculated by taking into account elastic and inelastic encounters for different values of the parameter E/N, are given in Fig. 6.5.

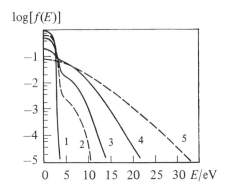

Fig. 6.5. EEDF in dc discharge plasma of N_2 dependence on electric field strength E at E/N ratio (in 10^{-16} cm^2) of 1.1 (1), 3.4 (2), 5.6 (3), 14 (4), and 27 (5).

Collisions between electrons lead to Maxwellization of the EEDF, the low-energetic part of EEDF being Maxwellized first with increasing degrees of ionization, and the whole distribution doing so at degrees of ionization of 10^{-3}–10^{-2}.

Under the experimental conditions, all plasma parameters are found to be interrelated. Indeed, even simplified models of a gas discharge plasma (for example the Schottky theory) give an unambiguous relation between the electron temperature, ionization potential, gas pressure, and characteristic size of a discharge system.

6.4 THERMODYNAMICS OF PLASMOCHEMICAL SYSTEMS

Let us consider the conditions when a thermodynamic approach can be realized in an analysis of the phase, structural, and chemical transformations of a low-temperature plasma. Such an approach may be used only for an equilibrium low-temperature plasma, regarding it as a macroscopic system whose spatial dimensions and lifetime are sufficient for its parameters to be measured. The latter condition means that the relaxation times of the system are small compared with the typical periods of change in the external conditions and with the duration of measurement.

At high temperatures, in addition to dissociation of complex structures, there occur process resulting in the formation of new types of species and compounds which do not exist at ordinary temperatures (for example C_3, C_4, Al_2O, AlO, etc.).

When reliable thermodynamic constants are available, it is possible and, in most cases, even necessary to estimate the optimum conditions for carrying out reactions, such as pressure and temperature ranges, ratio of reactants, and energetic characteristics of processes as well as to evaluate the expected yields of desired products and by-products. In addition, it is possible to determine the transport and optical properties for multicomponent systems.

Information on the composition of by-products, that is, the purity 'index' is sometimes of vital importance. For example, if the manufacturing processes of carbides of prescribed composition are accompanied by formation of tiny amounts of carbon, metal, or carbides of another composition, then these products are generally unsuitable for commercial use.

Calculated thermodynamic equilibrium compositions, if compared with experimental values, make it possible to know:

(i)   whether the experimental compositions are at equilibrium or are formed at intermediate stages of chemical transformation,

(ii)   whether the conditions chosen for carrying out the experiments (pressure, temperature, and energy consumption per unit mass of product) are optimum,

(iii)   whether the conditions of the quenching stage (see section 7.1.3) are optimum, and

(iv)   what is the contribution of radicals and other transient species reacting in the process of chilling to the yields of products obtained.

Gas-phase plasmochemical reactions usually occur for periods from tenths to millionths of a second. This time can be insufficient to establish chemical equilibrium for all the reactions. In particular, much more time is required for formation of a condensed phase or complex molecules demanding multiple encounters of certain transient compounds or radicals. Therefore, to describe gas-phase plasmochemical reactions it is sometimes more correct to consider not the true equilibrium composition but the quasi-equilibrium one, regardless of formation of a condensed phase or any other relevant compound.

In any case it is always worthwhile to carry out a thermodynamic analysis of a plasmochemical process.

6.4.1 Thermodynamic considerations of case systems
Molecular nitrogen. At high temperatures the system contains the following

components: molecular nitrogen N_2, atomic nitrogen N, and singly, doubly, and triply-ionized nitrogen atoms N^+, N^{2+}, and N^{3+} as well as electrons e^-. The equations to describe equilibrium are as follows:

$$n_{N_2} + n_N + n_{N^+} + n_{N^{3+}} + n_{e^-} = 1 \tag{6.4}$$

$$n_N + 2n_{N^{2+}} + 3n_{N^{3+}} - n_{e^-} = 1 \tag{6.5}$$

$$K_1 = n_N^2 / n_{N_2}, \qquad\qquad N_2 \rightleftharpoons 2N \tag{6.6}$$

$$K_2 = n_N + n_e / n_N, \qquad\quad N \rightleftharpoons N^+ + e^- \tag{6.7}$$

$$K_3 = n_{N^2} + n_e / n_{N^+}, \qquad N^+ \rightleftharpoons N^{2+} + e^- \tag{6.8}$$

$$K_4 = n_{N^{3+}} + n_e / n_{N^{2+}}, \qquad N^{2+} \rightleftharpoons N^{3+} + e^-. \tag{6.9}$$

where n_i is the concentration of the i-th component.

Results of calculations on the dependence of the equilibrium composition of a nitrogen plasma on temperature are shown in Fig. 6.6. At 4000–8000 K, the dissociation of molecular nitrogen is observed to occur extensively. Ionization begins at a temperature of 6000 K. At $T \approx 15 \times 10^3$ K the major components of the plasma are the N^+ ion and electrons. The N^{2+} ion becomes the dominant component only for $T \approx 2 \times 10^4$ K.

Knowing the equilibrium compositions, the temperature dependences of the enthalpies have been calculated for almost all the plasma gases. It has been

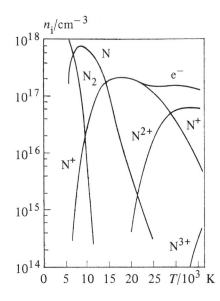

Fig. 6.6. Equilibrium composition of nitrogen plasma as a function of temperature ($P \approx 0.1$ MPa).

shown that the minimum temperature of a low-temperature plasma (*ca.* 10^4 K), at which the degree of ionization needed for maintaining its stationary state is realized, is much higher than the maximum possible temperature of a flame (*ca.* 3000 K). By using monoatomic gases, it is possible to obtain even higher temperatures, but enthalpies are less than those of diatomic gases at the same temperatures. In diatomic gas plasmas the enthalpies can be as large as 800–1200 kJ mol^{-1} owing to dissociation of the gases. Thus, the temperatures in plasmas are considerably higher than any of the values required for dissociation of a chemical compound. Also, the high temperature of plasmas makes it possible to use a significant part of their energy for induction chemical processes. For example, if a reaction takes place near 2500 K, then its realization in the propane–oxygen flame draws only 20% of the energy of the source in contrast to 90% in the case of nitrogen plasma at 10 000 K.

C–H System.
Let us scrutinize this system in more detail and define what valuable information can be obtained from thermodynamic calculations: it will be of value in considering plasmochemical processes of production of unsaturated compounds, carbon black, pyrocarbon, and reducing gases for metallurgy and chemistry.

Equilibrium compositions. Calculations on the equilibrium mixture of solid carbon and hydrogen at temperatures from 2000–5000 K involving 12 components in the gas phase and condensed carbon C_8 were carried out. The vapour pressure of carbon was fixed for every temperature which prescribed the C/H ratio in the gas phase. Up to 3000 K acetylene is the major hydrocarbon product, while at 3000 K the concentrations of radicals $\cdot C_2H$, $\cdot C_3H$, and $\cdot C_4H$ become nearly equal to that of acetylene, and the amounts of these radicals increase at higher temperatures to reach their maximum at 3400 K.

The yield of C_2H_2 in the gas produced by plasma pyrolysis (after quenching) will depend on the temperature and C/H ratio. On increasing the C/H ratio from 0.25 to 2.5, the C_2H_2 yield increases from 12% to 42% as the temperature changes from 3200 to 4000 K. The energy consumption for the preparation of 1 kg of acetylene is high and is pressure-dependent. Thus, it takes 19.8 kW h at *ca.* 0.1 MPa and *ca.* 3200 K and 11.8 kW h at 1.5 MPa and 3400 K.

The calculated values for the acetylene concentration agree satisfactorily with experimental data obtained with plasma devices using a high intensity arc on graphite and hydrogen or on graphite and methane. However, they differ considerably from the experimental compositions obtained in the plasma pyrolysis of hydrocarbons. This is related to the fact that a condensed phase cannot be formed within the period of reaction.

Quasi-equilibrium compositions. Assume carbon to be absent in the condensed phase. Then the gas phase at 1000–2000 K may contain 22 components (the principal products of pyrolysis are hydrogen, methane, acetylene, and toluene, while the total concentrations of other compounds do not exceed 1% vol.).

The amount of toluene can reach a value of 15% (at 1400 K). The condition for acetylene production can be realized at 1800–2000 K at low dilutions with hydrogen, when it amounts to 30–35% vol. Then the fraction of acetylene, ethylene, and propylene equals 73–83% wt, and the power consumption reaches 5.4–6.4 kW h per kg of acetylene, ethylene, and propylene.

However, the experimentally obtained compositions did not include benzene and toluene at such high concentrations. Indeed, the characteristic times of formation of benzene from $C_2H_4 + C_2H_2$ are 0.1 to 1 s at 1000–1500 K, hence for a time of 10^{-4}–10^{-3} s the formation of these compounds is negligible.

Consider the quasi-equilibrium compositions, which include neither carbon in the condensed phase nor aromatic compounds. Acetylene, ethylene, and propylene of total concentration approaching 22% vol. are simultaneously present in these compositions at 1000–2000 K. The maximum concentration of acetylene is observed at 1800 K and is equal to 29.5% vol.

The effects of pressure and C/H ratio on the composition are different. An increase in the C/H ratio leads to an increase in acetylene concentration, its largest value being close to the maximum possible quantity obtained on the assumption that acetylene and molecular hydrogen are the only compounds formed. Changing pressure within 0.01–0.1 MPa at temperatures above 1800 K barely affects the concentration of acetylene. In the temperature range 1000–1800 K the yield of C_2H_2 decreases abruptly with increasing pressure.

The acetylene ratio depends only on temperature unless the latter is below 1800 K: at such a low temperature it is also determined by the C/H ratio and by pressure.

The use of this dependence to demonstrate the dependence of the process on temperature and the C/H ratio is based on the possibility of controlling the process: as regards the manufacture of products of a desired composition with a varying C/H composition of feedstock materials, it is achieved through a corresponding change in the power supply. The power consumption per kg of acetylene produced also depends on the temperature and C/H ratio and passes through a minimum at 1500–1800 K. As the C/H ratio decreases, the power consumption is shifted to that of the low-temperature region. Given below are the values of the minimum power consumption (in kW h kg^{-1}) for the production of acetylene from different hydrocarbons:

CH_4	C_2H_6	C_3H_8	C_4H_{10}	C_5H_{12}
6.7	5.5	5.1	4.9	4.7

The calculated quasi-equilibrium concentrations were repeatedly compared with the experimental compositions obtained in the plasmochemical pyrolysis of different hydrocarbons. It was found that in the temperature range 1600–1800 K (the acetylene-producing regime of the pyrolysis) the reaction temperatures (or, more accurately, the temperatures of the reaction products before quenching) and the gas-phase composition in the pyrolysis of methane (C/H = 1:5.16) and petrol (C/H = 1:3.8) fitted the calculated values for the relevant C/H ratios within determination error.

Thus, some stages of the gas-phase plasmochemical processing of hydrocarbons are accurately described with quasi-equilibrium compositions, disregarding the possibility of forming a condensed phase and aromatic hydrocarbons.

6.4.2 Thermodynamic considerations of systems with a solid phase

Let us ascertain the thermodynamic features of production of high-melting silicon carbide SiC by the decomposition of methyltrichlorosilane (MTCS) in an argon–hydrogen plasma. (Silicon carbide is widely used in the electronics and electrical engineering industries).

Calculations were conducted on the decomposition of CH_3SiCl_3 at 0.1 MPa and different dilutions of the reactant in the temperature range 1000–3200 K. As many as 39 components with three of them in the condensed phase were taken into account. It was found that the maximum yield of a given product was achieved at 1400–1700 K, and was strongly affected by the feedstock composition. Dilution with hydrogen causes both an increase in the SiC yield up to 97% and its very weak dependence on temperature over a wide (1400–2400 K) interval. The latter feature seems very important since appreciable temperature gradients can be met in the plasma.

The target product SiC_k can be contaminated by carbon black C_k, silicon, and silicon chlorides. Silicon appears as a product at considerably higher temperatures than the temperature of maximum SiC_k yield. Carbon black is deposited at 1400–2400 K (Fig. 6.7). Strong dilution of the feed by hydrogen causes the carbon black to be formed over a narrower temperature interval.

Dilution of the feed with hydrogen strongly decreases the amount of silicon chlorides (SiCl, $SiCl_2$, $SiCl_3$, and $SiCl_4$) owing to formation of HCl.

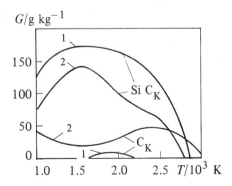

Fig. 6.7. The yield of SiC_k and C_k as a function of temperature (0.1 MPa): 1—[CH_3SiCl_3] = 0.65, [H_2] = 0.17, [Ar] = 0.18; 2—[CH_3SiCl_3] = 0.78, [H_2] = 0.011, [Ar] = 0.21 (in mass fractions).

Thus, the optimum conditions for manufacturing SiC from MTCS are as follows: 1400–1600 K, feed mixture $CH_3SiCl_3:H_2:Ar = 0.65:0.17:0.18$.

6.5 CHEMICAL REACTIONS IN TURBULENT PLASMAS

6.5.1 Mixing of gases with different energies

Chemical reactions at high energies (temperatures) are in many cases carried out in one of the following three ways:

(i)   by 'pumping' with energy or heating of a premixed reactant mixture,
(ii) by mixing of a cold component with a hot one (for example, in a plasma jet),
(iii) by mixing two reactant gases (one of them being the plasma gas) having different energies (temperatures).

The second and third ways include mixing of the reactants as a necessary stage of the process. This process occurs within the finite period when the chemical reaction can take place in a nonuniform nonisothermal system at an appreciable rate depending on time and space, τ_{mix} almost always being nearly equal to τ_{rxn} under the conditions of nonequilibrium reacting systems.

The process of establishing equilibrium in velocity space is known to be fundamentally different from establishing equilibrium in configuration space. The velocity distribution in any element of coordinate space approaches monotonically and very quickly (in $ca.$ 10^{-9} s or the order of the mean period between collisions at a pressure of 0.1 MPa) to the local Maxwell distribution. This distribution is never fully attained since external forces and the existence of flows prevent this [389].

The establishment of equilibrium in the distribution in coordinate space takes place nonmonotonically and requires a rather longer time. This does *not* mean that the equilibrium is established in two well-defined consecutive stages; firstly Maxwellization and then establishment of equilibrium in the coordinate space of type $\exp\{-\beta U(r)\}$ (where β is a constant and $U(r)$ the potential of an external field. In reality, the two processes are interrelated and the velocity distribution becomes strictly Maxwellian only upon achieving complete equilibrium.

If a chemical reaction is also involved, the process of establishing equilibrium will be more complicated because of the endo- or exothermicity of the reaction and the breeding (emergence and combination) of particles in every unit of the volume. Thus, according to data obtained with spectral, probe, and shadow methods [91], τ_{mix} is longer than, or nearly equal to, 10^{-5} s on mixing a plasma jet of argon ($T_{in} = 300$ K) and methane ($T_{in} = 300$ K), while the formation time of acetylene on decomposing methane is $\tau_{rxn} \approx 10^{-4}$. It is evident that some part of the reaction needing study and definition proceeds under complicated nonequilibrium conditions (unlike conventional chemistry, where $\tau_{mix} < \tau_{rxn}$).

Therefore, the problem of describing the mixing of reacting gases is a constituent part of nonequilibrium chemical kinetics. To solve the problem it is necessary to study it experimentally under conditions of different gas flows and to consider a system of equations of hydrodynamics and chemical kinetics, also making use of similarity theory [91] or of various stochastic methods.

The rate of gas-phase reactions is known to be determined first of all by the effective frequency of collision of reactant molecules. When a reactant mixture is not homogeneous[†], the reaction rate depends both on the degree of inhomogeneity of the given mixture and on the rate of relaxation of its properties to those of the homogeneous mixture. Thus, the relation between physical and chemical kinetics, whose concepts are generally used in describing the behaviour of relaxing systems, can be manifested here to a great extent. A rather simple example of a system wholly different from a chemically-reacting homogeneous mixture is given by the totality of the semi-infinite volumes of two substances where a bimolecular chemical reaction occurs under isothermal conditions parallel to the mutual diffusion (mixing) of the substances. The closer the characteristic time of diffusion τ_{mix}, the larger the difference between the reaction rate in such a system and that of the same reaction in a homogeneous mixture. In the limiting case of $\tau_{rxn} \ll \tau_{mix}$ the rate of the chemical reaction is obviously determined by the rate of the physical process of mixing the reactants ('very fast' chemical reaction).

For some plasmochemical processes in which a given product is obtained in the so-called kinetic region, it is extremely important to solve the problem of controlling the mixing of the plasma and reactants [126]. A similar problem also arises in calculations on combustion processes in turbulent flows and in determination of the parameters of ballistic tracks of bodies moving at high velocity in gases and liquids.

Consideration of the general case of simultaneous occurrence of chemical reactions on mixing two (or more) turbulent flows (one of which can be, as normally happens in industrial plasma chemistry, a dust-laden jet) is so complicated a problem that its complete solution remains elusive.

6.5.2 Effect of turbulent fluctuations

It should be borne in mind that local hydrodynamic fluctuations of temperature, pressure, density, and concentration of the substance constituting the medium take place even under thermal equilibrium. Strong fluctuations in these properties are observed in turbulent media [388]. A medium where a reaction occurs can experience the affects of adventitious fields: acoustic, electromagnetic, or

† Reactants are considered as mixed homogeneously if, for however small a volume of mixture (but sufficient for statistical purposes), the ratio of the average amount of reactant molecules in the absence of reaction is equal to the ratio of the initial quantities of reactants taken for preparation of the mixture.

radiation fields. Fluctuation in pressure is also possible and is taken into account by the steric factor in the equation for k [389–391].

It may often be assumed that the occurrence of a chemical reaction does not appreciably affect the hydrodynamic properties of the medium. This is true if, for example, the reaction takes place in dilute solution, does not involve solvent molecules, and the heat of reaction is low. Then the fluctuations in properties of the medium are external with respect to the system of reactions under consideration [392].

The existence of fluctuations in the properties of a medium, and the action of adventitious fields, cause the parameters involved in the kinetic equations to acquire certain random terms: external noise with pre-determined statistical characteristics thus appear in the equations of chemical kinetics . With the external noise taken into account, these equations become stochastic differential equations. If, as usually happens, the noises appear in the coefficients for different combinations of variables describing the chemical system, they are named multiplicative noises.[†]

It is especially important to include the effect of fluctuations when describing plasmochemical reactions occurring in a turbulent medium. So long as the turbulent fields are random, each may be associated with a certain system of multidimensional probability density distributions. Since the turbulent fields can be related statistically with each other, it seems to be natural to assume that there also exist mutual probability density distributions of these fields. If this is so, then, knowing the function of the mutual probability density distribution (MPDDF) $P(\Omega)$, the average value of any function of a turbulent field $f(\Omega)$ may be defined as an integral of the form

$$\langle f \rangle = \int \ldots \int P(\Omega)\, d\Omega \qquad (6.10)$$

where Ω is the collection of field variables.

Eq. (6.10) gives the so-called theoretico-probabilistic average, while values determined experimentally are usually averaged by either time or space. Thus a problem arises on the correspondence of these averages.

The probabilistic approach has been used extensively in statistical physics. The problem of the correspondence of different averages arising is then solved with the use of the so-called ergodic hypothesis, that is, with the assumption that the ensemble averages (theoretico-probabilistic averages) coincide with the time averages when the time interval over which the averaging is performed increases infinitely.

With the probabilistic approach, the problem of studying the processes of turbulent mixing of reacting flows can be reduced to that of determination of the MPDDF of accidental fields.

---

† Unlike additive noises, which are involved in stochastic differential equations as an additional summation but not the coefficient as a combination of variables of the system.

By the method of determination of MPDDF, calculations on the plasma pyrolysis of methane in a cylindrical reactor have been carried out [393–396]. The results fit the experimental data satisfactorily (Fig. 6.8). From consideration of the curves in Fig. 6.8 it becomes clear that neglect of the pulsations (fluctuations) of the hydrodynamic field leads to a decrease in the calculated degree of conversion of methane by almost a factor of 2 with respect to the experimental figure (and to a decrease in the acetylene yield by more than 4 times at a cross-section Z/D of 3, where Z is the distance from the inlet to a given cross-section of the reactor and D is the internal diameter of the reactor; the ratio Z/D is called the pass). However, taking account of the pulsations made it possible to obtain satisfactory agreement (within experimental error) of the experimental and calculated concentration distributions of methane and its decomposition products.

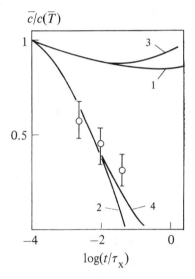

Fig. 6.8. Dependence of $\bar{c}/\bar{c}(\bar{T})$ on t/τ_x for unimolecular (1, 2) and bimolecular (3, 4) reactions ($\alpha^2\beta = 10$): 1, 3—$t_f/\tau_x = 10^{-3}$; 2, 4—$t_f/\tau_x = 0.1$ ($\alpha = E_a/RT$, $t_f = q/\varepsilon$, $\beta = (\bar{T}')^2/\bar{T}^2$, and $\tau_x = 1/k(\bar{T})c_0$; where q is the turbulence energy and c_0 the concentration of the reactant at zero time). Circles represent experimental data.

Evidently, it is insufficient to know only the average values for such fluctuating quantities as temperature and the concentrations of reacting components for complete description of the complex processes of chemical transformation under conditions of nonequilibrium, nonisothermicity, and turbulence, even in those cases where the influence of chemical reaction on the hydrodynamic properties of the system may be neglected [90, 95, 397]. The necessity to take into account the fluctuations of temperature and reactant concentrations and their mutual correlation is caused by the fact that the average rate of an elementary step of a chemical transformation for nonisothermal turbulent mixing of the reactants does not obey the Arrhenius law for the mean values of these quantities. Moreover, the existence of fluctuations results in a significant change in transport coefficients, the values of which in these cases are determined not only by the properties of the reacting gases but by the properties of the flow itself [397].

From the physical point of view, the reason for this appreciable dependence of the average rate of chemical transformation on the *fluctuations* in the determining parameters, in addition to their *mean values*, is that 'normal' turbulent mixing of the chemically-reacting flows of gases or plasmas (implying establishment of the turbulent profiles of the flow rate and temperature) means the lack of attainment of complete molecular homogeneity, which is the sole factor determining the true rate of chemical reaction [91, 394–396, 398].

Even for isothermal processes in initially homogenized multicomponent systems, the rate of any elementary step is strongly affected by fluctuations in the concentrations of the reactants, and neglecting these may lead to considerable disagreement between the calculated and experimental data [96, 139]. In nonisothermal turbulent processes, in addition to this, the fluctuations in temperature and flow rate also play an important role. The existence of these fluctuations and the correlations between them make the true rate of a chemical transformation depend nonmonotonically on pulsation in temperature.

6.5.3 Turbulent transport

According to modern theories, transport phenomena in a turbulent flow are related mainly to the turbulent motion of large eddies which exist sufficiently long [91, 388]. The transport of various entities by these eddies is, to a certain extent, analogous to molecular transport in a rarefied gas [97]. It is this similarity which was meant by Chung [399, 400] when he suggested consideration of the process of turbulent mixing from the generalized theory of Brownian motion [401]. Under these conditions the substance transferred is assumed to be a passive scalar impurity (that is, the influence of this impurity on the turbulence is negligible).

On the basis of known laws of turbulent transport, Chung believes that there are eddies of different size existing simultaneously in a turbulent flow. The statistical properties of units of the liquid are determined by the joint influence of these eddies. At a large enough Reynolds number, a statistical separation is observed in turbulent flow between statistically-nonequilibrium low wave numbers representing vortices of large size, and equilibrium high wave numbers representing eddies of small dimensions. The majority of the observed features of turbulent flow are related to the behaviour of large eddies. In addition, the rate of molecular scattering determined by small eddies is actually controlled by the rate of degeneration of large eddies.

Such a structure of a turbulence field at high Reynolds numbers can be described by the stochastic Langevin equation with relevant interpretation of its different terms [91, 96, 97]. Using the Langevin equation and some additional simplifying assumptions, Chung managed to derive an equation of the Fokker–Planck type for the distribution function of units of liquid in the phase space. In addition to the fact that this equation gives a rather complicated description of the turbulence field, it is statistically closed; hence, all equations for moments which can be obtained from this equation are also closed. Moreover, it becomes unnecessary to find different moments as independent functions, since the moments of all orders are now connected via the distribution function of units of the liquid. Consequently, the equation for moments in the Chung theory have another, somewhat different, meaning compared with equations of this type in classical statistical theories or ordinary phenomenological theories of turbulent transfer. These equations are needed only to facilitate solution of the Fokker–Planck equation, *cf.* the methods used in the solution of the Boltzmann equation in the kinetic theory of gases [402–405].

In classical statistical theories of turbulence, the turbulence field is described not by the Navier–Stokes equation but by a set of equations for moments of different order. Chung has shown that the equations for moments obtained from the Fokker–Planck equation constructed in this way are essentially the same as those derived from the Navier–Stokes equation (at least to second-order moments) [406].

6.6 ROLE OF MULTIPARTICLE INTERACTIONS AND EXTERNAL FIELDS IN A PLASMA

Let us estimate quantitatively the role of multiparticle interactions and external fields in plasmas. As indicated in Chapter 3, Eq. (3.46) for k_i is true only for binary collisions, and the reaction cross-section corresponds to collisions of totally uncorrelated pairs. In other words, the so-called statistical independence of molecules is assumed before collision, the condition for the latter being $N/a^3 \ll 1$ (where a is the interaction radius of the molecules and N is the density of molecules). For dense neutral gases $N/a^3 \approx 1$, and the assumption about collisions of uncorrelated pairs is invalid in plasmas with long-range Coulombic interactions. Calculations using Eq. (3.46) can produce incorrect results in these cases.

The application of an external field either electric, magnetic, or gravitational, can affect the molecular structure and hence the nature of the molecular forces of colliding particles. As a result, the cross-sections and distribution functions of the reactants and products can change. The cross-sections change slightly in weak fields since the interaction Hamiltonian remains almost unchanged, but the distribution function of the translational degree of freedom, for example for electrons or ions, may vary strongly in an external electric field [407].

The effect of a strong electric field is more profound. If the product of the field strength and the characteristic molecular size is of the order of magnitude of the molecular energy, then the field can influence the electronic wave functions of the molecules and, consequently, the distribution of the electrons in the reactant molecules, the molecular structure of the reactants, and the intermolecular potentials. So long as the size of the molecule is less than 1 nm and the molecular energy corresponds to *ca.* 1 eV, fields of 10^8 V cm^{-1} are needed to produce such effects. Such strong electric fields can be realized in outer space or in high-intensity laser beams. Under these conditions the cross-sections change because of the marked change in the interaction Hamiltonian and the distribution functions deviate significantly from their equilibrium values. The implementation of Eqs (3.46) and (3.47) in such cases can lead to appreciable errors.

6.7 DIAGNOSTICS OF LOW-TEMPERATURE PLASMAS

The diagnostics of low-temperature plasmas consist of the measurement of the concentration of particles and determination of the form of their energy distribution. A laboratory plasma is, in the thermodynamic sense, an open system, so that considerable gradients of all its parameters are observed. Thus, another task of the diagnostics is the determination of the spatial distribution of these quantities. In the case of a nonstationary plasma (turbulent, pulsed, etc.) it is necessary to study the time-dependence of the parameters measured, and in the diagnostics of plasma flows to measure their dynamic properties. The diagnostics of two-phase plasma systems are more complicated since determination of the parameters of a second phase, that is, temperature, velocity, and size distributions of particles, is also necessary.

The techniques of plasma diagnostics currently considered as most reliable are spectral and optical methods, mainly because their application does not disturb the plasma under study (as frequently happens with different probe techniques). The principle difficulty in using spectral and optical methods lies in interpreting the resulting data, that is, they are largely determined by the validity of the model for the plasma under investigation. Indeed, the spectral methods are based on measurements of the emission or absorption characteristics of the plasma from line or continuous spectra. If the emission intensity of a plasma at a state of local thermal equilibrium (LTE) is determined by relatively simple relationships, then, for interpretation of data on the emission intensity of a nonequilibrium plasma, it is usually necessary to use more complex schemes of the excitation and deactivation of the corresponding quantum transitions of the emissive particles in the plasma [89, 126].

6.7.1 Contact-free (spectral) methods
Determination of the form of the energy distribution in a plasma.
Energy introduced into a plasma is distributed in some way between its particles. If the plasma were a closed system the energy distribution would approach a Maxwell–Boltzmann type equilibrium distribution with time. A laboratory

plasma is always an open system, and hence some nonequilibrium energy distribution can be stationary in it. Energy exchange through collisions of particles in the system frequently occurs very rapidly via one or several degrees of freedom, so that quasi-equilibrium energy distributions of particles can be observed for these degrees of freedom:

$$N_i = N_0 \exp[(\varepsilon_0 - \varepsilon_i)/kT]$$ (6.11)

where N_i is the concentration of particles in the i-th energy state, ε_i is the energy of the i-th state, and T is the temperature.

Such a distribution is established most rapidly for translational and rotational degrees of freedom.

The translational temperature of atoms (or molecules) can be determined from the halfwidth $\delta\lambda$ of a spectral line emitted by these atoms and broadened by the Doppler effect:

$$\delta\lambda = 7.17 \times 10^{-7} \lambda \sqrt{(T/M)}$$ (6.12)

where λ is the wavelength of the spectral line and M is the atomic mass.

From Eq. (6.12) we obtain (in K)

$$T = 1.95 \times 10^{12} M(\delta\lambda/\lambda)^2.$$ (6.13)

The accuracy of this method is not high when using spectral lines emitted by light atoms at high temperatures. The effect of other possible mechanisms of line broadening (the Stark effect, resonance broadening, etc.) should also be taken into account.

The rotational temperature of molecules is determined by studying the distribution of the rotational-line intensity of molecular bands in spectra of diatomic molecules. The condition for applying the method is adherence to the Maxwell velocity distribution and the Boltzmann rotational and vibrational level population distribution of the molecules. The rotational temperature is believed to coincide with the gas temperature. The temperature is found from the slope of the straight line

$$\ln \frac{I_{j'j''}}{S_{j'j''} v^4} = -\frac{E_{rot}}{kT}$$ (6.14)

where $I_{j'j''}$ is the intensity of the rotational lines (j' and j'' are the quantum numbers of the upper and lower rotational levels, respectively), $S_{j'j''}$ are the Henle–London intensity factors, v is the frequency of emission $E_{rot} = Bhcj'(j' + 1)$ is the energy of the upper rotational level, and B is the rotational constant for a given vibronic state.

An acceptable accuracy of the method can be achieved with spectral instruments of high resolution (a reciprocal dispersion of more than 0.1–0.2 nm mm^{-1}). However it is possible to use devices with very low resolving power where the rotational lines merge and form a continuous intensity distribution at the exit.

The vibrational temperature of molecules is determined from the relative intensities of separate bands of diatomic molecules, belonging to the same system of bands, based on the slope of the corresponding plot:

$$\ln \sum_{v''} \frac{I_{v'v''}}{v^4} = c_i - \frac{g(v')hc}{kT},$$ (6.15)

where $I_{v'v''}$ is the integral band intensity (v' and v'' are the quantum numbers of the upper and lower vibrational levels, respectively, for the transition $v' \rightarrow v''$), and $g(v')$ is the vibration constant for the given electronic transition.

Temperature of the free electrons of a plasma
The electrons in a plasma are usually assumed to have a Maxwell velocity distribution: this is perhaps true when the concentration of electrons is sufficiently large. Then data on measurements of the intensity of continuous plasma radiation may be used for determination of the electron temperature. The emission coefficient of the plasma continuum is defined by the equations

$$\varepsilon_\lambda = \text{const } \xi(v, T_e) \frac{N_e N_i}{\sqrt{(kT_e)}} \exp\left[-\frac{h(v - v_g)}{kT} \right] (v \geq v_g)$$ (6.16)

$$\varepsilon_\lambda = \text{const } \xi(v, T_e) \frac{N_e N_i}{\sqrt{(kT_e)}} \qquad (v < v_g)$$ (6.16a)

where N_e and N_i are the concentrations of electrons and ions, respectively, T_e is the electron temperature, v_g is some boundary frequency in the observed spectrum, and $\xi(v, T_e)$ is the factor taken as calculated by Schluter [122] for noble gases.

The electron temperature is determined from the slope of the function $\ln [\varepsilon_\lambda/\xi(v, T_e)]$ plotted against the emission frequency v.

The mean energy of the electrons (or the electron temperature in the case of a Maxwell distribution) can be evaluated by the use of plasma radiation in the microwave region of the spectrum. To do this, it is necessary to measure the radiative power P of the plasma received by a matching waveguide receiver:

$$P = AT_r\Delta f$$ (6.17)

where A is the absorption coefficient of the plasma, T_r is the effective radiation temperature (for a Maxwell velocity distribution of the electrons $T_r = T_e$ and for other types of distribution T_r is related to the mean electron energy), and Δf is the transmittance of the receiver.

The temperature of the electrons can be found from measurements of the halfwidth of scattering of laser radiation in the plasma on the assumption of the limiting case of Thomson scattering of radiation by free electrons. In this case the parameter α of scattering theory is much less than unity [149], and the envelope of the scattering line for a Maxwell velocity distribution of the electrons has a Gaussian form with a halfwidth of

$$\Delta\lambda = 4\sqrt{(2\ln 2)}\,\frac{\lambda}{c}\sqrt{\frac{kT_e}{m}}\sin\frac{\theta}{2}$$

where λ is the wavelength of the probing radiation, m is the mass of the electron, and θ is the scattering angle.

In general, this method makes it possible to find the electron energy distribution function.

The electron temperature of a plasma described by the so-called corona model [89] can be determined from the relative intensity of two spectral lines from the same element. These lines have to be chosen such that the excitation functions differ strongly (for example, a combination of singlet and triplet lines).

The temperature of a plasma residing in the state of local thermal equilibrium (LTE) can be determined from the absolute intensity I of a spectral line emitted by an optically thin layer of the homogeneous plasma:

$$I = \frac{1}{4\pi}P_n^m N_0 \frac{g_m}{g_0^{(i)}}\exp\left(-\frac{E_m}{kT}\right)h\nu l \tag{6.18}$$

where P_n^m is the probability of the transition, N_0 is the concentration of particles in the ground state, g_m is the degeneracy of the state m, $g_0^{(i)}$ is the overall partition function of the emitting atom, E_m the energy of the upper level, and l is the thickness of the emitting layer in the direction of the observer.

Tungsten ribbon filament lamps and the anode crater of a dc arc between pure graphite electrodes are used as a standard emitter. In the diagnostics of inhomogeneous plasmas, the integral Abel transformation is recommended for the determination of the emission coefficient of a spectral line or continuum.

Another method of determination of the temperature of an LTE plasma is based on measurements of the relative intensities of lines emitted by a given type of atom. The relative transition probabilities and energies of the upper levels have to be known for these lines. Then from Eq. (6.18) we obtain

$$\ln\frac{I_i}{P_n^m g_m \nu_{i_i}} = -\frac{E_{m_i}}{kT} + \text{const}, \tag{6.19}$$

and the temperature is found from the slope of the above function plotted versus E_{m_i}. Obviously, the wider the range of E_{m_i}s measured, the greater the accuracy.

For evaluation of the temperature of an LTE plasma, the Larentz method can be used, which is based on the phenomenon that the temperature-dependence of the spectral line emission coefficient at constant pressure has a maximum at some temperature [408].

Determination of concentration of particles in plasmas

The concentration of particles in LTE plasma can be found by calculations using a system of equations including the Saha equation, the Dalton equation, quasi-neutrality condition, and the equation of conservation of the initial composition. To make sure that the conditions of LTE are fulfilled in a plasma under investigation, it is necessary to determine the temperatures and concentrations of the different components of the plasma. The concentration of particles in an excited state is determined from the absolute intensity of the corresponding spectral lines (see Eq. (6.18). The concentration of particles in the ground state can be found either by measurements of the absorption of radiation from an external source by these particles or by interferometric methods. Unfortunately, the latter technique enables only the total concentration of heavy particles to be determined if the components of the plasma are unknown. Use of absorption methods is limited by purely instrumental considerations since most resonance lines of atoms and molecules, which are usually of interest to plasma chemists, lie in the vacuum UV region.

Techniques for the measurement of concentrations of the free electrons of plasmas are more numerous and diversified than those for particles in the ground state. First of all, a method based on the broadening of the spectral lines of hydrogen in a plasma must be mentioned; this is caused by the linear Stark effect in the microfields of charged particles, and the theory has been developed by Griem [408]. The most precise method is by comparison of the envelope obtained experimentally with a set of theoretically calculated values [409]. With reduced accuracy, the relationship between the electron concentration and the spectral halfwidth can be used, for example, for the H_β-line:

$$\ln N_e = 1.452\Delta\lambda + 14.465 \qquad (6.20)$$

where N_e is the concentration of electrons cm^{-3} and $\Delta\lambda$ the halfwidth of the H_β-line in Å.

Measurements of the width or shift of isolated spectral lines caused by the quadratic Stark effect can also be used for determination of the electron concentration. Unfortunately, there is no general equation in this case, but data calculated from these parameters for atomic lines of different elements are found in the literature.

The electron concentration can be calculated by the method of successive approximation by temperature from Eq. (6.16a) for the coefficient of continuous plasma radiation. The absolute intensity of this radiation must be measured in this case. The main difficulty arising in such measurements is in evaluation of the contribution to the continuous radiation intensity made by impurity particles in the plasma.

The determination of the electron concentration in a plasma can be carried out by using the Inglease–Teller rule which states that the principal quantum number m_{max} of the last Balmer line observable in the spectrum is related to the electron concentration as follows:

$$\log N_e = 23.26 - 7.51 g m_{max}. \tag{6.21}$$

The associated error in this determination possibly does not exceed an order of magnitude.

In studying a plasma in the LTE state, the concentration of electrons can be obtained from the relative intensities of lines generated by atoms in different states of ionization (for example, of an atom and its first ion):

$$\frac{I_0}{I_1} = \frac{v_0}{v_1} \frac{N_0}{N_1} \frac{P_0 g_0 g_1^{(i)}}{P_1 g_1 g_0^{(i)}} \exp\frac{E_1 - E_0}{kT}. \tag{6.22}$$

Combination of Eq. (6.22) with the Saha and Dalton equations and with the condition of quasi-neutrality enables the temperature and concentration of electrons to be calculated.

A convenient method for measurement of the electron concentration is interferometry. This makes possible the determination of the refractive index n which is described by the equation:

$$n_e^2 \approx 1 - f_p^2 / f^2, \tag{6.23}$$

where $f_p = 8.98 \times 10^3 N_e$ s^{-1} is the plasma frequency, and f is the frequency of the probe radiation, for free electrons.

By using SHF-generators and optical emitters (especially lasers) as the probe radiation, it has become possible to measure N_e over a relatively wide region (from 10^{11}–10^{17} cm^{-3}). The use of lasers as sources in interferometers makes their design much simpler and increases their sensitivity.

The concentration of electrons can also be determined from measurements of the absolute intensity of laser radiation scattered by the free electrons of a plasma. The calibration of instruments is carried out by using Rayleigh scattering by neutral molecules or atoms. So long as the ratio of Thomson (by free electrons) and Rayleigh scatterings is independent of the concentration of scattering particles and can be calculated with sufficient accuracy, the ratio of the intensity of light scattered by the plasma to that scattered by the molecules of some gas under the same experimental conditions is given by the equation

$$I_e / I_M = N_e \sigma_T / N_M \sigma_R \tag{6.24}$$

where I_e and I_M are the intensities of light scattered by the electrons and gas molecules, respectively, and σ_T and σ_R are the cross-sections of Thomson and Rayleigh scattering, respectively.

By measuring I_e/I_M for a known concentration of molecules N_M, the concentration of electrons N_e is easily obtained with Eq. (6.24).

Small concentrations of electrons ($\leq 10^{13}$ cm^{-3}) are measured by the use of SHF-resonators. This technique also enables the effective collision frequency of electrons with heavy particles to be determined.

To control some plasmochemical processes it is necessary to measure the velocity and temperature of finely-dispersed particles present in a plasma flow. A laser Doppler velocity meter has in this case a considerable advantage over other techniques of velocity measurement (for example, measurements from particle tracks). However, there is no such reliable method as this for determination of the temperature of particles. To find the temperature of a particle in a plasma flow it is usually necessary to solve a system of equations of motion and heat balance, making numerous assumptions about the conditions of motion and heating of the particle.

The methods of laser spectral diagnostics of plasmas (laser-induced fluorescence, optogalvanic spectroscopy, Raman and Rayleigh scattering, and ground state multiple absorption) are undergoing intense development, but have yet to become routine.

Laser Doppler velocity meters (LDVMs) and some methods of determination of the size distribution of particles based on the scattering of laser radiation have found a somewhat wider use in studying plasma flows.

6.7.2 Contact methods

Together with spectral and optical methods, contact methods are extensively used in the diagnostics of low-temperature plasmas. They consist of placing specially-constructed solid objects (made of metal or dielectric) into a plasma and measuring the energy and particle fluxes upon them. Such objects can be thermocouples, heated wires, and different types of probe, including samplers. The quantities thus measured are the concentrations of electrons and heavy particles in the ground and excited states and their energy distributions. The conditions for applying some contact methods are indicated in Table 6.2.

The wide application of contact methods is due, firstly, to their simplicity and to the possibility of obtaining the local characteristics of the plasma. Hitherto, they have given the only means of obtaining information on the distribution of electric fields in plasmas, on the electron energy distribution in nonequilibrium plasmas, etc. [93].

In spite of the simplicity of the instrumental design and the means of obtaining the primary data, the interpretation of the results appears to be complicated and, frequently, not unambiguous. Consider, for example, a thermocouple designed and used for the measurements of gas temperatures in equilibrium systems. On turning to measurements in nonequilibrium systems, there arise a number of problems due, in particular, to the fact that the surface heat effects are related to energy transfer onto the surface not only from the translational degrees

Table 6.2. Contact methods for diagnostics of reacting plasmas [91, 93]

Method	Quantities measured	Type of plasma
Electric probe	$N_e, T_e, \bar{\varepsilon}, E$, EEDF[†]	DC glow discharge, HF and SHF low-pressure discharges
	Rate of growth and VCC of dielectric film	Low-pressure electric discharge
Cooled electric probe	N_i, T_e, E, fluctuations of N_i and T_e	Ambient pressure electric discharge
Dielectric probe	N_e, T_e, rate of growth of conducting films	Low-pressure electric discharge
Cooled calorimetric probe	Enthalpy of plasma	Ambient-pressure electric discharge
Cooled sampler probe	Chemical composition of stable products	Ambient-pressure electric discharge
Thermocouple	T_g, heat effects on surface	Low- and medium-pressure dc discharge
Heated wire	T_g, gas flow rate, accommodation coefficient	The same

† Abbreviations. E is the electric field strength, T_g is the gas temperature, VCC are the voltage-current characteristics, and EEDF is the electron-energy distribution function.

of freedom. In general, it is necessary to take into account energy transfer from the internal degrees of freedom. Under these conditions, although the measurement task will be complicated, by combining several techniques it is possible to acquire information on such quantities as the concentrations of excited and atomic species, accommodation coefficients, etc.

Another example is given by the electric probes commonly used for measurements of parameters of the electron component of a plasma. The results are extremely sensitive to the purity of the probe surface. As a consequence, this gives the possibility of using the probes in studies on film deposition in nonequilibrium plasmas.

It should be noted that, being a perturbing technique, contact methods can change the properties of the plasma studied significantly. The perturbations introduced by contact methods are typified, first of all, by geometrical perturbations: a body placed in a plasma disturbs the spatial distribution of the electric and thermal fields and of the particle concentrations.

Thus, when using contact methods it is necessary to minimize their influence on the plasma. Moreover, a careful analysis of the information obtained by these methods is needed in each case. In addition to other methods of studying plasmas, the contact methods can provide a large body of data on processes occurring in plasmas.

7

Introductory industrial high-energy chemistry

Industrial chemistry, as a science, deals with the techniques of processing of raw materials required in the manufacture of chemically-based consumer goods and their means of production [410, 411]. Its tasks include the design and selection of the conditions, schemes, and types of industrial process and auxiliary operations as well as the structural design of equipment and the selection of materials for machines and facilities.

The fundamentals of industrial high-energy chemistry and engineering must include, in addition to the principles general to industrial chemistry, those particular features necessary in the implementation of the conditions inherent to high-energy chemistry. The overwhelming majority of nonequilibrium and quasi-equilibrium processes of HEC are characterized by their low wastage. Materials obtained in these processes generally possess unusual structures and unique properties.

The high rates of HEC processes require highly sensitive detection methods with good time resolution.

Solution of the problems associated with industrial high-energy chemistry is a crucial element of the scientific approach to industrial management on the eve of the 21st century [412].

7.1 PROCESSING OF PLASMAS

A typical scheme of any industrial process includes

(i)   the mechanical and physical operations of treatment of the raw materials for their subsequent use in chemical processes,

(ii)   chemical reactions yielding a collection of products, and

(iii)   physical and/or chemical procedures needed for conservation, separation, and purification of a desired product.

In principle, a scheme of processing a plasma does not differ from that considered above. A complete scheme for an industrial plasmochemical process includes the stages of plasma generation, of chemical transformations in the plasma, and of chilling (quenching) of the products, the latter two stages coinciding in time and/or space for some of the processes.

7.1.1 Low-temperature plasma generators

Substances are made in their plasma state in specially designed reactors called plasma generators. One chooses the type of plasma device, depending on the work in hand, but in any case the generator chosen must satisfy the kinetic and thermodynamic properties of the process concerned.

Applied plasma chemistry employs different types of generator (termed a torch, plasma reactor, or plasmotron):

(i)   arc generators,
(ii)   high-frequency (HF, or radiofrequency) generators of two kinds, namely high-frequency generators with inductive coupling (HF–IC) and with capacitive coupling (HF–CC), and
(iii)   super-high frequency (SHF, or microwave) generators,
(iv)   glow-discharge reactors,
(v)   corona-discharge devices, etc.

From the early 1960s d.c. and a.c. arc plasmotrons operating at mains frequency were those most widely used in research and industrial processes. This was probably the time when the term 'plasmotron' was invented. Such a generator consists of electrodes, a discharge chamber, and a plasma gas feed port (Fig. 7.1). The gas passes through an arc striking between the anode and the cathode and flows in the form of a plasma jet through a hole in the anode, that is, a nozzle. The arc column is stabilized in space by the chamber walls or by injection of a gas tangentially to the walls (eddy gas feed). In the latter case the colder gas is centrifuged to the reactor walls and rotates the plasma spot in the anode nozzle, thus preventing melting or any other significant erosion of the electrode.[†] In powerful plasmotrons (up to 50 MW) the plasma spot is moved by a magnetic field created along the nozzle axis by a solenoid.

The arc length in self-adjusted arc plasmotrons depends on the dynamics of the gas flow and on by-passing in the anode nozzle (Fig. 7.1a). The larger the axial component of the gas velocity, the longer the arc, but any extension of the

---

† Electrode erosion means taking away material from the body of the electrode by various complex thermal, electrical, chemical, and mechanical processes in the near-electrode space caused by drawing of electric current.

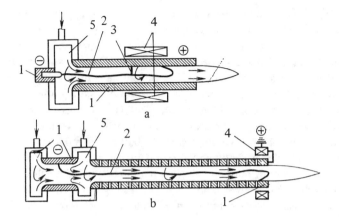

Fig. 7.1. (*a*) Gas-stabilized plasmotron with self-adjusted arc length and (*b*) segmented
anode plasmotron with constricted arc length: 1—electrodes; 2—arc column; 3—
channel by-passing breakdown; 4—magnetic coils; 5—eddy chambers.

arc increases the potential at some of its cross-sections with respect to the anode
walls, and at a certain cross-section gas breakdown occurs. Thus the arc is
shortened, and this process, known as by-passing (shunting), is repeated with a
frequency of several kHz.

Control of the arc length in a constricted arc plasma reactor is provided by
using segmented electrodes, so that only definite segments are connected to the
power supply (Fig. 7.1b), or by the use of an overexpanded nozzle. A stall eddy
zone is formed behind the nozzle, which is exactly the region where arc by-
passing takes place.

Multi-arc plasmotrons, where the jets are mixed in one chamber, thus enabling the power of a single device to be increased, have been designed recently.

Electrodes are made from high-melting materials (tungsten, molybdenum) or
from copper and steel supplied with cooling. The electrode lifetimes are determined by erosion and can reach several hundred hours.

Arc plasmotrons are characterized by their voltage–current characteristics
(these, in particular, determine the requirements of the power supply unit) and
their thermal efficiency. The latter is defined as the ratio of the amount of energy
taken from the nozzle per unit time to the arc power. These quantities are either
found experimentally or calculated on the basis of a system of criteria obtained
from similarity theory, dimensional theory, and equations describing physical
processes in the arc.

Self-adjusted arc plasmotrons have falling   voltage–current characteristics
(the voltage drop across the arc decreases with increasing current) and require
special power supply units or are connected into a circuit in series with a so-
called ballast resistor.

Constricted-arc plasmotrons show upward voltage–current characteristics. They operate steadily without a ballast resistor. The efficiency of this type of plasmotron with a power of a few MW to tens of MW approaches 80–90%.

The temperature and the velocity of the gas in plasmas jets are distributed nonuniformly along and across the axis, for example in an argon jet flowing out to the atmosphere. The variations in temperature can be as large as 5×10^4 K mm^{-1} along the axis. Most of the gas (ca. $\frac{2}{3}$) is consumed in the peripheral areas. The maximum temperature along the axis of the jet varies from 10 to 50 $\times 10^3$ K (the mean mass temperature is ca. 10 000 K) for monoatomic gases and from 4 to 6×10^3 K for diatomic gases). The gas velocity at the nozzle tip depends on the gas flow rate, arc power input, and nozzle diameter; it can vary from several m s^{-1} to several km s^{-1}. Argon, nitrogen, oxygen, air, water vapour, ammonia, natural gas, carbon monoxide, carbon dioxide, halogens, etc. can be used as plasma gases.

The plasma of an arc reactor is always contaminated to some extent by electrode materials because of erosion. If this is unacceptable, then HF–IC reactors, HF–CC reactors, SHF plasmotrons, discharge plasma reactors, etc. have to be used. Though the power of these plasma generators is less than that of arc reactors (it is ca. 0.5–1.0 MW for HF reactors and ca. 0.1 MW for SHF devices), and the efficiency does not exceed 0.6, their lifetime is much longer, up to several hundred thousand hours.

Reliably working HF plasmotrons appeared in the 1960s, based on vortex gas stabilization (they were invented by G. I. Babat as early as 1942). These plasmotrons function on the following principle: having a conductivity both finite and independent of its parameters, a plasma (created by means of a special device operated by switching on a plasmotron) placed in a HF field gives rise to vortex currents in its skin layer. The latter currents provide heating of the plasma gas.

A high-frequency plasmotron with inductive coupling, for example designed for the treatment of powder materials (Fig. 7.2), consists of an inner discharge tube and an outer tube made of quartz. An inductor is installed in the lower part of the tubing. The plasmotron is fed with argon in three streams: one enters the upper end of the inner tube tangentially, to stabilize the discharge and to cool the reactor walls, a second stream passes between the tubes, and the third carries the powder material through water-cooled tubing. Very complex dynamic phenomena are observed in plasmotrons of this type. Gas streams inside and outside the plasmoid are directed oppositely to the axial component of the plasma flow. Therefore the reactants are introduced into a specially-designed axial flow. Since the major part of the gas goes through the cold near-wall region, the mean mass temperatures in such plasmotrons are low. The maximum values of the temperature in argon plasmas are found on the axis, and they range from 9 to 11×10^3 K for rather widely varying pressure (from 0.05 to 0.3 MPa), frequency (from 0.3 to 300 MHz), gas flow rate (from 3 to 300 dm^3 min^{-1}, power (from

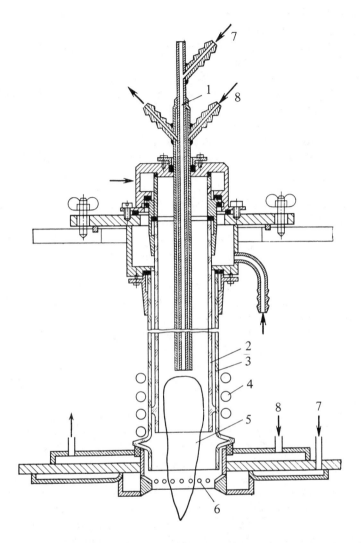

Fig. 7.2. High-frequency inductively-coupled plasma reactor: 1—powder inlet; 2—inner discharge chamber; 3—outer chamber; 4—inductor; 5—radiant plasmoid; 6—quenching port; 7—plasma gas inlet; 8—coolant water inlet.

1 to 200 kW), and discharge tube diameter (from 1.2 to 12 cm). The radial temperature gradients in the central part of the plasmoid are not more than 100 K mm^{-1}.

The maximum temperatures in molecular gas plasmas do not exceed 7–8×10^3 K under the same conditions as in the argon plasmas. The radial position of the temperature maximum is determined by the ratio of the skin layer to the diameter of the plasmoid.

A high-frequency capacitively coupled plasmotron consists of a discharge tube supplied with circular outer electrodes. Even under atmospheric pressure, a plasma generated in this plasmotron can be nonequilibrium. For example, in argon the electron temperature can exceed the temperature of heavy particles by an order of magnitude.

The outflow velocity of the plasma jet in a HF plasmotron is less than the velocities of plasma jets in arc plasmotrons and is larger than 100 m s^{-1}, while the radial distribution of the mass flow rate has no axial dip typical for arc plasmotrons.

The fraction of the radiant energy of the plasma increases drastically with increasing pressure and temperature; in a HF–IC xenon plasma it reaches 80% at a pressure of 4 MPa. The emission intensity is of importance for determining the efficiency of the plasmotron, and its spectral composition is crucial in considering protective shielding. The spectral composition of the emission depends on the nature of the plasma gas and on the pressure and temperature. Thus, at identical temperatures the argon plasma exhibits intense emission in the UV-region, the neon plasma in the IR-region, and a xenon+argon plasma emits light spectrally similar to solar radiation.

A low-pressure HF plasma is nonequilibrium. Departure from equilibrium is also possible for an atmospheric pressure plasma in its peripheral regions at the boundaries of the skin layer and in the zone of gas entry into the plasmoid body (where the temperature gradients are large and the electron concentration is low).

The absorption of energy by a plasma is enhanced by increasing the frequency of the electromagnetic field. Therefore, the energy input in the SHF range can already be effective at a relatively low plasma temperature (*ca.* 4000 K), and this is actually used in SHF plasmotrons.

In principle, the physical processes occurring in HF and SHF plasmotrons are identical, but the designs of the devices are different. Three types of SHF plasmotron are currently used, namely resonance, coaxial, and waveguide plasma generators. A resonance plasmotron is a closed toroidal or cylindrical metal chamber. Electromagnetic energy is introduced through an opening in the wall and it is absorbed in the plasma initially created by an ignition circuit if its conductivity is larger than that of the chamber walls. Depending on the type of oscillation, the plasma has the form of a filament or hollow cylinder coaxial with the resonator. A coaxial SHF plasmotron does not differ from a HF plasma torch [89], but its electrode erosion is less at higher plasma energy densities. In a waveguide SHF plasmotron (Fig. 7.3) the plasma is generated by a travelling wave in the waveguide. Such a plasmotron consumes up to 85% SHF energy in production of the plasma, and its efficiency can be as high as *ca.* 60%.

A low-pressure stationary SHF plasma and intermediate-pressure (60–130 Pa)

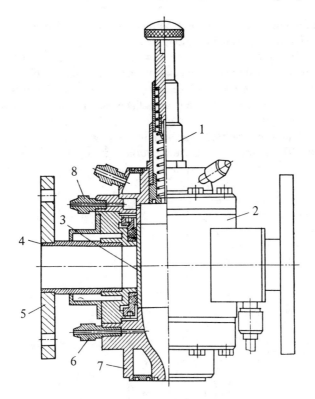

Fig. 7.3. SHF plasmotron: 1—discharge ignition port; 2—reactor body; 3—silica tube where plasma is generated; 4—rectangular waveguide; 5—energy feed flange; 6—reactant feed connection; 7—nozzle; 8—plasma-gas feed connection.

plasmas are nonequilibrium, for example in the pulsed 60 Pa nitrogen plasma at a power of 1 kW, the gas is barely heated, yet has an appreciable degree of ionization and a mean electron energy of 3 eV.

The departure from equilibrium in atmospheric-pressure SHF plasmas is determined by the nature of the gas. Thus, in an argon plasma doped with hydrogen, the gas temperature is 4500 K at a power of 800 to 1200 W, and the population-level temperature[†] of argon and hydrogen is 7000 K with an electron concentration of 10^{14} cm^{-3}. The nitrogen plasma is at equilibrium under these conditions.

To obtain a nonequilibrium plasma, a stationary glow of a corona discharge is usually employed. The corona discharge is formed in highly-nonuniform electric fields near bends in the electrodes where the ionization and excitation of gas molecules take place. From this region (corona) charged particles drift out to the

† The population temperature (energy) is defined by the population of the group of atomic quantum levels considered.

rest of the interelectrode space. The corona discharge device is a metal tubular chamber with a thin wire situated along its axis.

In glow discharge devices (Fig. 7.4) several dark zones and several radiant zones are formed, completely occupying the cross-section, the dimensions of the zones depending on pressure and the distance between the electrodes. A spatial charge is localized basically in the near-surface cathode area which also has the maximum concentration of charged particles. In the positive column zone, the energy distribution of the electrons in molecular gases is far from the equilibrium (Maxwell) distribution and the electron concentration changes along the radius of the cylindrical dielectric tube (in accordance with a zeroth order Bessel function).

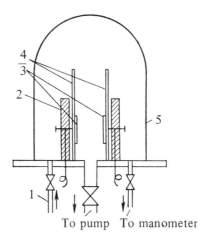

To pump   To manometer

Fig. 7.4. Set-up for glow-discharge deposition of films: 1—injection of gaseous reactants; 2—electrode holders; 3—flat electrodes; 4—sheath plates restricting plasma volume; 5—working chamber.

In laboratory studies some types of apparatus for plasma generation are used which utilise different types of radiation (X-ray, γ-ray, or optical laser radiation).

7.1.2 Chemical plasma reactors
Gas-phase reactors
A plasma reactor for gas-phase processes has two parts: a mixer and the reactor itself. In the mixer a homogeneous mixture of reactants is formed at a temperature determined by the kinetic and thermodynamic features of the process. This mixture is injected into the reactor, some chemical processes already taking place in the mixer [91, 92, 392–396, 398]. In most cases the reactor has separate inlets for flows of reactant and plasma. Under these conditions it is necessary, first of all, to mix the plasma with the reactants in such a manner that their

molecules are in direct contact, that is, mixed on the molecular scale. However, as shown above, plasmochemical processes occur under condition when the characteristic times for chemical reactions and for mass- and energy-transfer processes approach each other. As a result, a significant part of the chemical transformation can proceed on mixing. One must consider not only this part but get an idea as to how a plasmochemical process becomes equilibrium. The mixing time is defined as

$$\tau = L_m / v \qquad\qquad (7.1)$$

where L_m is the length of the mixing path and v is the linear velocity of the mixture in the reactor tube.

According to an approximate model of mixing gaseous flows in a cylindrical channel, the length of mixing (the distance from the gas inlet to a cross-section with the established concentration profile of the injected gas) is equal to $2L_m$ (accurate to $\pm 15\%$). The minimum diameter of the mixer D_{min} has to be 2.5–3.5 times the inlet diameter of the injected gas. Then for $L_m = 3.5$ mm and $v \approx 300$ m s^{-1}, the mixing time τ is about 20–30 µs.

Such a model describes the mixing of the plasma flow with the injected gas stream on the macroscopic scale. This means that so-called gas globules are mixed, each of them consisting of molecules and ions of either plasma only or of injected gas only. But we have already said that for chemical reaction to occur it is necessary that the mixing has been carried out on the molecular scale.

Turbulent mixing of the plasma and reactant flows have been studied by using fast chemical reactions methods and 'physical process' methods [91]. When choosing a relevant chemical reaction or a physical process, the following requirements were taken into consideration: the characteristic time for the reaction had to be less than the time of turbulent mixing, and no secondary reactions were allowed.

With the fast chemical reactions methods of decomposing N_2O, the mixing of an argon plasma with cold argon containing 15% N_2O was studied in plasma reactors both with co-current and radial feeds of the cold gas.

The reactor with the radial feed was found to provide better conditions for mixing (mixing time $\tau \approx 50$ µs in a 3.5 mm channel); the length of the zone for complete mixing up to the molecular scale level L_{mn} is 2.3 times greater than the length of establishing the concentration profile L_m.

Cylindrical type reactors are usual in plasma processing. Their characteristics are similar to those of the ideal plug-flow reactor. Any differences are caused by three factors:

(i) Intensive cooling of the reactor walls results in radial temperature and flow-rate gradients. The degree of conversion in near-wall streams decreases because of decreasing temperature. To obtain a desired (previously calculated) degree of conversion, it is necessary to increase the length of the reactor. However, this cannot be done if the given product is transient. The problem is that any extension of the reactor length, that is, increasing the reactant's residence time, decreases the selectivity of the process since the reaction rates of the high-temperature axial part of the flow are large. Therefore, to carry out reactions of this type it is necessary that the difference between the wall temperature and that of the reaction zone in a plasma reactor is as small as possible.

(ii) Turbulent diffusion at the reactor axis. Its effect is estimated from a quantity reciprocal to the Peclet number: $P^{-1} = D_{eff}/vL_m$, and the larger the quantity, the more profound the effect (D_{eff} is the turbulent diffusion coefficient).

(iii) The type of radial profile of the eddy flow rate. Deviation of the mean residence time for a gas in a reactor from the value calculated with the axial gas velocity can be as large as 30%, which needs to be taken into account in calculating the reactor volume.

Modelling of the plasma reactor
It has now become clear that the fields of turbulence and temperature must be identical. It is fairly easy to provide the same turbulence characteristics in reactant feed streams and in a large mixer. The properties of turbulence fields and temperature fields in plasma jets depend on the type of plasmotron used. Frequently a plasma jet is initially eddied to make the turbulence and temperature fields more uniform.

If the fields of turbulence and temperature in reactor mixers of different size are identical, then the length of the mixing zone on the molecular scale is directly proportional to the mixer diameter. The mean mixing time is found from $\tau_m^{(1)} = L_m^{(1)}/v_1$ and $\tau_m^{(2)} = L_m^{(2)}/v_2$, where v_1 and v_2 are the mean plasma flow rates for radial feeds of the components in mixers of different size. The degree of conversion will be the same provided that $\tau_m^{(1)} = \tau_m^{(2)}$, when $D_1/v_1 = D_2/v_2$.

If the output of one reactor is n times larger than that of the other, then $n\rho_1 v_1 s_1 = \rho_2 v_2 s_2$, where ρ_1, ρ_2, and s_1, s_2 are the densities of the mixture and the reactor cross-sections, respectively. But $\rho_1 = \rho_2$ since the temperature fields in the reactors are identical and the pressures are the same, hence $n v_1 D_1^2 = v_2 D_2^2$.

The reactor diameters and corresponding flow rates are related to each other by the following equations:

$$D_2 = D_1 n^{1/3} \quad \text{and} \quad v_2 = v_1 n^{1/3} \tag{7.2}$$

which can be easily obtained from the two formulae considered above.

Let us now consider the reactor modelling itself. A homogeneous mixture at the reactor inlet is at the reactor temperature T_r. The residence–time distributions of the mixture in the reactors have to be identical. For a plug-flow reactor and plasma reactors with a wall temperature close to T_r, which are quite similar, this requirement is reduced to $\tau_\rho^{(1)} = \tau_\rho^{(2)}$. Then we find

$$L_\rho^{(1)} \big/ v_1 = L_\rho^{(2)} \big/ v_2 \tag{7.3}$$

and

$$L_\rho^{(2)} = L_\rho^{(1)} n. \tag{7.4}$$

Reactors for heterogeneous processes

There are three types of plasmochemical process involving substances in the condensed phase. In heterogeneous processes of the first type, the reactants are injected into the reactor in the gaseous state and the products are in the condensed state. In heterogeneous processes of the second type, both reactants and products are in the condensed state. The third type of process is when the reactants are in the condensed state and the products formed are gaseous.

The participation of substances in the condensed state in a plasmochemical process complicates the reactor design and the operational procedure for some of the stages. In heterogeneous processes of the first type, the operational conditions of the mixer and its design are the same as for homogeneous processes. For the second and third types of process, the mixer must provide, in addition to the functions specified above, a uniform distribution of powder in the plasma jet, while the reactor has to be connected to a feed port to receive a uniform supply of condensed-state material.

A strong mutual influence of chemical and heat and mass transfer processes is observed in heterogeneous plasmochemical reactions [347]. Some mathematical models have been proposed describing the behaviour of particles in plasma jets.

The results of calculations have made it possible to study the dynamics of changes in the gas and particle temperatures, their velocities, the heat transfer coefficient, the radii of the particles, and their degree of vaporization as a function of the initial temperature, particle size, the mass flow rate of the powder, the thermophysical properties of the plasma gas, and the distribution of the reactant along the length of the reactor where the heterogeneous processes take place.

The characteristic time required for heating to the melting points and vaporization temperatures, as well as for inducing phase transitions, are proportional to the square of the particle size. It also appeared to decrease dramatically with increasing temperature of the plasma T_p and its 'thermal conductivity potential'

$$S = \int_{H_o}^{H} \frac{\lambda}{C_p} \, dH \tag{7.5}$$

where λ, C_p, and H are the thermal conductivity, heat capacity, and enthalpy of the plasma, respectively.

For instance, a spherical particle of Al_2O_3 with a radius of 50 µm is melted in a nitrogen plasma at 10^4 K for 1.2 µs, and it is completely vaporized in 2 µs. But at $T_p \leq 4250$ K this particle cannot be vaporized, because of the radiation losses.

Experimental studies on the behaviour of solid particles in a plasma have shown that the particles and the gas move with different velocities in the reactor: the gas proceeds as if flowing around the particles. A powder placed in the plasma decreases the gas temperature and makes the distribution of the parameters in terms of the cross-section of the jet more uniform. The powder induces eddies in the jet, if it was laminar before powder injection, and vice versa, it decreases the turbulence if the jet was initially turbulent. This is essential when selecting the length of the reactor.

Kinetic calculations on many plasmochemical processes lead to the conclusion that reactions in the gas phase occur for a 10^{-4}–10^{-3} s, whereas the period of heterogeneous reactions is longer by a few orders of magnitude, which is in agreement with experimental data. Thus, the parameters of reactors for carrying out heterogeneous processes of the second and third types are restricted by the rate of transition of the components to the gas phase.

Design of plasma reactors
Those most extensively used for carrying out equilibrium gas-phase plasmochemical processes are jet-type plasma reactors. Depending on the technique of mixing the plasma flow with the feedstock stream, they are distinguished as co-current flow and countercurrent flow reactors. However, the combined feeding of reactors is possible in which part of the reactants is fed in co-current mode and the other part in countercurrent mode.

Reactors of this type are fed with stock materials by means of crosscurrent flow, co-current vortex flow, and countercurrent vortex flow through different number of openings or by means of transport through porous walls in accordance with some distribution law.

The front part of the reactor accommodates the feedstock inlets, and this part serves as a mixer. The lower part of the reactor has openings for injection of a quencher; this part of the reactor acts as an antechamber for the quenching port (see section 7.1.3).

The feedstock in co-current vortex and countercurrent vortex reactors (Fig. 7.5) is introduced tangentially to the plasma jet. At this point the material moves in one direction with the plasma jet (co-currently) or in the opposite direction (countercurrent feed). A conical shape of the internal part of the reactor improves the mixing process. Furthermore, a condensed phase is more easily isolated and its deposition on the cooled internal surface of the reactor walls is suppressed in reactors of this type.

Industrial plasma chemistry generally uses the co-current vortex scheme of reactant feed.

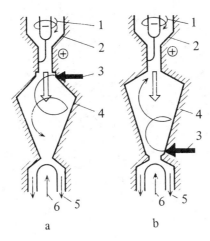

Fig. 7.5. Schematic diagram of co-current vortex (*a*) and countercurrent vortex (*b*) types of plasmochemical apparatus: 1—coolant feed; 2—plasmotron body; 3—reactant feed; 4—reactor; 5—product outlet; 6—quencher feed.

Countercurrent reactors are most frequently used when the feedstock is injected into the plasma jet in the sprayed liquid state or in the form of gas-fluidized solid particles. Mixing of the streams is intensified by feeding by different specific pulses. Under these conditions a plasma is fed through an expanded cone nozzle at a supercritical pressure drop or through a converging nozzle at a subcritical pressure drop. The feedstock is injected through a diffuser. Expansion of the flow and its decrease in velocity take place in the latter. Additional mixing occurs on the backward motion of the mixture in an annular gap between the diffuser wall and the flow core.

To increase the residence time in the plasma, treatment of dispersed materials reactors with countercurrent plasma jets is applied. Particles of powder in a flow of carrier gas are introduced into the plasma jet at the exit end of the plasma-torch nozzle. They are accelerated by the jet, pass the zone of jet encounter, and are decelerated by the contrary jet. The latter carries them to the former jet. Thus the particles are fluidized in the high-temperature zone.

Heterogeneous plasmochemical processes are frequently carried out in fluidized-bed reactors where the period of particle contact with the plasma is much longer (Fig. 7.6). A reactor of this type is conical (1), having a condensed-phase feeder at the top and an electric arc plasma torch (4) at the bottom. The jet of plasma is introduced into the bottom of the reactor through the nozzle (5). Gaseous products of the process heat the powder feed and leave the reactor after passing through the separator (6). Liquid products drain off the reactor walls and nozzle walls to the hopper (7).

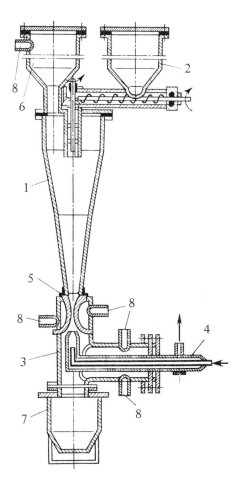

Fig. 7.6. Plasmochemical apparatus (reactor with plasmotron) for fluidized bed treatment of dispersed materials: 1—conical body; 2—feeder (device for condensed-phase injection), 3—reducer; 4—plasma torch; 5—nozzle; 6—separator; 7—hopper; 8—water cooling.

In the high-temperature part of the reactor which is the entry zone of the plasma jet, the temperature field is nonuniform, and this is a disadvantage of this type of reactor.

A more uniform temperature field is obtained in a reactor where powder-laden plasma jets from several plasma torches are introduced tangentially to the mixer chamber walls. Accelerated by the plasma jet to considerable velocities, solid particles are held in the mixing chamber by centrifugal force to be distributed by their fractional composition. Melted unreacted particles are deposited on the walls and drain off at the bottom cone to the hopper where they are cooled and granulated to be removed.

7.1.3 Quenching of products of plasma processing

Thus, at the exit of the plasma reactor we obtain a mixture containing some desired products of a chemical process. It is then necessary to remove them from the high-temperature (energy) zone in such a manner that their concentration is either not reduced or is kept within prescribed limits. This is achieved by means of quenching. The objective of quenching in this case is to chill the reaction products as quickly as possible to prevent decomposition in the intermediate temperature range. The quenching rate is defined as $\partial T/\partial t$, and it usually lies in the range 10^5–10^8 K s^{-1}.

In general, the quenching of products is needed when the plasmochemical processes of interest are at equilibrium or quasi-equilibrium. In nonequilibrium processes where the temperature of the heavy particles is relatively low (< 1000 K), quenching either becomes completely unnecessary or, very often, its conditions are milder.

If there is a need to stabilize radicals or compounds unstable at room temperature, then they must be frozen out on cold walls at cryogenic temperatures after extensive dilution with an inert gas. Thus, noble gas fluorides such as XeF_4 and KrF_4 were obtained in this way from a nonequilibrium glow-discharge plasma.

There are two types of quasi-equilibrium gas phase plasmochemical reaction when the composition of the products obtained depends on the quenching conditions to a decisive extent. The first type includes reactions yielding consecutively a series of intermediate products. The conversion of methane to carbon and hydrogen provides an example. The desired hydrocarbon product here is acetylene which is formed at one of the intermediate stages. Naturally, not only the quenching rate but the instant at which the reduction of temperature starts is important in such reactions; if it is delayed at the beginning of quenching by 0.002 s then the acetylene concentration falls from 15.5 to 10% by volume.

A characteristic of reactions of the second type is that the substance obtained is the final product of reaction occurring exclusively at high temperature: this product is reasonably stable only at room temperature. This type of reaction is exemplified by, for example, the thermal process of formation of nitric oxide in air. It is important in this case to provide the required quenching rate and to start the quenching at not too early a point; that is, when equilibrium has not been established. For instance, a decrease in the quenching rate of nitrous gases in the thermal production of nitric oxide in air from 10^8 to 10^7 K s^{-1} reduces the concentration of NO from 9.6 to 6.4%. The quenching conditions, that is, the initial moment and the temperature dependence of the temperature decay rate dT/dt, are determined by the kinetics of the process. Disturbing this dependence in one temperature region cannot be compensated by increasing the quenching rate in another region.

Consider the character of quenching of nitric oxide shown in Fig. 7.7 as calculated from kinetic data on NO decomposition [90]. It is seen that the

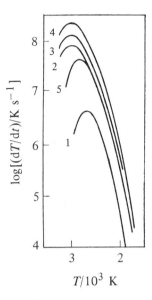

$$T/10^3 \text{ K}$$

Fig. 7.7. Chilling curves of nitroso gases: $1—T_{in}$ = 3000 K, P = 0.01 MPa; 2—
3300 K, 1 MPa; 3—3300 K, 2 MPa; 4—3300 K, 5 MPa; 5—3100 K, 2 MPa.

required quenching rate reaches its maximum value at a temperature close to the
initial temperature of quenching and then decreases drastically. However, it is
important in some cases not only to remove the heat from the system rapidly but
to use it with maximum efficiency; this is often necessary to make the process
commercially viable.

The available procedures and quenching conditions makes it possible to solve
in practice some problems associated with the commercial and materials science
aspects of plasma processing, namely, to conserve the major part of the target
product in a plasmochemical process and to obtain certain (prescribed) structures and properties of substances prepared as solids (ultradisperse powders,
films, etc.). Thus, quenching is one of the most important methods for optimization and controls plasma processing.

The quenching conditions determine both the structure and properties of the
condensed products of plasmochemical processes. By adopting one or other rate
of quenching in a low-temperature plasma, it is possible to obtain substances of
both stoichiometric and nonstoichiometric composition and any transient species
of equilibrium or nonequilibrium nature. Moreover, by using different quenching conditions, for example in plasma reduction processes, it is possible to
prepare metals in the form of powders with different ranges of particle size,
filament aggregate, and ingot. Thus, the appropriate choice of partial pressure of
the metal vapour and degree of supersaturation (which is achieved by change in
the mass flow rates of the gas and powder as well as by change in

temperature at the entrance to a quenching port) enables ultradisperse tungsten powders of spherical form to be obtained. The appropriate choice of initial timing and quenching rate has provided the possibility of limiting their particle dimensions within 40–50 nm. The time of chilling the particles is restricted, on one hand, by the dimensions and thermophysical properties of the material and, on the other, by the characteristic time of a change in structure or, generally, in the thermodynamic state due to decreasing temperature. The minimum possible time of quenching for the finely dispersed oxides TiO_2, ZrO_2, and Nb_2O_5 is 10^{-6}–10^{-4} s.

Considering the principal techniques of quenching, the commonest is cooling by means of heat exchangers, for example with water-cooled tubes. This procedure is used for chilling gases with a temperature up to 400 K. The quenching rate depends on the design of the heat exchanger, the gas-flow conditions, temperature, and the physical characteristics of the flow. It can reach *ca* 10^7 K s^{-1} at the entry of the heat exchanger and is about 10^6 K s^{-1} on average.

A widely applied method for quenching products is flooding with streams of liquid (water or reactant) or gas. Thus a gas containing nitrogen oxides at *ca.* 1 MPa and 3000 K can be chilled with water jets at a rate of 10^8 K s^{-1}.

The same quenching rates are also attainable in a fluidized bed reactor and on mixing the products with a cold gas.

The reactants themselves can be used as a quencher. This improves the output of a process and its parameters, and sometimes leads to a change in industrial procedure to make it commercially more worthwhile, for example quenching of the products of plasma pyrolysis of hydrocarbons by the hydrocarbons themselves increases the yield of the desired product, gives the possibility of fine control over its composition, and decreases the power consumption per unit mass of product.

The quenching of hot gases can be carried out by means of an aerodynamic method, using a de Laval nozzle providing a quenching rate as high as 10^8 K s^{-1}. However, the cooling rate decreases with increasing temperature, whereas it should be maximal in many cases near the temperature at which chilling of the products begins. Moreover, since the gas temperature drops to its initial value on slowing down the gas flow, the de Laval nozzle should be used in combination with other devices designed for heat exchange.

7.2 RADIATION PROCESSING

The objective of industrial radiation chemistry as a science is the study and development of procedures and devices for the implementation of radiation-induced physicochemical processes in manufacturing consumer goods, and in improvement of the performance of materials and articles as well as in the solution of environmental problems [413]. Working out the process of radiation-chemical manufacture of a product has the following basic steps:

(i)   adoption of a scheme of the process and determination of its technical parameters with a pilot plant,

(ii)   selection of the constructional materials and their radiation corrosion testing,

(iii)   choice of the radiation source and design of the radiation-processing unit,

(iv)   elaboration of the procedures for isolation of a given product and its by-products,

(v)   mathematical simulation of the operation of the radiation-processing facility and optimization of the process as a whole,

(vi)   estimates of the industrial economics and comparison with other available procedures for manufacture of the desired product,

(vii)   evaluating safety measures for the personnel operating the radiation plant.

The importance of radiation chemistry and radiation processing is emphasized, in particular, by the fact that seven International Conferences on Radiation Processing have already been held (in 1976, 1978, 1980, 1982, 1984, 1987, and 1989) [414–416]. The total value of products manufactured by radiation processing in the USA in 1980 was about 10^9 dollars, and has since increased more than two-fold [416] (the production output in the USA was estimated as 200 million dollars in 1969 [424]). Combined approaches to radiation processing, for example a combination of radiation treatment with biocatalysis [423], have been sought in recent years.

The principal feature of radiation processing is the use of ionizing radiation for the induction of a chemical process. The following advantages of ionizing radiation as an industrial strategy [414] should be noted. The highly energetic efficiency of ionizing radiation, making radiation processing an energy-saving technology, stimulates interest against the background of increasing costs of traditional sources of energy. Consider one example. The curing of coatings by a conventional procedure consumes only 2% of the energy for the process itself while the rest is dissipated. However, if the curing is carried out by using low-energy electron beams, then the electrons are completely absorbed in the cured layer, that is, there is no energy-waste.

Ionizing radiation has considerable powers of penetration, that is, mm in solids for electron beams (processes in films) and tens of cm for gamma-radiation (processes in equipment and reactor units).

The amount of radiation used in materials processing can be measured easily, and it is possible to 'dose' radiation in both time and space.

Ionizing radiations do not contaminate the product during processing, while in chain processes involving chemical initiators, the final product always contains the residue and conversion products of the initiator.

In Chapter 1 were listed the types of ionizing radiation generated by means of different radiation sources. The basic types of radiation facility used in industry are gamma-sources and electron accelerators [413, 417–421]. Their total power in industry worldwide was about 15 MW at the end of 1980, with nearly 1 MW due to the gamma-sources. This total power has since more than doubled. To illustrate the growth in power one might note that in 1972 the total power was only 750 kW [414].

7.2.1 Sources and facilities for ionizing electromagnetic radiation

The most widely used sources of high-energy (hard) electromagnetic radiation are radioisotopes, fissile heavy elements, and X-ray tubes.

Of the various radionuclides, only a few are used in the different types of radiation facility (see Table 7.1) [424].

Table 7.1. Radionuclides used as gamma-emitters [425]

Isotope	Half-life	Quantum energy /MeV	Type of irradiation facility
^{60}Co	5.3 y	1.17 (1)† 1.33 (10)	Gamma source installation
^{24}Na	14.9 h	1.37 (1) 2.75 (1)	Gamma loop of fast breeder (draft)
^{137}Cs + ^{137m}Ba	33 ± 2 y	0.661 (1)	Gamma source installation
^{116m}In	54 min	0.137 (0.03) 0.406 (0.25) 1.085 (0.51) 1.274 (0.75) 1.487 (0.21) 2.090 (0.25)	Gamma loop of slow reactor (acting)

† Indicated in parentheses is the number of gamma-quanta of given energy released per decay event.

There are a large number of gamma-irradiation units of different design in use today (Table 7.2). They fall into two large classes: devices with a (i) movable and (ii) fixed radiation source (emitter, or container with a radioactive material), respectively. In the case of a movable emitter, the radiation source, unless in use, is placed in storage. When the object to be irradiated is ready, the source is moved into the operations room. Set-ups of this type make it possible to treat objects of practically any shape and size. If a facility has a fixed radiation source, then the object, after appropriate preliminary procedures, is transferred to the working (storage) chamber where it undergoes irradiation. The dimensions of the object to be irradiated and its configuration are determined by the size and shape of the working chamber, which is often not particularly large (for example in the first four setups listed in Table 7.2 the volumes of the working chamber are 4.4, 1.2, 4.4, and 0.3 dm^3, respectively).

Table 7.2. Gamma source installations for radiation research purposes [425]

Type	Radiation source	Shielding	Irradiator	Maximum dose rate/Gy s^{-1}
RKh-γ-30	^{60}Co	Dry	Immobile	4.4
MRKh-γ-100	"	"	"	5.5
RKhM-γ-20	"	"	"	2.5
LBM-γ-1M	^{137}Cs	"	"	0.55
K-20 000	^{60}Co	"	Movable	11
K-120 000	"	"	"	46
K-200 000	"	"	"	70
K-300 000	"	"	"	40
UKP-100 000	"	Pond	Immobile	50
UK-120 000	"	"	Movable	28
MGU-30 000	"	"	Immobile	60 to 70
GUG-120 000	"	"	Movable	18
UGU-200 000	"	"	"	4.5

As regards protective shielding from the biological action of gamma-radiation, irradiation plants are distinguished as having 'dry' shields (lead, concrete, and cast iron) or 'pond' shields (the radiation source is stored or used under water).

An example of a large-scale radiation research facility is the universal radiation plant K-300,000 [420] with a concrete shield (Fig. 7.8). The top chamber with a working table is designed for irradiation, and the bottom chamber for storage of the isotope source. The set-up has 16 channels where the ^{60}Co compound is placed. The ^{60}Co preparations are transported through the channels by means of a hydraulically operated hoist. Different numbers of samples can be raised to the operation room which enables the mean dose rate to be varied. Fig. 7.9 gives a general view of the working table of this facility so that the channel pipes for transportation of the ^{60}Co preparations are clearly seen.

7.2.2 Electron accelerators. Ion-beam machine sources

Accelerators of charged particles are devices accelerating electrons or ions in an electric field. There are two types of accelerator differing principally in their design: (i) linear accelerators, where the charged particles travel in a rectilinear trajectory, and (ii) cyclic machines where the motion of the particles is circular. In accordance with the type of electric field used, accelerators are divided into high-voltage machines, where the direction of the electric field does not change with time while the particles travel, and resonance devices where the acceleration of the particles is provided by an alternating high-frequency electric field. The principal units of an accelerator are a high voltage resonator, a source of

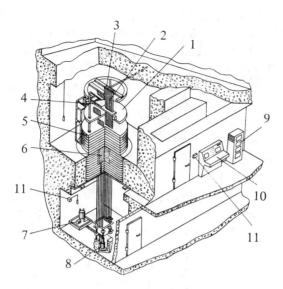

Fig. 7.8. Universal gamma-irradiation facility K-300 000 [420]: 1—working table; 2—emitter channels; 3—collector; 4—radiation-chemical research unit; 5—storage chamber; 6—uptake passage; 7—hydraulically-operated lift; 8—hydraulic cylinder; 9—research unit control; 10—central control panel desk; 11—radiation meters.

charged particles (ion source), and a specially designed chamber where acceleration of the charged particles takes place. The accumulation of energy by a particle in a resonant accelerator takes time depending on the mass and charge of the particles being accelerated; accordingly they work periodically in the so-called pulsed mode. Some types of high-voltage accelerator, such as the van de Graaf machine and the avalanche accelerator, can operate with a continuous flow of charged particles. Most accelerators are used for accelerating both electrons and protons, deuterons, helions, or positive ions of heavier elements.

Continuous acceleration in the resonant machines (cyclic and linear accelerators) is achieved by pushing the charged particle which is always in the accelerating phase of the electric field. A cyclic accelerator (cyclotron, synchrotron, or synchrophasotron) provides the required acceleration energy by multiply-repeated journeys of the particle along the circumference of the device, while the linear-type device gives energy to the accelerated particle through the application of a high-frequency electric field along a linear periodic system of electrodes.

Many designs of accelerator have been made [417, 419, 426]. Radiation chemistry employs electron accelerators with electron energies which do not exceed 10 MeV, thus avoiding induced radioactivity in irradiated samples. Pulse accelerators are used mainly in researches in radiation chemistry (they were the basis for the development of the ns and ps pulse radiolysis techniques indispensable to investigation of transient species [177, 178]. Radiation processing is

Fig. 7.9. Working table of universal gamma source K-300 000.

carried out with high voltage 0.3–1.5 MeV electron accelerators with a power up to 50 kW. The service properties of USSR-made electron accelerators are given in Table 7.3.

An electron accelerator is a very powerful source of energy. The electron beam in an accelerator of conventional design is not more than 3 cm in diameter. With an accelerator power of 10 kW, the 1 MeV electron beam creates a dose of 2.5 MGy s^{-1} in water (the equivalent dose rate of absorbed energy is 2.5 kW g^{-1}). Any substance will simply be vaporized at such a rate of energy input, as, for example, the exit window of the accelerator. To avoid this, a powerful electron beam scans a certain range of sample surface are so that the

Table 7.3. Electron accelerators produced in the USSR [425]

Commercial name	Electron energy/MeV	Mean power /kW	Operating conditions
ELT-1.5	0.7	15	Semiperiodic, 50 Hz
ELIT-0.8	0.8	0.8	Pulsed, $\tau^\dagger = 1$ μs
ELIT-1V	1.1	4.5	Pulsed, τ = 2.5 μs
ELIT-2	1.5	10	Pulsed, τ = 3.3 μs
ELV-1	0.4 to 1.0	20	Semiperiodic, 50 Hz
ELV-2	0.8 to 1.0	20	"
ELV-3	0.5 to 0.7	50	"
ELV-4	1.0 to 1.5	50	"
Elektron-III	0.7	7	"
Elektron-IV	0.5	10	"
Avrora	0.5	20	"
Avrora-II	0.5	25	"
Avrora-III	0.4	30	"
Avrora IV	0.3	25	"
U-10	3	1.0	Pulsed, $\tau = 2.5$ μs, 400 Hz‡
U-12	5	0.6	Pulsed, $\tau = 2.2$–3.0 μs, 400 Hz
U-13	10	0.6	Pulsed, $\tau = 3.0$ μs, 400 Hz
U-16	1.4 to 2.0	1.2	Pulsed, $\tau = 2.5$ μs, 400 Hz
U-27	10	5	Pulsed, $\tau = 3$–5 μs, 430 Hz
U-33	3	11	Pulsed, $\tau = 3$–5 μs, 430 Hz
LUE-8-5	8	5 to 7	Pulsed, $\tau = 2.8$ μs, 31–500 Hz
LUE-8-5V	8	5	Pulsed, $\tau = 2.8$ μs ., 31–500 Hz
LUE-15-10	13	≤10	Pulsed, τ = 5.5, 2.5, and 0.5 μs; 150, 300, and 600 Hz
LUE-215-15V	13	<10	Pulsed, $\tau = 5.5$ μs; 150 Hz
ELU-4	3 to 5	5	Pulsed, $\tau = 4.5$ μs; 200 Hz
ELU-6	4 to 7	5	Pulsed, $\tau = 4.5$ μs; 200 Hz

† Pulse duration
‡ Pulse frequency

equivalent dose rate in the sample decreases to reasonable values and avoids any undesirable heating of the object.

7.3 PHOTOCHEMICAL PROCESSING

Photochemical reactions are used on the industrial scale for syntheses, for materials processing, for recording of information (images), and in printing arts and microlithography as well as in quantum electronics.

In some cases photochemical methods provide relatively simple procedures for the production of compounds which are barely obtainable by other methods (see examples in Chapter 8). Nevertheless, the realization of large-scale photochemical processing technologies in industry is rare because of the large energy cost of light (the cost of 1 mole of photons is *ca.* $1 (US)) and technical difficulties. Much more important are other aspects of photoprocessing which use the ready availability of optical excitation and the precise measures of intensity and radiation dose for different purposes such as the homogeneous initiation of bulk processes (which is very important for, for example, manufacturing of the optically-uniform polymers used in optics) and, on the other hand, the induction of processes in thin surface films or in high-definition drawings. These 'fine' photochemical technologies are used universally in printing arts, in microelectronics for the production of integrated circuits, in quantum electronics for modulators, and for the production of complex optoelectronic circuits.

From the standpoint of the utilization of light energy, photochemical processes used or planned for use in industry are divided into two types, differing in principle: endoergonicand exoergonic processes.

The light energy in endoergonic chemical processes is consumed in the production of molecules possessing energy higher than that of the reactant molecules. The energy of the absorbed photon is partly transformed in these reactions to that of the reaction product, and the maximum quantum yield cannot exceed unity. The energetic yield, that is, the fraction of absorbed energy stored in the reaction products, is usually less than unity because of the conversion of a part of the photon energy into thermal energy in relaxation processes and because of competing side-reactions which decrease the quantum yield of reaction. The issue of the energetic field is especially important when the conversion of solar to chemical energy is concerned. Another example of an endoergonic process is the pumping of photochemical lasers whose efficiency depends in a complicated way on the fraction of energy dissipated in relaxation processes. It should be noted that the efficiency of action of all systems of this type has an optimum at a certain ratio of rates of relaxation processes, providing an irreversibility to chemical reactions of excited species and preventing their recombination to the original ground state.

In exoergonic processes, the light energy is used only in initiating a process or in facilitating the overcoming of an activation barrier. In chain processes, the quantum yield can exceed unity by many times, and can even achieve very high values (10^3–10^9). Most industrially-significant processes are of this type. By using relevant sensitizers, it is possible to induce reactions with light of practically any wavelength, and also IR.

The efficiency of photochemical synthetic processes is described by their chemical yield (the degree of conversion of the most valuable reactant to the desired product), the quantum yield, and the energetic yield of the product. The latter is expressed in moles (or kilograms) of product per kilowatt-hour of light or electric energy consumed:

$$\eta = 0.003 \, \lambda \Phi \beta \alpha$$

where λ is the wavelength in nm, Φ is the quantum yield, β is the efficiency of the light source, and α is the fraction of light absorbed by the reactant.

Light sources

Gas discharge lamps, filament lamps, and sometimes lasers are used as sources of UV, visible, or IR radiation. The principal characteristics of these sources are their emission intensity, spectral composition, and directivity.

Low-pressure discharge lamps produce a spectrum composed of narrow lines. Lamps of this type are used as sources of monochromatic radiation, with selection of a required line being made by means of special glass, liquid, or gaseous light filters. The use of interference filters in photochemical procedures is undesirable since they transmit to a considerable extent (up to few per cent) over a very wide spectral region.

Depending on the composition of the gas or vapour, different spectra can be obtained. Spectral lines are strongly broadened with increasing vapour (gas) pressure. The spectra of lamps operating at pressures above 10 kPa is a continuum with broad peaks (Fig. 7.10). The most widely used sources are mercury and xenon lamps. Superhigh-pressure (about 3 MPa) xenon lamps produce an almost continuous spectrum beginning from 200 nm with several maxima in the visible and UV-regions.

All discharge lamps need a special power supply unit to limit the current (a choke if a.c. power is used). The high-pressure and superhigh-pressure lamps are compact and have a high intensity, but their lifetime is short (a few hundred hours), and they require cooling and also a special protective cap in case of possible explosion.

Filament lamps are more convenient for operation but they a low efficiency; their spectrum has a maximum in the IR region (900–1000 nm). At present, halogen tungsten filament lamps with quartz bulbs are most frequently used, where the halogen additives provide regeneration of the eroding tungsten filament, enabling an increase in its temperature and thus enabling the emission in the UV region to be enhanced.

The properties of some discharge and filament lamps are reported in Table 7.4. All these lamps give nondirectional divergent light. To provide the more complete use of light energy for synthetic purposes, submerged lamps are used (the lamp is surrounded by the reaction mixture). As the working temperature for most of the lamp is high, a water-cooled jacket is placed between the lamp and the reaction mixture.

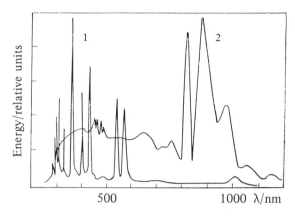

Fig. 7.10. Emission spectrum of high-pressure mercury lamp (1) and high-pressure xenon lamp (2).

Table 7.4. Approximate photon flux of some radiation sources

Light source	Electric power	Wavelength/nm	Photon flux/E h^{-1}
Tungsten filament lamp	500 W	300–400	0.01
		400–500	0.1
		500–600	0.25
		600–700	0.5
		700–800	0.7
		800–1000	2
Tungsten–halogen filament lamp	500 W	300–400	0.05
		400–500	0.2
		500–600	0.5
		600–700	0.7
		700–800	0.9
		800–1000	2
Xenon lamp	500 W	300–400	0.5
		400–500	0.8
		500–600	0.8
		600–700	1.0
		700–800	0.9
		800–1000	2.5

Table 7.4. (continued)

Light source	Electric power	Wavelength/nm	Photon flux/E h^{-1}
Mercury lamps			
Heraeus TNN 150/150	150 W	254	0.35
		313	0.01
Heraeus TQ-150	150 W	240–270	0.06
		302	0.02
		313	0.04
		334	0.005
		366	0.07
		405–408	0.04
		436	0.06
		546	0.08
		577–579	0.08
Heraeus TQ 40055.200	40 kW	240–270	18†
(industrial)		302	5†
		313	7
		334	1
		366	20
		405–408	10
		436	18
		546	30
		577–579	33
He–Ne laser,§ (100 mW)		633	0.002
N$_2$ laser (1 ns, 2 MW peak power, 100 Hz)		337	0.002 0.005‡
(10 ns, 1 MW)		337	0.03‡
Ar–ion laser,§ (40 W)		458–514	0.6
XeF exciplex laser (15 ns, 5 MW, 100 Hz)		351	0.1 0.2‡
KrF exciplex laser (15 ns, 15 MW, 100 Hz)		249	0.2 0.5‡
Ruby laser,§ (1 mW)		694	0.02
(1 ms, 0.5 MW, 0.02 Hz)		694	0.2 2000‡
(20 ns, 1000 MW, 0.001 Hz)		694	0.0004 120‡
(2 ps, 5000 MW)		694	0.06‡

Table 7.4. (continued)

Light source	Electric power	Wavelength/nm	Photon flux/E h^{-1}
Nd–YAG laser,§ (20 W)		1060	0.7
(10 ns, 50 MW, 1 Hz)		1060	0.02
			50‡
Nd glass laser (50 ns, 200 MW)		1060	100‡
Dye laser,§ (1 W)		200–1100	0.004–0.04
(10 ns, 1 MW)		200–1100	0.01–0.1‡
(5 ps, 1000 MW)		200–1100	0.01–0.05‡
CO_2, laser,§ (200 W)		10600	11

† Lamp without jacket.
‡ Photon dose in microeinstein per pulse.
§ Continuous wave laser (light output power).

7.4 ERGOMETRY

7.4.1 Units used in ergometry

A chemical or physical effect produced in a system under different types of radiation is some function of the amount of energy introduced into the system, of the nature of this energy, and of the rate of energy input. These three characteristics need to be determined independently, thus an auxiliary branch of high-energy chemistry has arisen, called ergometry. The task of ergometry is the development of techniques for the measurement of the energy introduced (absorbed) and the rate of energy input (absorbed dose rate) as well as the determination of its spatial distribution in the system. This field of science has been developed independently in each of the branches of HEC. It is known as dosimetry in radiation chemistry and actinometry in photochemistry.

In accordance with the Système International d'Unites (SI), high-energy chemistry uses the following basic quantities to characterize the incident energy W expressed in joules or electron volts (1 J = 6.24×10^{18} eV) (Table 7.5).

(1) Energy flow P equal to

$$P = \Delta W / \Delta t; \quad [P] = 1 \text{ J s}^{-1} = 1 \text{ W}; \quad \text{dimensions of } P = L^2 M T^{-3}.$$

(2) Energy density (radiation intensity) Ψ as an energy flow per unit area

$$\Psi = \Delta P / \Delta S; \quad [\Psi] = 1 \text{ W m}^{-2}; \quad \text{dimensions of } \Psi = m T^{-3}.$$

If energy is transferred not by a field but by particles (electrons, phonons, photons, etc.) then the following quantities are introduced.

(3) Flux of particles Φ as the number of particles passing through some surface within a period Δt:

$$\Phi = \Delta N / \Delta t; \quad [\Phi] = 1\ s^{-1}; \quad \text{dimensions of } \Phi = T^{-1}.$$

(4) Flux density of particles J as the flux per unit area;

$$J = \Delta \Phi / \Delta S; \quad [J] = 1\ s^{-1}\ m^{-2}; \quad \text{dimensions of } J = T^{-1}L^{-2}.$$

For energy absorbed in a medium, two characteristic quantities are used: absorbed energy (dose) D as the amount of energy absorbed by unit mass, unit volume, or unit area, and absorbed dose rate (dose rate) $\dot{D}$ as the energy absorbed per unit time. Different units are used for these quantities in different branches of HEC.

7.4.2 Dosimetry in radiation chemistry

Numerous publications dealing with the dosimetry of ionizing radiations are available [5, 6, 427–430], so we shall consider only the basic units [431].

(1) Radiation dose (absorbed dose, or just dose) D is the amount of energy transferred to matter referred to unit mass:

$$D = \Delta W / \Delta t; \quad [D] = 1\ J\ kg^{-1} = 1\ Gy\ (\text{gray});$$

$$\text{dimensions of } D = L^2 T^{-2}.$$

(2) Dose rate of radiation (absorbed radiation dose rate, or just dose rate) $\dot{D}$:

$$\dot{D} = \Delta D / \Delta t; \quad [\dot{D}] = 1\ Gy\ s^{-1};$$

$$\text{dimensions of } \dot{D} = L^2 T^{-3}.$$

In addition to the gray, an off-system unit eV kg^{-1} is used. Other off-system units eV g^{-1}, rad, and kW h ton^{-1} used previously are equal to 1.6×10^{-16}, 0.01, and 3.6×10^3 Gy, respectively. For electromagnetic radiation the following quantities are used besides the ones mentioned above.

(3) Exposure dose of photon radiation (exposure dose of γ- or X-rays) X as the sum of the electric charges Q of all ions of either charge created by electrons released in irradiated air per unit mass of air provided that the ionizing power of the secondary electrons is realized completely:

$$X = \Delta Q / \Delta m; \quad [X] = C\ kg^{-1}; \quad \text{dimensions of } X = M^{-1}TI.$$

Earlier, the roentgen (R) was used as a unit of this quantity, which corresponds to formation of 2.08 ion pairs in 1 cm^3 of dry air at 0°C and 0.1 MPa. For 3 MeV-photons it corresponds to 8.69×10^{-3} Gy.

(4) Exposure dose rate of photon radiation $\dot{X}$:

$$\dot{X} = \Delta X / \Delta t; \quad [\dot{X}] = 1\ C\ kg^{-1}\ s^{-1} = 1\ A\ kg^{-1};$$

$$\text{dimensions of } \dot{X} = M^{-1}T.$$

Table 7.5. Conversion factors for energy units

Unit	Mean energy of particle/eV	Energy per unit quantity of matter	
		/kJ kmol^{-1}	/kcal mol^{-1}
1 eV	1	9.65×10^4	23.05
1 kJ kmol^{-1}	1.04×10^{-5}	1	23.9×10^{-5}
1 kcal mol^{-1}	4.34×10^{-2}	4.19×10^3	1
1 K (1)	1.29×10^{-4}	12.48	2.98×10^{-3}
1 K (2)	0.862×10^{-4}	8.32	1.99×10^{-3}
1 cm^{-1}	1.24×10^{-4}	12.0	2.86×10^{-3}
1 nm	0.807×10^{-3}	77.87	0.0186
1 Hz	0.414×10^{-14}	3.99×10^{-10}	9.53×10^{-14}

Notes: (1) 1 eV corresponds to an absolute temperature of 7733 K if the average kinetic energy of the molecules or atoms (in translational motion $mv^2/2 = 3kT/2$).
(2) 1 eV corresponds to a temperature of 11 600 K if considering the energy of an electronic transition from one state to another and the value for this energy is placed in the Boltzmann exponential factor $\exp(A/kT)$, that is, on the basis of the equation $mv^2 = kT$.

There are only two cases when the dose in a sample can be calculated *a priori*: upon complete absorption of a known flux of ions or electrons with known energy by the system, and if using α- or β-isotopes of known composition and activity. In all other cases one is obliged to determine the dose by some effect caused by the action of radiation on the system treated, which may be heating, induced conductance, coloration, luminescence, chemical transformation, etc.

A physical or chemical system employed for dose determinations is conventionally called *a dosimeter* (physical or chemical depending on the type of effect developed). A direct technique for determination of the amount of absorbed energy is calorimetry provided that there are no chemical processes occurring in the working body of the calorimeter (metals) or chemical equilibrium has been established (some solutions). Calorimetric dosimetry, together with the other two above methods of direct calculation of dose, is regarded as an absolute technique of dosimetry. Other methods are relative since for each of them it is necessary to find an energy conversion factor for the effect measured.

By using a dosimeter at some point of a system or in place of the system as a whole, the absorbed dose D_d is determined. This dose differs from the dose in the system under investigation D_s because of the difference in chemical composition of the dosimetric system: $D_s = kD_d$, where k is a coefficient depending on the type of radiation, its energy, and the chemical composition. For example, with electromagnetic radiation of arbitrary energy, the conversion factor for a point source is calculated from the equation:

Mean energy of particle in ensemble		Wave number/cm^{-1}	Wavelength /nm	Frequency of radiation quantum/Hz
/K (1)	/K (2)			
7733	11600	8066	1240	2.42×10^{14}
80.1×10^{-3}	120.2×10^{-3}	83.6×10^{-3}	0.1285	2.51×10^{9}
335.4	503.4	349.9	53.8	1.05×10^{13}
1	1.50	1.043	0.160	3.13×10^{10}
0.667	1	0.695	0.107	2.08×10^{10}
0.959	1.44	1	0.154	3.00×10^{10}
6.24	9.36	6.508	1	1.95×10^{11}
3.20×10^{-11}	4.80×10^{-11}	3.34×10^{-11}	0.513×10^{-11}	1

$$k = \frac{1 - \exp[-(\mu/\rho)_s \rho_s \, x]}{1 - \exp[-(\mu/\rho)_d \rho_d x]}$$

where μ/ρ are the mass absorption coefficients for electromagnetic radiation (see section 1.1.3), ρ_s and ρ_d are the densities of the system under study and of the dosimetric system respectively, and x is the thickness of the sample (the equation is valid if the sample thickness does not exceed one half-layer of extinction).

At moderate photon energies (0.5–3 MeV) where Compton absorption is predominant, the simpler equation $k = n_s/n_d$ (where n_s and n_d are the electron densities of the system under study and of the dosimeter system respectively) may be used. With ion or electron beams which have a range larger than the thickness of the irradiated layer, the conversion factor is determined by the ratio of the stopping powers S of the two systems:

$$k = S_s/S_d ,$$

but k is equal to unity if the range is less than the thickness.

Parallel to determination of the dose it is also important to know its spatial distribution. This can be obtained for long objects by dosimetry at separate points with subsequent interpolation of the dose field throughout the volume.

The physical methods of dosimetry are rather diverse, including calorimetric, ionization, luminescence (scintillation), radiothermoluminescence, chemiluminescence, optical, and activation techniques. Calorimetric methods are usually employed for calibration of other dosimeters as well as for measurements of energy in mixed radiation. Ionization methods are used for radiation detection and determination of the dose field in long specimens. Different types of optical and luminescence method are used for determination of the dose at separate points and for survey of a field of dose, and activation techniques are used for dose measurements in neutron fluxes.

Chemical dosemeters are those most used in radiation chemistry. Of these, the Fricke dosimeter serves as a secondary standard when finding the radiation yield in any other system since the radiation yield of conversion of Fe^{ii} to Fe^{iii} ions in the dosimetric system has been determined very carefully for all types of radiation.

Chemical dosimeters have been the subject of exhaustive studies (see, for example, [34]), and some of them are described in Table 7.6. The systems can be gaseous, liquid, or solid. Gaseous dosimeters give the average value of the dose over the entire irradiated volume. They are convenient when the dose in set-ups with a sophisticated configuration needs to be known. Solid-phase dosimeters are useful for dose measurements at separate points. Film dosimeters can be applied in dose-field measurements. A variety of chemical dosimetric systems make it possible to choose a dosimeter closest to the system of interest in chemical composition as well as to operate the dosimeter in the same physical state as the subject of study. Such circumstances reduce the error in dose determination on transferring from the dosimeter to the real system. One type of chemical dosimeter is photographic film, which is convenient for radiation detection and for dose-field evaluation, but it produces a large error.

Normally, three dose ranges can be distinguished which are examined with individual dosimetric systems:

(i)   the biological regime (< 10 Gy),
(ii)   the research regime ($10–10^4$ Gy),
(iii) the industrial regime ($> 10^4$ Gy).

7.4.3 Actinometry

Chemical actinometers, photoelectric cells, and thermocouples are used for measurements of light intensity. In photochemistry it is necessary in most cases to express the intensity in quantum units, as the number of photons per unit time, but not in energetic or photometric units. The off-system unit, the einstein (E), equals one mole of photons (6.02×10^{23} photons), and is frequently used. The number of photons is most conveniently measured by means of chemical actinometry which enables the intensity of the light to be determined by using some standard chemical reaction with a constant quantum yield independent of or weakly dependent on wavelength and temperature. Although chemical actinometry is rather time-consuming (the measurement of one dose takes about

Table 7.6. Chemical dosimetry systems for ^{60}Co γ-radiation [6]

System	Detected effect or quantity	G	Dose range/Gy
Methyl violet leuco-base solution in methyl ethyl ketone	Colour change	2.0	0.15–15
Doped lithium fluoride	Radiothermo-luminescence	—	$0.01–10^3$
Aqueous solutions or emulsions of chlorohydro-carbons	Decrease in pH	—	0.02–5000
Iron(II) sulphate (Fricke) dosimeter	Fe^{3+} ions	15.6	$10–10^3$
Cerium(IV) sulphate dosimeter	Ce^{4+} disappear-ance	2.32	$10^2–10^6$
Cyclohexane	H_2 formation	5.0	2×10^4
Polystyrene solution in CCl_4	Change in viscosity	—	10^6
Aqueous solutions of carbo-hydrates (glucose, etc.)	Change in angle of polarization	—	$10^4–4 \times 10^6$
3M Aqueous methanol solution	Ethylene glycol formation	3.2	$100–10^6$
Methanol	Ethylene glycol formation	3.0	$100–10^6$
Polyisobutylene solution in heptane or CCl_4	Change in viscosity	—	$10–10^8$
Cellulose triacetate film	Colouring	—	$5000–10^6$
20–40% Chlorobenzene solution in ethanol	Change in pH	5.9	$100–2 \times 10^5$
Gaseous methane	H_2 formation	5.5–6.9	5×10^7
Gaseous N_2O	N_2 formation	9–12	$1000–3 \times 10^7$
Air	NO_2 formation	1.4–1.45	1.5×10^7

Note: In those cases when the yield is not determined, the system must be calibrated with a standard.

half an hour to a few hours), calibration by standard light sources is unnecessary, and it is possible to carry out the measurements in the same vessel to be used for the chemical reaction under study. However, chemical actinometry is an integral method and does not provide the possibility of monitoring the intensity of measured light. It is reasonable to use the chemical actinometry in cases when measurements of light intensity are to be carried out rather infrequently. Otherwise, the time consumed for calibration required by other methods is repaid only with many determinations.

The sensitivity of photocells is strongly dependent on wavelength, so their application requires special calibration for different wavelengths and is admissible in photochemistry only if monochromatic light is used. Thermocouples are not very sensitive, and they measure the intensity of light in energy units. Because of this, the determination of the number of photons appears to be possible only for monochromatic light and requires the use of a wavelength-dependent conversion factor (1 E = $1.196 \times 10^5/\lambda$ kJ, where λ is the wavelength of radiation, in nm).

In recent years, so-called luminescent transformers have been adopted which transform any wavelength of light to a certain standard spectrum. This gives the possibility of using a photocell or photomultiplier in combination with a luminescent transformer to measure the number of photons in a nonmonochromatic light beam. Different luminophores in the form of a solution or powder, as well as luminescent glasses can be used as luminescent transformers. The following features are important: that complete absorption of the incident light occurs, that there is constancy of the quantum yields of luminescence over a wide range of excitation wavelengths, and the stability of the luminophore is guaranteed. Most popular is a solution of rhodamine C in ethylene glycol (8 mg dm^{-3}) which can be used over a quite large wavelength range (250–600 nm). To obtain absolute values for light intensities using photocells, calibration with a light source of known intensity or with a chemical actinometer is needed.

Chemical actinometry is based on determination of the amount of substance formed or destroyed in a standard photochemical reaction. The rate ω of formation or consumption of the substance is proportional to the rate P_a of photon absorption ($\omega = \Phi P_a S/V$) and the change in molar concentration $\Delta[A]$ of this substance over a period of time Δt is proportional to the dose $Q = P_a \Delta t$ of the absorbed radiation

$$\Delta[A] = \Phi P_a \Delta t\, S/V = \Phi Q_a S/V,$$

where V is the total volume of the illuminated gas or solution, S is the cross-section of the light beam measured and P_a is the rate of light absorption, in $E\, m^{-2}\, s^{-1}$).

To avoid solving the problem on specifying the ratio of absorbed to transmitted light (which is strongly wavelength-dependent) it has become common practice to make the concentration and thickness of the solution (or gas pressure) sufficiently large to ensure the complete absorption of light. Various chemical and physical techniques are used for determination of the amount of substance formed or decomposed in a photochemical reaction, the most suitable being spectral methods. Some systems utilized as chemical actinometers are shown in Table 7.7.

Table 7.7. Chemical actinometers

Photochemical reaction	Wavelength range/nm	Quantum yield
Ferrioxalate $[Fe(C_2O_4)_3]^{3-} + h\nu \rightarrow Fe^{2+} + \frac{5}{2}C_2O_4^{2-} + CO_2$	250–480	0.9–1.25
Reinecke's salt $[Cr(NH_3)_2(NCS)_4]^- + H_2O + h\nu \rightarrow$ $[Cr(NH_3)_2(NCS)_3(H_2O)] + NCS^-$	315–600	0.32–0.27
Malachite green leuco-cyanide $[(CH_3)_2NC_6H_4]_2C(CN)Ph + h\nu \rightarrow$ $[(CH_3)_2NC_6H_4]_2CPh^+ + CN^-$	250–340	1.0
Uranyl oxalate $(UO_2)C_2O_4 + 4H^+ + h\nu \rightarrow U^{4+} + CO_2 + 2H_2O,$ by-products CO and HCOOH	200–410	0.61–0.49
$Cl_2 + H_2 + h\nu \rightarrow 2HCl$	280–380	15–16
$2HBr + h\nu \rightarrow H_2 + Br_2$	180–250	1.0
$CH_3COCH_3 + h\nu \rightarrow C_2H_6 + CO$	250–320	1.0
$2NOCl + h\nu \rightarrow 2NO + Cl_2$	365–635	2.0

One of the most extensively used systems is the tris(oxalato)ferrate actimometer which absorbs radiation in the UV and blue regions of the spectrum and has a high quantum yield. The scheme of reactions occurring in the actinometric system is rather complicated and involves radical stages. The quantum yield changes somewhat with wavelength, which requires corresponding corrections to be made. The concentration of Fe^{2+} ions formed is measured spectrophotometrically with using 1,10-phenanthroline which gives a deeply coloured complex with iron(II).

To measure the intensity of visible light (450–600 nm) it is convenient to use an aqueous solution of Reinecke's salt $K[Cr(NH_3)_2(NCS)_4]$ where ligand exchange of thiocyanate by water takes place under the action of light. The concentration of thiocyanate ions thus formed is measured spectro-photometrically utilizing their reaction with ferric ions. The rather low extinction coefficient of Reinecke's salt of 30 to 100 M^{-1} cm^{-1} in the 390–590 nm wavelength region makes it necessary to employ a relatively high (*ca.* 0.1 M) concentration of the salt to ensure the complete absorption of light.

8

Some applications of high-energy chemistry

8.1 PLASMOCHEMICAL PROCESSES

Applied plasma chemistry covers a wide range of processes of significance to various industries including chemical, metallurgic, electronic, radioengineering, and electrical engineering. Moreover, the areas of application of plasma processes has continuously expanded in recent years, in particular, to the surface treatment of solids and to chemical and instrumental analysis. Many technologies of plasma processing are already used in industry, and others are being tested for economic viability. Consideration of the results of thermodynamic and kinetic calculations and experimental studies on plasmochemical processes show that low-temperature plasma processes have real potential for use on the industrial scale in those cases where high product yields are obtained under essentially nonequilibrium conditions. In nonequilibrium plasmas, unique compounds are formed, high-purity materials such as semiconductor materials are produced, the equilibrium is displaced to the high-temperature side, reaction rates increase dramatically with increasing temperature which enables the significant miniaturization of equipment, the number of stages in the production line is reduced, and widely available, cheap raw materials can be used. It is virtually impossible to consider all examples of plasma processes published in the scientific and patent literature, and we shall simply illustrate the ideas discussed in the preceding chapters by referring to one plasmochemical process for each of the different applications.

8.1.1 Synthesis of organic compounds
Many aspects of the industrial plasma pyrolysis and oxidation of hydrocarbons, and of the selective synthesis of valuable compounds, have already been elaborated. Let us consider the plasmochemical manufacture of a reaction mixture for the production of vinyl chloride $CH_2 = CHCl$.

From analysis of thermodynamic calculations on the C–H–Cl system, it follows that the composition of the mixture $C_2H_2:HCl = 1$ (*ca* 20% vol.) required for the production of vinyl chloride can be prepared at 1500–1800 K and a C:Cl ratio of ≥ 2. Under these conditions the acetylene yield (in terms of carbon) must reach 80–90% by mass, and the power consumption should be relatively low (nor more than 2 kW h kg^{-1} mixture). These conclusions have been confirmed using pilot plants in the USSR and elsewhere.

In kinetic calculations on the degradation of chlorohydrocarbons it is assumed that the reactants are decomposed to HCl, CH_4, C_2H_4, and H_2 in a relatively short time (*ca.* 10^{-6} s), with respect to the period taken by a chemical reaction. If they also contain aromatic groups, C_2H_2 is also formed. Then these hydrocarbons are decomposed in accordance with the scheme suggested by Kassel for methane degradation. Values for the initial ratio $CH_4:C_2H_4:C_2H_2$ are obtained on the assumption that the C–C and C–H bonds in the chlorohydrocarbon are ruptured at different sites. In the calculations (a system of equations of chemical kinetics and jet hydrodynamics was solved) the quenching of the products at a rate of 5×10^6 K s^{-1} was also taken into consideration, which was 'switched on' at different distances from the reactor. The results of calculations fit the experimental data rather well.

Experimentally, the mixture with equal concentrations of C_2H_2 and HCl was obtained by the pyrolysis of petrol (end b.p. 165°C) in chlorine-containing plasma jets or by pyrolysis of chlorohydrocarbon mixtures in a H_2 jet. In the first case a mixture composed of 17% to 20% of both C_2H_2 and HCl was formed at 1550–1700 K in 50 to 150 µs. In the second case the mixture was produced within *ca.* 300 µs. These processes when incorporated into the production line of vinyl chloride simplify the procedure and significantly decrease (up to 40%) the cost price of the product [200, 432–434].

The behaviour of hydrocarbons in a nonequilibrium plasma was extremely interesting. Isomerization, elimination and polymerization reactions yielding hydrocarbons with five- and six-membered rings are characterized by their high conversion and selectivity; these reactions are used in preparative chemistry.

8.1.2 Synthesis of inorganic compounds

In the field of plasmochemistry of inorganic compounds, the plasma oxidation and reduction of various substances, ores, and minerals, their decomposition, production of high-melting compounds (nitrides, carbides and intermetallides), as well as exotic reactions such as the formation of noble gas compounds, have been studied.

Consider the production of titanium nitride by injection of $H_2 + TiCl_4$ into the SHF nitrogen plasma. According to calculations on the equilibrium composition of the heterogeneous Ti–N–H–Cl system, the maximum concentrations of titanium nitride are obtained at 1000–1700 K. Its yield depends on the dilution of $TiCl_4$ with nitrogen and hydrogen and it increases from 40 to 100% with increasing N/C and N/Ti ratios up to 5 and from 1 to 100, respectively. Experimentally, 100% conversion was observed in 1–10 ms and at a rate of quenching of 10^4 K s^{-1}; a powder of composition $TiN_{0.8}$ was obtained with properties wholly different from those of titanium nitride prepared by conventional means. The power consumption was 18 kW h kg^{-1} which is close to the thermodynamically predicted value. If a mixture of $TiCl_4 + BCl_3 + H_2$ is injected into a nitrogen plasma, then powders composed of TiN, TiN + TiB_2, or TiN + TiB_2 + B are formed, depending on the mass ratio of the reactants.

8.1.3 Manufacture of powders

The residence of crystalline particles in a plasma results in melting and changes in the microstructure of the particles to form 0.1–0.5 µm aggregates of the highest mechanical strength due to heating, phase transitions, and accompanying deformation phenomena. The remelted particles become rounded, and pores and microspheres develop inside them because of boiling. The material becomes refined: impurities with high vapour pressures volatilize (for example the content of silicon and zinc in a tungsten powder decreases several times, and that of manganese and lead by an order of magnitude). Its chemical composition may also change, thus the treatment of tungsten carbide WC decreases the carbon level in the product and new substances W_2C and W are formed. Depending on the composition of the plasma gas, oxides and nitrides are formed on the surface of the particle. A sublimed or vapour-phase-deposited material is condensed to particles with dimensions of a few tens of nm, which are near-critical values. The point is that with such small particles, the fraction of surface atoms increases considerably, which changes their conditions of thermodynamic equilibrium. The structure and certain properties of such finely-dispersed powders differ markedly from those of the bulk material.

Indeed, the phase composition itself depends on the particle size. Thus, if the particles are larger than 100 nm, regions with different phases are observed. Particles of 10–100 nm in diameter have a crystal structure either metastable or wholly irrelevant to the bulk material. The surface layer of ultradisperse powders is enriched with high-temperature modifications. As an example, let us consider the properties of finely-dispersed titanium nitride, the production of which has already been described. The product is represented by cubic single crystals of titanium nitride with a mean particle size of 50 nm. X-ray studies have shown that there are statistical distortions of the TiN lattice. Such distortions must cause a change in the thermo-e.m.f., which is a structurally-sensitive property. The thermo-e.m.f. of finely-dispersed TiN at 300 K is 10 µV K^{-1}, which is twice that of sintered TiN at 1600 K. Vacuum annealing at 1000 K results in an increase to 25 µV K^{-1}. However, further increase in the annealing temperature causes sintering of the particles, which leads to a decrease in the thermo-e.m.f. down to the value typical of monolithic specimens. The thermo-e.m.f. of vanadium nitride with a mean particle size of 50 nm prepared in a SHF plasma changed in the same manner on annealing [437].

The phenomenon of superconductivity is also affected by the particle dimensions: it disappears if the particles are 2.5–10 nm in size. The superconductivity of plasma-processed titanium nitride powders was found to depend on the compacting pressure, namely an increase in pressure increased the fraction of superconductive volume in the sample, increased the transition temperature, and decreased the temperature range of the transition. This temperature is 5.2 K for a bulk sample under a pressure of 2 GPa.

8.1.4 Preparation of films

There are a wide variety of industrial applications of plasma-deposited thin films, mainly in the production of microelectronic devices. Different dielectric films are used in manufacturing thin-film hybrid circuit elements, for encapsulation of integrated circuits, and as diffusion masks in manufacturing transistors. In addition, these films are used as protective corrosion-resistant coatings and for preparation of semipermeable membranes.

Plasma-deposited films can be crystalline or amorphous. Their thickness ranges from fractions of a micrometre to a hundred micrometres. The rate of film growth should be such that the separate stages of the process can be controlled, that is, desired layers can be created and dopants determining the electrical properties can be introduced. Glow-discharge and low-pressure (13–1300 Pa) HF discharge plasma reactors are used for film deposition. The growth of a film on the cathode is observed in a glow-discharge film-deposition reactor (see Fig. 7.4) on excitation of the plasma by direct current at a current density of a few mA cm^{-2}. However, if the discharge is excited by an a.c. field with a frequency of 50 Hz–2 MHz, the film is formed on both electrodes.

The set-up for film deposition in a HF discharge plasma shown in Fig. 8.1 provides the possibility of exciting the plasma in one of the reactants injected into a chamber, which causes decomposition of the other reactant and deposition of a film.

Organic polymer films are formed from organic compounds of different types, saturated, unsaturated, and aromatic hydrocarbons and organometallic compounds, as well as from a mixture of CO, H_2, and N_2. Their composition is varied by changing the composition of the gas phase. It is possible to obtain films with their composition varying by depth. Dilution of the monomer with inert gases changes the properties of the plasma and, correspondingly, the structure of the film. Thus, a film obtained in the presence of argon has a higher degree of crosslinking than that made without argon.

Plasma-deposited polymer films are characterized by the following properties: they are amorphous, nonporous, and crosslinked, and have considerable thermal stability, a high melting point, and low solubility; they are rapidly oxidized in ambient atmosphere. Such films are formed under low pressures and at

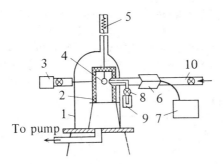

Fig. 8.1. Apparatus for glow-discharge film deposition. 1—chamber cover; 2—
furnace; 3—vacuum gauge; 4—support; 5—quartz balance; 6—HF power input
(waveguide) ; 7—radiofrequency generator; 8—reactant feed; 9—thermostat; 10—
oxygen feed.

high current densities. Polymers prepared at high pressures and low current
densities have a lower molecular mass, they are softer and more soluble. The
thickness of plasma-deposited polymer films does not usually exceed 2 μm.
Such a polymer contains many unsaturated groups.

Polymer films are characterized by low conductivity, small dielectric losses,
and high breakdown voltage. In high-intensity electric fields, films with a thickness of 5–250 nm have a nonlinear conductivity. The resistance of films *in
vacuo* does not change, but increases in air. The photoconductivity of the films
depends on the polymer composition.

The mechanism of plasma polymerization is rather complicated and still far
from being clear.

Inorganic films are dense, amorphous materials having significant hardness.
The following films have been prepared experimentally: of silicon by deposition
onto a cold surface in a silane vapour plasma, of silicon oxides in a silane +
oxygen plasma or by decomposition of tetraethoxysilane in an oxygen plasma,
of silicon nitride from a plasma of a silane–ammonia mixture or mixture of
$SiCl_4$, NH_3, N_2, and N_2O taken at different ratios, of silicon carbide in a silane-
ethylene plasma, of titanium dioxide from a mixture of $TiCl_4$ and CO_2, and of
metals from plasmas containing the corresponding metal carbonyls.

The electrical properties of these films vary rather widely. Taking silicon
nitride films as an example, their permittivity lies between 4 and 11, depending
on the SiH_4 concentration, their dielectric strength between 10^6 and 10^7 V cm^{-1},
tan $\delta \approx (6-7) \times 10^{-4}$, specific resistance 5×10^{-14} ohm cm, and refractive index
1.95–2.1. Their IR absorption falls in the 1.4–12.2 μm region of the spectrum.
These films are suitable as passivating coatings or protective masks; they can
also be used in metal-nitride-oxide-semiconductor structures.

The preparation of diamond-like films by decomposition of acetylene in a
13.5 MHz plasma (13–1300 Pa and a voltage up to 100 V) on a heated support

(to 620–800°C) of pure silicon or of silicon coated with oxide film has been reported recently.

Oxide films are also formed in an oxygen plasma on electrode surfaces or on the surface of metals placed in the plasma.

8.1.5 Preparation of semipermeable membranes

Membranes which are used in the separation of solutions of salts, organic compounds, and gaseous mixtures, are formed by plasma deposition of thin polymer films on porous supports. Compared with conventional methods, the plasmochemical preparation of such membranes has some advantages since it is possible to use hydrocarbons of different types and to choose any porous base of arbitrary shape; moreover, crosslinked polymer materials are formed which have the maximum possible density and do not change their dimensions during their service life.

The film deposition is carried out under reduced pressure in a prescribed and controlled atmosphere; as a result, pure films without traces of moisture are obtained.

Membrane films are prepared by two methods: by polymerization of hydrocarbons or by destruction of polymers. More frequently, pyridine derivatives have been used as the monomer and millipore filters with a pore size of 25 nm as the porous support. The process is carried out in an electrodeless HF or SHF plasma at a pressure of 15 to 2300 Pa and a power of 30–150 W. After 60–3000 s of plasma treatment of vinylpyridines, a several-hundred micrometre thick film is formed on the filter. The filtration efficiency of such membranes can be as high as 95%. Their selectivity depends on the composition of polymer film and on the service temperature of the membrane.

8.1.6 Surface treatment of metals and alloys

The treatment of metal surfaces is usually carried out in gas-discharge devices such as that shown in Fig. 7.4. The article to be treated is placed in a vacuum chamber containing a gaseous mixture of prescribed composition under a pressure of 15–1500 Pa. A voltage (from some hundred to *ca.* 1500 V) is applied to the object serving as the cathode and to the chamber walls acting as the anode. The bombardment of the cathode surface with ions, and the ion-recombination on the surface, results in heating of the article to 350–700°C. The required temperature for plasma treatment, depending on the structure and composition of the material, is established by attenuation of the power input. The time of treatment is varied from 10 min to 20 h, depending on the material and the depth of the modified layer required.

Another type of apparatus for film deposition is shown in Fig. 8.2. The power of the commercial device can be as large as 300 kW (1000 V, 300 A). Depending on the shapes of the objects to be treated, vacuum chambers of different

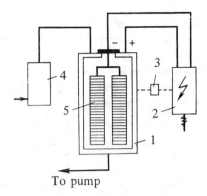

Fig. 8.2. Schematic diagram of apparatus for surface modification of metals in a glow discharge: 1—vacuum chamber; 2—power supply unit; 3—control unit; 4—gas feed system; 5—treated article.

configuration and size are used. These chambers can reach 2 m in diameter and 12 m in length.

The glow-discharge nitridation (ion nitridation) technique has been studied in most detail and implemented in particular countries (Germany, USA) to treat objects as varied as balls for ball-point pens and rolls for rolling mills.

A section of a metallographic specimen of plasma-treated iron has three distinct zones; the surface zone (*ca.* 30 μm in depth), the diffusion zone (300–3000 μm), and that of the material itself. The surface zone consists mainly of γ-Fe$_4$N and ε-FeN nitrides without clearly-displayed grain boundaries. It can be non-porous and possess high hardness ($H_v \approx 1000$–1500) and be corrosion-resistant. The diffusion zone provides a smooth transition from the surface zone to the bulk of material. In this zone carbon nitride strips are frequently formed along residual austenite grain boundaries as a result of decarbonization of the grains and the grain boundary reactions of iron with carbon and nitrogen diffusing from the surface zone. The wear resistance and hardness of the material of the diffusion zone are higher than those of the bulk material and lower than those of the surface layer.

Ion nitridation occurs 2–2.5 times faster than gaseous nitridation. However, the type of dependence of the layer depth on the time of treatment is similar, which is, perhaps, some indication of the same mechanism of formation of the layer. Gas-phase nitridation proceeds as a result of the diffusion of atomic nitrogen in the metal. However, the increase in the diffusion rate of atomic nitrogen during glow-discharge nitridation is probably related to degradation of the oxide films on the metal surface as well as to the formation of large numbers of dislocations in the material due to ion bombardment and plasma radiation.

The hardness of the plasma-treated surface is almost unaffected by changes in pressure from 130 to 1300 Pa, in current density from 0.5 to 2.0 mA, and by the form of the discharge current. This position makes it possible to study the process with laboratory apparatus under the conditions of a normal glow discharge and to design commerical devices operating under pulse conditions in the region of abnormal glow discharge.

Data on the effect of the $N_2:H_2$ ratio (ranging from 1:9 to 9:1) in the plasma on the performance of the surface treatment are contradictory. As the temperature of the material being treated increases to 500–650°C, the depth of the modified layer usually increases (for example, from 0.1 to 0.7 mm for a 6 h treatment of chromium-molybdenum-aluminium alloyed steel). But in some cases, as in the plasma treatment of cast iron, the layer depth decreases on increasing the temperature of the metal above 600°C. Perhaps this is due to the development of thermoelectron emission from the cathode (a metal article) so that the fraction of the ion current in the total discharge current is reduced. As a result, the flux of ions to the metal surface decreases.

The quality of the plasma treatment surfaces is also determined by the nature of the steel. Depending on the nature of the alloying element and its concentration, the modified layer changes its mechanical properties. For instance, the hardness of the layer decreases on alloying steel with elements in the following sequence: Al, V, Mo, Mn, Si, and Ni (Fig. 8.3).

Glow-discharge plasma carburizing, siliconizing, aluminizing, boronizing, and titanizing are also described in the literature.

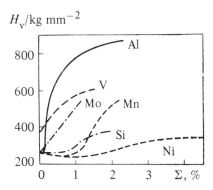

Fig. 8.3. Hardness H_v of modified layer as a function of type and concentration of alloying element.

8.1.7 Plasma treatment of polymeric materials

The treatment of polymers in a nonequilibrium plasma results in changes in their wettability, molecular mass, and the chemical composition of the surface layer (1–10 μm). The wettability determines the adhesion, ability to maintain paints, and the possibility of gluing with other materials. The change in molecular mass affects the permeability, melting point, solubility, and cohesion. Grafting of oxygen-containing and certain other functional groups to the polymer molecules

located at the surface leads to an increase in the wettability. The change in molecular mass is related to the formation of crosslinks in the polymer chains induced by the action of plasma-induced UV-emission.

The treatment of natural polymeric materials such as wool, cotton, and leather by conventional methods aims to change their surface roughness, to improve their adhesion to various coatings, and, sometimes, to remove a surface layer.

The treatment of natural polymeric materials in a corona discharge increases the cohesion of the fibres to each other and to synthetic materials. Thus, after oxygen-plasma treatment, the adhesion of cellulose plates toward each other and to other polymer films increases five- to seven-fold.

The surface of the treated fibres becomes more hydrophilic and rough after treatment. Materials treated in a chlorine-containing plasma possess the best adhesive properties. The treatment strongly decreases the shrinkage of wool fibre, increases the wettability of cotton fibre, destroys the sheath of wool fibre, and changes its electrostatic properties.

A description of a corona discharge pilot plant for treatment of wool and mohair in a plasma of air–chlorine mixture has been published. Its capacity is 18 kg of fibre per hour, the duration of the treatment in 2 s, and the power is 4.5 kW.

8.1.8 Application of low-temperature plasmas in chemical and instrumental analysis

At present, the basic steps of preparation of samples for chemical and instrumental analysis can be carried out in a nonequilibrium plasma, including low-temperature (from -196 to $300°C$) oxidation of organic, biological, and polymeric materials; removal of any organic matrix from an inorganic sample for analysis, the formation of gaseous products in plasma reactions for further gas chromatographic or mass spectral analysis, and purification of sample surfaces for rehabilitation of their optical properties, are also possible.

A plasma used for the above purposes must contain oxidizing or reducing agents.

The plasma pretreatment of samples for analysis is usually carried out in HF or SHF plasmas at a pressure of 60–200 Pa for a few seconds to several hours, depending on the structure and chemical composition of the samples. Samples can be treated in the form of powders, flat sections, films, fumed or porous materials, and low-volatile liquids.

A low-temperature plasma combined with IR-spectroscopy is applied for trace analysis of metals in foods, drugs, and biological samples. Oxidation of biological samples in a HF discharge plasma for 1.5 h results in almost complete precipitation of radionuclide traces from Na to Au, present in the samples, in a boat in which the sample has been placed. If the target elements are highly volatile, they can be confined by adding some other elements to the matrix, which can form involatile compounds with the elements of interest; for example, when mixed with Cu, As or Hg, Se is left in the boat in practically the same initial amount after plasma treatment of its sample.

A nonequilibrium plasma is applied in pretreatment for analysis of inorganic materials in geochemistry and metallurgy and for analysis of sea water and soils. In these cases the organic constituents should be removed and the phases separated at a relatively low temperature to avoid degradation of the material's structure.

In the last decade nonequilibrium plasmas have been widely used in sample preparation for microscopic studies. The techniques of plasma ashing and of plasma etching of surfaces for subsequent production of replicas are realised in this way.

The use of plasmas in the pretreatment of dust-like particles in the analysis of inorganic substances is especially promising. For example, solid products of the combustion of fuels in relatively small concentration are collected on filters made from organic materials. As a result of oxidation in a nonequilibrium plasma, the organic compounds making up the filter and free carbon are removed, while the inorganic particles are retained and their concentration is much increased. Moreover, the size of the particles, their morphology, and composition do not change.

The removal of carbon from matter to be analysed without changing the composition and structure of its residual inorganic constituents is carried out in chemical and crystallographic studies on the mineral components of coals.

Replicas for the investigation of crystal boundaries and of their orientation are obtained in a plasma in the following way: after etching, the metal surface is oxidized in the plasma. It has been found that oxide films grow epitaxially on metal surfaces treated in this way.

8.2 RADIATION-INDUCED PROCESSES

Radiation processing is a rather diversified field of the industrial application of radiation. We cannot give a very detailed consideration to the different aspects of radiation technologies, but shall confine ourselves to just two key issues: (i) the radiation-induced modification of polymers, and (ii) radiation synthesis will be outlined. Other areas of radiation processing are described comprehensively elsewhere [413–417, 422, 436–439].

8.2.1 Modification of polymers

The radiation-induced modification of polymers and oligomers is undertaken for several reasons: radiation crosslinking, manufacture of composite materials, radiation curing, and graft polymerization.

Radiation crosslinking

This is also known as radiation vulcanization [440–442]. It leads to the creation of a three-dimensional structural net resulting in an increase in mechanical strength and in thermal stability, as well as in the development of other useful properties. For example, polyethylene diffuses into a formless mass on heating to 100°C, but radiation crosslinked specimens retain their shape at this temperature. If polyethylene is doped with certain substances before irradiation, the product resists temperatures above 200°C for a long time and serves as a good electrically insulating material. An important property of irradiated polyethylene is its 'shape memory'. Based on this effect, thermally-cured polyethylene films have been formulated which enable excellent insulating and packing materials to be obtained. Radiation curing of natural rubber, of polybutadiene and of other elastomers, is used in the production of tyres, insulating materials, and other rubber goods. The radiation technology of the production of poroplasts widely used in automobile and light industries has been developed. The output of radiation crosslinked poroplast exceeds 25 thousand tons a year in the USA alone.

Production of composite materials

Composite materials have been used extensively in recent years in various fields of human life and activities. Radiation processing provides the opportunity of producing a new class of composite materials on the basis of low-cost porous raw materials coated with a monomer which is to be polymerized and grafted to the support under the action of radiation [415]. In this way it is possible to manufacture very wear-resistant and attractive parquetry from birch, aspen, and poplar wood and other polymer–wooden materials, polymer–concrete products (compared with normal concrete items, their compressive and tensile strengths increase four- to five-fold, their modulus of elasticity 3.5-fold, their flexural modulus of elasticity 1.5-fold, their attrition resistance 3- or 4-fold; their stability on freezing and thawing, and their corrosion resistance also increase), and modified tufa (such a tufa-polymer resembles marble in its facing properties). Radiation-processed glass-reinforced plastics have improved structural and insulating properties.

Organic semiconductors can be prepared from linear polymers by the combined action of radiation and high temperature [443].

Cure of coatings

The most developed application of radiation chemistry is the curing of paints on articles made of various materials [444]. The idea of the method is that both polymerization of an oligomer and its bonding to the surface is caused by low-energy electrons penetrating through the entire cured layer (*ca.* 0.5 mm). The process occurs rapidly without heating (taking a few minutes in an inert atmosphere), while the thermochemical procedure requires a drying oven operating at 60–70°C for several hours. Radiation-cured coatings suffer ageing in air to a slower extent than heat-cured ones. There is no need to polish surfaces after radiation curing.

Graft polymerization

This is a method of changing the surface properties of a polymer or other solid material while retaining the properties of the support [360]. In contrast to other methods, the radiation procedure enables grafting to be carried out in a variety of chemical systems, at any temperature, and without chemical initiators and catalysts contaminating the end-product. The high penetrating power of ionizing radiation enables the process to occur uniformly throughout the volume of the material. Table 8.1 lists some materials modified by radiation-induced graft polymerization. Some other materials modified by nonequilibrium gas discharge plasma are included in the same table. As can be seen, radiation-induced graft polymerization makes it possible to endow materials with quite different properties.

Table 8.1. Radiation-induced graft polymerization treatment of materials [360]

Substrate	Grafting procedure	Effect achieved	Implementation
Cotton fabrics	Direct liquid phase	Germ resistance, crease resistance	Industrial scale production
Medical gauze	Direct liquid phase	Hemostatic properties	Industrial pilot plant
Linen fabrics	Direct vapour phase	Rot resistance	Pilot plant
Cotton and polyamide fabrics	Direct liquid phase with local irradiation	Structural and colour effects	Experimental production
Cotton-polyester fabrics	Direct liquid phase	Crease resistance	Industrial-scale production
	Post-grafting from vapour phase (glow discharge)	Improvement of washing-off properties	Pilot plant
Polyamide fabrics and knitted fabrics	Direct liquid phase	Hydrophilization	Industrial-scale production
Synthetic polymer films and fabrics	Direct liquid phase	Chemical and thermal resistance, resistance to light	Laboratory plant
Poly(vinyl chloride)	Direct gas phase	High impact strength	Pilot plant

Table 8.1. (continued)

Substrate	Grafting procedure	Effect achieved	Implementation
Butadiene-styrene thermoelastoplast	Direct liquid phase	Adhesion to metals	Pilot plant
Polyalkenes, organosilicon polymers, polyurethanes	Liquid-phase grafting	Biocompatibility, thrombosis resistance	Laboratory plant
Polypropylene fibres	Direct vapour phase	Ion-exchange properties	Industrial-scale production
Polyethylene (powder)	Post-grafting from vapour phase (glow discharge) Post-grafting from liquid phase	High-catalytic activity	Industrial pilot plant
Polyethylene films	Direct gas phase	Improvement of adhesive properties	Pilot plant
Polyalkenes	Direct liquid phase	Grafting of acrylic acid	Laboratory plant
Polyalkene films	Direct grafting from sublimed monomer vapours	Membranes with enhanced radiation stability	Laboratory plant
Teflon (secondary)	Direct liquid phase	Possibility of using as filter	Laboratory plant
Teflon	Direct liquid phase	Possibility of using for water purification membranes	Laboratory plant
Fluoropolymer and polyalkene films	Liquid-phase grafting	Ion-exchange properties (separation membranes)	Experimental production

Table 8.1. (continued)

Substrate	Grafting procedure	Effect achieved	Implementation
Rubber goods	Direct vapour-phase plasma treatment	Decrease in friction coefficient and in adhesion, increase in wear resistance	Industrial-scale production
Natural leather		Improvement of service properties and durability	Laboratory plant
Kaolin	Direct gas phase	Decrease in density of suspensions, reduction of costs of papermaking	Laboratory plant
		Prolonged action	Laboratory plant
Fertilizers	Direct gas phase	Ion-exchange properties	Laboratory plant
Mineral sorbents and supports	Direct gas plasma		
Silica fabrics and glass fibres	Direct gas plasma	Semiconductivity	Laboratory plant

8.2.2 Radiation synthesis

The idea of radiation-induced chemical synthesis has attracted much attention [439, 445–448] (Table 8.2). Typical features of the radiation-induced processes which differentiate them from traditional synthetic techniques are:

(i)   the possibility of carrying out a process at low temperatures which reduces degradation of the end-product and decreases the fire risk and explosion hazard,

(ii)   the relatively simple control of the process by variation of the dose rate,

(iii) the absence of residual catalysts and initiators or of their transformation products among the end-products.

Radiation-induced synthetic processes may be divided arbitrarily into three groups: long-chain $(G \geq 10^3)$, short-chain $(10 < G < 10^3)$, and non-chain $(G \leq 10)$ processes.

Table 8.2. Selected radiation-induced syntheses [445, 446]

Process	Type of process	Particular system[†]	Product
Cracking of aliphatic hydrocarbons	Chain	Heptane, 400–600°C	H_2, CH_4. and alkenes
Structural isomerization of hydrocarbons	Non-chain	Butene-1, vapour-phase, 23°C, in presence of aluminosilicates	*Cis/trans*-butene-2 mixture
Cis/trans-isomerization of hydrocarbons and their derivatives	Non-chain	*Cis*-butene-2 in benzene	*Trans*-butene-2
Carbonylation of hydrocarbons	Non-chain	CO-saturated aqueous solution of phenol	Salicylaldehyde
Carboxylation of hydrocarbons, acids, amines, etc.	Non-chain	CO_2-saturated aqueous methanol	Glycolic acid
Hydroxylation of hydrocarbons, alcohols, etc.	Non-chain	Aqueous solution of tyrosine	Dihydroxyphenyl-alanine
Oxidation of hydrocarbons, etc.	Chain	Heptane, 100°C	Peroxides, carbonyl compounds and alcohols
Chlorination of hydrocarbons	Chain	Kerosene, 20–60°C	Alkyl chloride mixture
Bromination of hydrocarbons	Chain	2,3-dimethylbutane	2,3-Dibromo-3,3-dimethylbutane
Iodination and fluorination of hydrocarbons	Non-chain	$C_2F_2Cl_3$ in benzene	Haloethylbenzene
Sulphochlorination of hydrocarbons	Chain	Synthine solution of SO_2 and Cl_2	Alkylchloro-sulphonate mixture

Table 8.2. (continued)

Process	Type of process	Particular system[†]	Product
Sulphoxidation of hydrocarbons	Chain	Solution of SO_2 and O_2 in kerosene	Mixture of sulphonic acids
Nitration of hydrocarbons	Non-chain	NO_2 solution in dodecane	Nitrododecane
Fischer–Tropsch synthesis of hydrocarbons	Non-chain	CO–H_2 mixture, 160–250°C	Hydrocarbons up to hexane
Destructive chlorination of hydrocarbons	Non-chain	Benzene or hexane with Cl_2	CCl_4
Addition of alkanes, carbon halides, PCl_3, dialkylphosphonates, silanes, alkylsilanes, chlorosilanes, alcohols, ethers, thioalcohols, amines, and aldehydes to C=C bond (telomerization)	Short-chain	CCl_4 and allyl alcohol	2,4,4,4-Tetrachlorobutanol and 2,6,6,6-tetrachlorohexanol
		Hexene-1 and PCl_3	$C_4H_9CHCH_2PCl_2$ $\quad\quad\mid$ $\quad\quad Cl$
		Allyl alcohol solution in methanol	1, 4-butanediol
		Acetaldehyde in hexene-1	$C_6H_{13}COCH_3$
Dimerization of different types of substance	Non-chain	Methanol	Ethylene glycol
		Acetone	Acetonylacetone
Condensation of substances	Non-chain	Heptane and cyanogen	Heptanonitrile
Alkylation of tin	Short-chain	Tin in alkyl halide	Dichloroalkyltin
Preparation of organophosphorous and organoarsenic compounds	Non-chain	Triphenylphosphine in fluorobenzene	$(Ph_4P)F$
		Trimethylphosphonate and iodobenzene	$PhPO(OCH_3)_2$
Nitrogen monoxide addition	Non-chain	NO–CCl_4 mixture	Cl_3CNOH

Table 8.2. (continued)

Process	Type of process	Particular system[†]	Product
HCN synthesis	Non-chain	Mixtures of N_2 or NH_3 with CH_4	Hydrogen cyanide
Alkylation of chloro- and alkylchlorosilanes	Chain	Chlorobenzene and $HSiCl_3$, 300°C	$PhSiCl_3$
Production of phosgene	Chain	$CO–Cl_2$ mixture	Phosgene

† Room temperature unless otherwise specified

Long-chain processes (chlorination, sulphoxidation, etc.) can be carried out with adequate efficiency with chemical photochemical, and radiation initiation. Thus, comparison of the chlorination of benzene by the different methods has shown that the product composition and chain length are identical for all of the above methods. The radiation procedure, however, has some advantages: it provides an equal rate of the process throughout the entire volume and prevents overheating of the product. The overheating in the chlorination of alkanes results in formation of dehydrochlorination by-products in the case of chemical initiation, while radiation-induced chlorination gives no such side-products.

Short-chain processes (telomerization, oxidation, etc.) induced by ionizing radiation are superior to the same procedures carried out by using light or a chemical initiator since the initiators are rather expensive and the radiative power of light sources is not very high so that a large number of sources must be used, while high-intensity radiation sources are readily available.

Non-chain processes, also known as power-intensive processes, often make it possible to considerably reduce the number of stages in fine-organic synthesis. This is because a particular group can be introduced into a compound at a given position in one step by using a radiation-induced process; this cannot be done by conventional methods. However, not one but several products are usually formed in these processes. This disadvantage can be eliminated or even be turned to advantage for small-scale production, especially for microscale synthesis, so long as the products can be isolated by efficient techniques such as preparative chromatography. Implementation of the procedures of radiation synthesis gives the possibility of necessary expansion of the range of manufactured products, and it is often acceptably profitable. Moreover, radiation chemistry gives the best chance of increasing the variety of radioactive tracer compounds.

The procedures of radiation synthesis have certain advantages over other techniques from the energetic, economic, and environmental points of view.

Their implementation does not require considerable amounts of process water, there are no wastes, and the isolation of the products and their purification is simpler. At the same time, an economic analysis of radiation-induced processes is always needed in order to compare them with other means, since only by economic gain can any technology take off.

8.3 PHOTOCHEMICAL PROCESSES

8.3.1 Photochemical synthesis

Among photochemical processes applied on a large scale in industry are the chlorination and sulphochlorination of aliphatic and aromatic hydrocarbons, sulphoxidation of higher alkanes, addition of chlorine to benzene, addition of hydrogen sulphide to alkenes, and nitrosation of cyclohexane occurring by radical and chain mechanisms.

The chlorination of C_{11}–C_{14} alkanes is used in industry for the manufacture of monochloroalkanes [449]. The reaction is carried out by means of irradiation of a reaction mixture flowing through Vycor glass tubes with a 7.5 kW medium-pressure mercury lamp. The primary process consists of the photodissociation of a chlorine molecule to atoms:

$$Cl_2 + h\nu \rightarrow 2\dot{C}l.$$

A radical chain reaction then develops:

$$\dot{C}l + RH \rightarrow HCl + R$$

$$R + Cl_2 \rightarrow RCl + \dot{C}l.$$

The temperature of the reaction mixture containing 15% mole chlorine is maintained below 40°C to prevent the occurrence of uncontrolled reactions and the formation of polychlorinated alkanes. The output of such a reactor is about 27 tons of chloroalkane per day.

Hexachlorocyclohexane is prepared by the photoaddition of chlorine to benzene [450]:

$$C_6H_6 + 3Cl_2 \rightarrow C_6H_6Cl_6.$$

This reaction also occurs by a radical chain mechanism. Its quantum yield in an industrial-scale process can be as large as 2500. Up to twenty photochemical reactors connected in series are used, each being supplied with two 40 W fluorescent lamps. The output of a set-up of this sort is about 270 kg of hexachlorocyclohexane per hour.

The photochemical chlorination of toluene makes it possible to produce benzyl chloride, benzylidene chloride, and benzotrichloride [451, 452]:

$$PhCH_3 + Cl_2 \xrightarrow{hv} PhCH_2Cl + HCl,$$

$$PhCH_2Cl + Cl_2 \xrightarrow{hv} PhCHCl_2 + HCl,$$

$$PhCHCl_2 + Cl_2 \xrightarrow{hv} PhCCl_3 + HCl.$$

These reactions also occur by a radical chain mechanism.

The photochemical sulphochlorination of alkanes is used in the production of alkane sulphonate surfactants obtained by the hydrolysis of primarily-formed sulphonic acid chlorides:

$$RH + SO_2 + Cl_2 \xrightarrow{hv} RSO_2Cl + HCl.$$

The sulphochlorination occurs by a chain mechanism. The initiating step is the photodissociation of chlorine to atoms. The chlorine atoms abstract hydrogen from a hydrocarbon, giving chain propagation:

$$\dot{R} + SO_2 \rightarrow \dot{R}SO_2$$

$$\dot{R}SO_2 + Cl_2 \rightarrow RSO_2Cl + \dot{C}l.$$

The quantum yield under industrial conditions is about 2000, and the power consumption is *ca.* 100 kJ (300 W h) per 1 kg sulphonyl chloride.

The one-stage production of alkane sulphonates is based on the photochemical sulphoxidation of alkanes [453, 457–459]:

$$RH + SO_2 + 0.5\,O_2 \xrightarrow{hv} RSO_3H.$$

The reaction occurs by a radical chain mechanism. The power consumption is 720 kJ per 1 kg alkylsulphonate. Medium-pressure mercury lamps with a power up to 49 kW are used. The output of a reactor is about 1600 tons of alkyl-sulphonate per year.

The photoaddition of hydrogen sulphide to higher alkenes gives mercaptans which used to be added to natural gas as odorants to detect gas leaks. Hydrogen sulphide dissociates under the action of UV with $\lambda < 317$ nm

$$H_2S + hv \rightarrow H\dot{S} + \dot{H}.$$

The anti-Markovnikov addition of $\dot{S}H$ radical occurs by a chain mechanism:

$$RCH{=}CH_2 + H\dot{S} \rightarrow R\dot{C}CH_2SH$$

$$R\dot{C}HCH_2SH + H_2S \rightarrow RCH_2CH_2SH + \dot{S}H.$$

The photonitrosation of cyclohexane is used in the production of cyclohexanone oxime which is converted to ω-caprolactam by the Beckmann rearrangement [460]. ω-Caprolactam is used in the production of Nylon-6. The photonitrosation is a radical non-chain reaction

$$NOCl + h\nu \longrightarrow NO + Cl,$$

The quantum yield for the oxime is *ca.* 0.8. This process is used in the industrial scale by the Torey Co. in Japan. Submerged medium-pressure mercury lamps (with thallim iodide added to improve the emission spectrum) of 60 kW power are used. The output is 24 kg of oxime per hour per lamp. The commercial reactor is equipped with a large set of these lamps adjusted in staggered rows to facilitate the effective absorption of radiation. A mixture of NOCl and HCl is passed through liquid cyclohexane from the bottom. Oxime hydrochloride formed precipitates and is isolated through an outlet pipe at the bottom of the reactor.

Photochemical reactions are also used in small-scale industrial syntheses, for example in the production of medicines, provided the target compounds can barely be prepared by other techniques. A typical example is the manufacture of vitamin D_3. The photolysis of 7-dehydrocholesterol, made from cholesterol, causes electrocyclic ring-opening, yielding pre-vitamin D_3 which is converted to vitamin D_3 by a thermal 1,7-hydrogen shift [461, 462].

The reaction occurs under the action of 280–287 nm mercury lamp with a quantum yield of 0.3. By using a submeged 40 kW mercury lamp, about 5 kg vitamin D_3 per hour is produced.

Photochemical *cis-trans* isomerization is used in the industrial-scale production of vitamin A. Vitamin A acetate obtained by the Wittig reaction consists of two types of stereoisomers: the all-*trans*-isomer and 11 *cis*-forms. Since only the *trans*-isomer is needed, the photoisomerization is carried out [463]:

To provide more complete use of the radiation from the mercury lamps, sensitizers such as zinc tetraphenylporphin or chlorophyll are applied. Other photochemical processes used in producing medicines can be found in Fischer's review [464].

8.3.2 Photopolymerization and light-stabilization of polymers

Radical polymerization is easily induced by light in the presence of various initiators, the most available and suitable agents being azobisisobutyronitrile, aromatic carbonyl compounds (for example benzophenone derivatives), and iodine. Light initiation enables polymerization to be carried out at lower temperatures and at an easily-controlled reaction rate. This makes it possible to produce materials with exactly prescribed properties. For commercial reasons, photo-induced polymerization is not practised very widely for the production of plastics but is used only for the curing of lacquers and for manufacturing optically uniform articles (low-cost lenses, Fresnel lenses, etc.) as well as for the fabrication of printing plates in photolithography. Most applications of photopolymerization are in the field of free radical polymerization of acrylic esters and their derivatives.

The properties of most polymers deteriorate under the action of light because of scission or crosslinking of the polymer chain or by other processes. The photodegradation of polymers is usually a sequence of photoinduced oxidation processes. Photoageing of polymers is caused by the presence of various functional groups, additives, and impurities, since most of the widely-used polymers do not absorb radiation at $\lambda > 300$ nm. Two methods are used to protect polymeric materials from photoageing: (i) the introduction of light-absorbing additives exhibiting high light-fastness due to the effective degradation of electronic excitation energy and thus shielding the material from deeper UV radiation, and (ii) the introduction of radical inhibitors (many radical inhibitors of the thermal ageing of polymers possess the necessary light fastness needed for this purpose).

Environmental problems have given rise to the concept of designing polymers capable of light-induced self-degradation. Thus, unsaturated ketone-alkene copolymers are very sensitive to exposure to light, which causes polymer chain-scission due to the Norrish type-II reaction. One can also add special sensitizers to the usual commercial polymers, thus ion(III) dithiocarbamate acts as a stabilizer at high concentrations but, at low concentration, the ferric ions formed via decomposition of the complex facilitate photo-oxidation. In this way it is possible to control the lifetime of polymeric articles; after this period has elapsed, a fast degradation of the article takes place.

8.3.3 Photoimaging
A photographic process consists in obtaining an image under the action of light which can be positive or negative. Photolithography is the manufacture of printing plates for relief, intaglio, or offset printing.

Silver halide photography
The essence of silver halide photography is the formation of latent image centres containing some silver atoms, under the action of light on silver halide crystals:

$$AgX + h\nu \rightarrow Ag + X$$

The subsequent development, which is treatment with a weak reducing agent, leads to the reduction of silver in those microcrystals which contain the latent image centres. Depending on the structure and size of the microcrystals, up to 10^{10} silver atoms can be formed per absorbed photon during development. In this way an intensification of the original latent image is achieved, which is the basis of the very high speed of silver halide photographic materials. By using different dyes as sensitizers (usually based on cyanines), it is possible to control the spectral speed (the spectral sensitivity of silver halide materials is in the range from UV–1.3 µm IR radiation). The dyes are strongly absorbed on the surface of AgX crystals. The dominant mechanism of sensitization is thought to be electron transfer from the excited sensitizer molecule to the conduction band of silver halide.

In silver halide colour photography, several layers of materials sensitive to different spectral regions are used where dyes of the corresponding colour are formed from the oxidation products of the developer during development, and the silver is subsequently removed after appropriate treatment.

Silver halide photography provides a unique speed but requires special procedures for the development and fixing of the image, and it consumes expensive silver. Meanwhile, other photographic techniques based on silver-free materials are being developed.

Silver-free photography
The first silver-free process widely applied some time ago for the copying of drawings and texts was based on the photodecomposition of iron(III) oxalates or citrates:

$$2Fe(C_2O_4)_3^{3-} + h\nu \rightarrow 2Fe^{2+} + 5C_2O_4^{2-} + 2CO_2.$$

The Fe^{2+} ions formed in the illuminated areas react with ferricyanide to give a negative image based on Turnbull's blue.

The more convenient diazotype process, which was widely used until the invention of xerography, is based on the decomposition of arenediazonium salts under the action of UV or short-wave visible light:

$$ArN_2X + h\nu \rightarrow ArX + N_2.$$

After making alkaline with gaseous ammonia, the undissociated salt remaining in the unexposed areas enters a coupling reaction with hydroxyaromatic compounds (the so-called azo-component) pre-existing in the light-sensitive layer, and forms a dye giving a positive image:

$$ArN_2X + HAr' \rightarrow ArN{=}NAr' + HX.$$

Such diazotype photo-materials show a high range of contrast and are used for the duplication of drawings and microfilms. It is possible to obtain images of any colour by choosing relevant diazo compounds and azo components. The quantum yield for the photodecomposition of diazonium salts is fractionally less than unity, and the speed of these materials is not high, *ca.* $10^{-2}\,m^2\,J^{-1}$ (the light-sensitivity is a quantity reciprocal to the dose needed to obtain an absorption different from that of background by a certain value, usually be 0.1 or 1.0). There is a possibility of increasing the light sensitivity of diazotype photomaterials by using photoinduced chain reactions to dissociate the diazonium salts, but then the shelf-life of such materials decreases strongly. Consequently, the chain decomposition of diazonium slats is not widely practised.

Photomaterials differing in their mode of action are based on light-induced polymerization in microcapsules containing reagents capable of forming dyes after mechanical or thermal destruction of the microcapsules. At the site of light irradiation, the microcapsules are polymerized but not destroyed, thus preventing escape of the agents and the formation of dyes. This method gives the possibility of a very wide choice of dyes. Examples of photomaterials for colour photography giving good colour rendition and light fastness have been worked out on this basis.

Photochromic materials

The photomaterials described above require special chemical treatment to obtain and fix the image. For many practical purposes, such as the rapid recording and optical processing of information, one must avoid any chemical operation, and such demands are met in photochromic materials [465, 466]. Photochromism is a reversible change in the absorption spectrum (colour) of a substance under the action of light owing to the occurrence of photochemical or photophysical processes. Photochromic processes are based on various endoergonic reactions such as photoisomerization, dimerization, and dissociation or photoinduced redox reactions which can be reversed either thermally or photochemically, for example the *trans-cis* photoisomerization of thioindigo:

Photochromic materials with a short relaxation time (0.1 µs to 1 s) are based on the development of light-induced triplet states of aromatic and similar compounds, for example the photoisomerization of spiropyrans to merocyanines:

Inorganic photochromic materials are prepared by the introduction of *ca.* 0.5% silver chloride or bromide in glasses. Silver atoms which absorb radiation in the visible region, are formed under the action of light. The glasses darken if exposed to bright light. The time for thermal relaxation of such glasses is a few minutes.

Photochromic materials have a low speed, *ca.* 10^{-2} m^{2}J^{-1}, which cannot be increased significantly.

Light-sensitive materials for the luminescent display of information, which are based on the creation of an image using luminescent substances, have a speed several orders of magnitude higher, since luminescence instrumentation makes it possible to detect less than 10^{-12} mol cm^{-2}, while to achieve an absorbance A of 0.1 one needs at least 10^{-9} mol cm^{-2} of a compound with an extinction coefficient ε of 10^5 M^{-1} cm^{-1}. Systems of this type are exemplified by stilbenes, capable of *cis-trans*-isomerization (where only the *trans*-isomer is luminescent) or spiropyrans which photoisomerize to luminescent merocyanines.

Certain light-sensitive systems, the refractive index of which changes on illumination, are used for recording holograms.

Photolithography and microlithography

To prepare relief and intaglio printing plates and integrated semiconductor circuits (microlithography), it is necessary to create a positive or negative relief image on the surface of a support or a semiconductor. For lithography it is sufficient to change the adhesive properties of the surface. Appropriate materials should be capable of giving a relief resistant to the corrosive media applied in chemical, electrochemical, or plasma etching of semiconductors (photoresists).

Photopolymerization is widely used to make printing plates. A viscous mixture containing monomers, oligomers, and photoinitiator is illuminated imagewise, and then the unpolymerized residue of the mixture is washed off. 'Dry' oligomer-base photoresists are also applied in microelectronics.

Photoprocessing of polymers based on changing their solubility or adhesive properties has been used in printing for a long time. A gelatin layer containing dichromate undergoes hardening under the action of light: first its swelling ability decreases and then its solubility in water is totally lost. The hardened layer is no longer wettable with water but can be wetted with a special paint used in lithography. Other natural or synthetic polymers are also suitable for this purpose. Irradiation of poly(vinyl cinnamate) results in crosslinking due to the photocyclization of the cinnamate groups yielding cyclobutanes, which leads to a decrease in the solubility of the polymer.

One can distinguish positive photoresists, where the stable coating is formed in the unirradiated area, and negative photoresists, where the coating appears on the illuminated areas. A widely-applied positive photoresist is made from phenol-formaldehyde resin containing a quantity of naphthoquinone diazide or its analogues. On illumination the elimination of nitrogen takes place and naphthoquinone diazide transforms eventually into indene carboxylic acid:

The regions containing the original diazide are not dissolved on treatment with aqueous alkali, but those containing the acid are dissolved and a positive image is thus formed.

Negative photoresists are usually based on the photopolymerization or photoinduced crosslinking of polymers. One type of negative photoresist met in practice is made of polyisoprene doped with a disubstituted diazide, which is transformed on exposure to a dinitrene, which crosslinks the polymer chains and makes the polymer insoluble. The poly(vinyl cinnamate) described above is also used.

The photoinduced depolymerization of thermodynamically unstable polyketone- and polyaldehyde-type polymers is used for making highly-sensitive photoresists which are convenient in that they do not require a wet treatment because the depolymerization products are removed *in vacuo* where subsequent operations with a semiconductor are carried out.

The photochemical reactions of compounds of low molecular mass are used to modify light-sensitive layers. Thus, the photolysis of naphthoquinone diazides produces carboxylic acids which are soluble in alkaline media and are readily removable. To provide the required mechanical properties, quinone diazide is added to different resins or linked to them chemically.

8.3.4 Photochemical lasers

Soon after the invention of lasers the applications of photochemical reactions for the production of population inversion and lasing systems was proposed [467]. The first photochemical laser employed the photodissociation of CH_3I, which produces excited iodine atoms, and stimulated emission of I ($^2P_{1/2}$) was observed. High-power photochemical lasers based on the dissociation of CF_3I were built for the ignition of nuclear fusion [468]. Photoinduced proton-transfer reactions are used in dye lasers for extension of the range of emission wavelengths [469]. In these lasers, the fast equilibrium in the excited singlet state supplies the population inversion and laser action from both (or even more) acid and base forms, which emit in different wavelength regions. Exciplex-formation reactions of excited Ar, Kr, and Xe atoms with halogens are used in lasers incorrectly referred to as 'excimer lasers' (the correct name should be 'exciplex laser') [470]:

$$Ar^* + F_2 \rightarrow ArF^* + F$$

$$ArF^* \rightarrow Ar + F + h\nu \ (193 \ nm).$$

Several types of chemical laser based on chemiluminescent reactions have been built [471]. In these, the population inversion of vibrationally-excited molecules is generated in fast, extremely exothermic reactions (for example the formation of vibrationally excited HCl in the reaction $H_2 + Cl_2$). An advantage of these lasers is their extremely large release of energy.

References

[1] Loudon, R. *The Quantum Theory of Light*. 2nd edn. Oxford: Clarendon Press, 1983.

[2] Bagdasarjan, Kh. S. *Dvukhkvantovaya fotokhimiya* (Two-Quantum Photo-chemistry). Moscow: Nauka, 1976.

[3] Starodubtsev, S. V., Romanov, A. M. *Vzaimodejstvie gamma-izlucheniya s veshchestvom.* Chast I (Interaction of Gamma-Radiation with Matter. Part I). Tashkent: Izd. Nauka UzbSSR, 1964.

[4] Starodubtsev, S. V. *Polnoe sobranie nauchnykh trudov.* Tom 2. *Yadernaya fizika.* Kniga 2. *Vzaimodejstvie izlucheniya s veshestvom* (Comprehensive Collection of Works. Volume 2. Nuclear Physics. Book 2. Interaction of Radiation with Matter). Tashkent: FAN, 1970.

[5] Aglintsev, K. K. *Dozimetriya ioniziruyushchikh izluchenii* (Dosimetry of Ionizing Radiations). Moscow: GIZ TTL, 1957.

[6] Pikaev A. K. *Dozimetriya v radiatsionnoj khimii* (Dosimetry in Radiaton Chemistry). Moscow: Nauka, 1975.

[7] Byakov, V. M., Nichiporov, F. G. *Vnutritrekovye khimicheski protsessy* (In-Track Chemical Processes). Moscow: Energoatomizdat, 1986.

[8] Mozumder, A., Magee, J. L. *Radiat.Res.* 1966, **28** 203; *J.Chem.Phys.* 1966 **45** 3332.

[9] Konovalov, V. V., Raitsimring, A. M., Tsvetkov, Yu. D. *Khim. Vys. Energ.* 1984, **18** 5.

[10] Popov, V. I. *Metody LPE spektrometrii ioniziruyushchikh izluchenii* (The Methods of LET-Spectrometry of Ionizing Radiations). Moscow: Atomizdat, 1978.

[11] Gusev, N. G., Mashkovich, V. P., Suvorov, A. P. *Zashchita ot ioniziruyushchickh izluchenii.* Tom 1. Fizicheskie osnovy zashchity ot izluchenii (Protective Shielding against Ionizing Radiations. Volume 1. Physical Aspects of Protection). Moscow: Atomizdat, 1980.

372   **References**

[12] Kaplan, I. G., Miterev, A. M. *Russ. Chem. Rev.* 1986 **55**, 377; *Adv. Chem. Phys.* 1987, **68**, 255.
[13] Mozumder A. *J. Chem. Phys.* 1969, **50** 3162.
[14] Santar, I., Bednar, J. *Intern.J.Radiat.Phys.Chem.* 1960, **1**, 133.
[15] Kowari, K., Sato, S., *Bull.Chem.Soc.Japan.* 1978, **51**, 741.
[16] Kaplan, I. G., Miterew, A. M., *Radiat.Phys.Chem.* 1986, **27**, 83.
[17] Vanshtein, L. A., Sobelman, I. I., Yurkov, E. A., *Sechenie vozbuzhdeniya atomov i ionov elektronami* (Cross-Section of Excitation of Atoms and Ions by Electrons). Moscow: Nauka, 1973.
[18] Eberson, L., *Acta Chem.Scand.* 1984, **38**, 439.
[19] *Average Energy Required to Produce an Ion Pair*. ICRU Report 31. 1979.
[20] Kaplan, I. G., *Khim.Vys.Energ.* 1983, **17**, 210.
[21] Kaptsov, N. A., *Elektricheskie yavleniya v gazakh i vakuume* (Electric Phenomena in Gases and in Vacuum). Moscow–Leningrad: GITTL, 1950.
[22] Lamola, A. A., *Energy Transfer and Organic Photochemistry*. N.Y.: Wiley, 1969.
[23] Ermolaev, V. L., Bodunov, E. I., Sveshnikova, E. B., Shakhverdov, G. A., *Bezyzluchatelnyi perenos energii elektronnogo vozbuzhdeniya* (Radiationless Transfer of Electronic Excitation Energy). Leningrad: 1977.
[24] Förster, Th., *Ann.Phys.* 1948, **2**, 55.
[25] Galanin, M. D., *Zh.Eksp.Teor.Khim.* 1951, **21**, 114, 126.
[26] Perrin, J., *Ann.Chem.Phys.* 1932, **17**, 283.
[27] Kalyazin, E. P., *Vestn.Mosk.Univ.* Ser. 2, Khim. 1982, **23**, 566.
[28] Förster, Th., *Naturwiss.* 1949, **36**, 186.
[29] Weller, A., *Prog.React.Kinet.* 1961, **1**, 187.
[30] Martynov, I. Yu., Demyashkevich, A. B., Uzhinov, B. M., Kuzmin, M. G., *Russ.Chem.Rev.* 1977, **46**, 1.
[31] *Energii razryva chimicheskikh svyazei, potentialy ionizatsii i srodstvo k elektronu* (Chemical Bond Energies, Ionization Potentials, and Electron Affinities), Ed. Kondratyev, V. N., Moscow: Nauka, 1974.
[32] Okabe, H. *Photochemistry of Small Molecules.* N.Y.-L.: Wiley-Inter-science, 1978.
[33] Kaplan, I. G., Miterev, A. M., *Khim.ys.Energ.* 1985, **19**, 208.
[34] Makarov, V. I., Polak, L. S., *Khim.Vys.Energ.* 1970, **4**, 3.
[35] De Heer, F. J., *Int.J.Radiat.Phys.Chem.* 1975, **7**, 137.
[36] Lias, S. G., Ausloos, P., *Ion-Molecule Reactions*. Washington: American Chemical Society, 1975.
[37] Terentyev, P. B., *Mass-spektrometriya v organicheskoi khimii* (Mass Spectrometry in Organic Chemistry). Moscow: Vysshaya Shkola, 1979.
[38] Virin, L. I., Dzhagatspanjan, R. V., Karatsevtsev, G. V. *et al. Ionno-molekulyarnye reaktsii v gazakh* (Ion-Molecule Reactions in Gas Media). Moscow: Nauka, 1979.
[39] Afanassiev, A. M., Kalyazin, E. P., *Khim.Vys.Energ.* 1982, **16**, 121.

[40] Saraeva, V. V. *Radioliz uglevodorodov v zhidkoi faze* (The Liquid Phase Radiolysis of Hydrocarbons). Moscow: Izd.Mosk.Univ., 1986.

[41] Bell, R. P., *The Proton in Chemistry*. 2nd Ed. Chapman and Hall, London, 1973.

[42] Pikaev, A. V., *Solvatirovannyi elektron v radiatsionnoi khimii* (The Solvated Electron in Radiation Chemistry). Moscow: Nauka, 1969.

[43] Christophorou, L. G., *Environmental Health Perspectives*. 1980, **36**, 3.

[44] Massey, H. S. W., *Negative Ions* (Russian Ed.). Moscow: Mir, 1979.

[45] Feng, D.-F., Kevan, L., *Chem.Rev.* 1980, **80**, 1.

[46] Raitsimring, A. M., Rapatskii, L. A., Samojlova, R. I., Tsvetkov, Yu.D., *Khim.Vys.Energ.* 1982, **16**, 394.

[47] Belloni, J., Billian, F., Delaire, J.-A., Delcourt, M. O., Marignier, J. L., *Radiat.Phys.Chem.* 1983, **21**, 177.

[48] Hunt, J. W., *Advances in Radiation Chemistry*, Burton, M. and Magee, J. L., N.Y.: Wiley, 1976, **5**, Ch. 3, 186.

[49] Pikaev, A. K., *Khim.Vys.Energ.* 1976, **10**, 107.

[50] Allen, A. O., *Drift Mobilities and Conduction Band Energies of Excess Electrons in Dielectric Liquids*, NSRDS–NBS. 1976, No. 58.

[51] Allen, A. O., *Yields of Free Ions Formed in Liquid by Radiation*, NSRDS–NBS. 1976, No. 57.

[52] Vannikov, A. V., Zhuravleva, T. S., *Khim.Vys.Energ.* 1980, **14**, 221.

[53] Salem, L. *Electrons in Chemical Reactions. First Principles.* N.Y.: Wiley-Interscience, 1982.

[54] Hart, E. J., Anbar, M., *The Hydrated Electron.* N.Y.-L: Wiley-Interscience, 1970.

[55] Zamaraev, K. I., Khairutdinov, R. F., Zhdanov, V. V., *Tunnelirovanie elektrona v khimii. Khimicheskie reaktsii na bolshikh rasstoyaniyakh* (Electron Tunnelling in Chemistry. Long Range Chemical Reactions). Novosibirsk: Nauka, 1985.

[56] Goldanskii, V. I., Trakhtenberg, L. I., Flerov, V. N., *Tunnel'nye yavleniya v khimicheskoj fizike.* (Tunnelling Phenomena in Chemical Physics). Moscow: Nauka, 1986.

[57] Bartczak, W. M., Kroh, J., Romanowska, E., Stradowski, C., *Curr. Top. Radiat. Res.* 1977, **11**, 307.

[58] Milinchuk, V. K., Tupikov, V. I., *Organic Radiation Chemistry Handbook.* Chichester: Ellis Horwood, 1989.

[59] Lam, R. J., Hunt, J. W., *Intern.J.Radiat.Phys.Chem.* 1975, **7**, 317.

[60] Tran-Thi Thu-Hoa, Koulkes-Pujo, A. M., *Radiat.Phys.Chem.* 1984, **23**, 745.

[61] Yakovlev, B. S., *Russ.Chem.Rev.* 1979, **48**, 615.

[62] Tezuka, T., Namba, H., Nakamura, Y., Chiba, M., Shinsaka, K., Hatano, Y., *Radiat.Phys.Chem.* 1983, **21**, 197.

[63] Kulik, P. P., Norman, G. E., Polak, L. S., *Khim.Vys.Energ.* 1976, **10**, 303.

[64] Kimura, T., Fueki, K., *J.Chem.Phys.* 1979, **70**, 3997.

[65] Biernbaum , V. M., Golde, M. F., Kaufman, F., *J.Chem.Phys.* 1976, **65**, 2715.

[66] Kay, B. D., Castelman, A. W., *J.Chem.Phys.*, 1983, **78**, (2), 4297.

[67] Nonhebel, D. C., Walton, J. C., *Free-Radical Chemistry. Structure and Mechanism.* Cambridge: University Press, 1974.

[68] Zubarev, V. E., Belevskii, V. N., Bugaenko, L. T., *Russ.Chem.Rev.* 1979, **48**, 729.

[69] Petryaev, E. P., Shadyro, O. I., *Radiatsionnaya khimiya bifunktsionalnykh organicheskikh soedinenii* (The Radiation Chemistry of Bifunctional Organic Compounds). Minsk: Izd.Universitetskoe, 1986.

[70] Palm, V. A., *Osnovy kolichestyvennoj teorii organicheskikh reaktsii* (The Foundations of Quantitative Theory of Organic Reactions). Leningrad: Khimiya, 1977.

[71] Pikaev, A. K., Kabakchi, S. A., *Reaktsionnaya sposobnost pervichnykh produktov radioliza vody* (Reactivity of the Primary Products of Water Radiolysis). Moscow; Energoatomizdat, 1982.

[72] Ross, A. B., Neta, P., *Rate Constants for Reactions of Inorganic Radicals in Aqueous Solutions.* NSRDS–NBS. 1979, No. 65.

[73] Belevskii, V. N., *Khim.Vys.Energ.* 1981, **15**, 3.

[74] *Konstanty skorosti gazofaznykh reasktsii* (Gas-Phase Reaction Rate Constants). Ed. Kondratyev, V. N., Moscow: Nauka, 1970.

[75] Silaev, M. M., Bugaenko, L. T., Kalyazin, E. P., *Vestn.Mosk.Univ.*, Ser. 2, Khim. 1986, **27**, 386.

[76] *Radiation Chemistry of Hydrocarbons*, Ed. Foeldiak, G. Budapest: Akad. Kiado, 1981.

[77] Pichuzhkin, V. I., Arshakuni, A. A. , *Khim.Vys.Energ.* 1975, **9**, 207.

[78] Denisov, E. T. *Konstanty skorosti gomoliticheskikh zhidkofaznykh reaktsii* (Rate Constants for Homolytic Liquid Phase Reactions). Moscow: Nauka, 1971.

[79] Baulch, D. L., *et al.*, *J.Phys.Chem.Rev.Data.* 1984, **13**, 1259.

[80] Jones, W. E., Macknight, S. D., Teng, L., *Chem.Rev.* 1973, **73**, 407.

[81] Buxton, G. V., Sellers, R. M. *Complication of rate Constants for the Reactivity of Metal Ions in Unusual Valency States*, NSRDS–NBS. 1978, No. 62.

[82] Bagdasaryan, Kh.S., *Russ.Chem.Rev.* 1984, **53**, 623.

[83] Melnikov, M.Ya., *Fizicheskaya Khimiya. Sovremennye Problemy*, Ed. Kolotyrkin Ya.M. Moscow; Khimiya, 1987, 48.

[84] Frankevich, E. L., *Khim.Vys.Energ.* 1980, **14**, 195.

[85] Buchachenko, A. L., *Fizicheskaya Khimiya. Sovremennye Problemy*, Ed. Kolotyrkin Ya.M. Moscow: Khimiya, 1980, 7.

[86] Anisimov, O. A., Molin, Yu.A., *Khim.Vys.Energ.* 1980, **14**, 307.

[87] *Comprehensive Chemical Kinetics.* Vol. 3. The Formation and Decay of

Excited Species. Ed. Bamford, C. H. and Tipper, C. F. H. Amsterdam-L.-N.Y.: Elsevier Publishing Company, 1969.

[88] Polak, L. S. *Neravnovesnaya khimicheskaya kinetika i ee primenenie* (Nonequilibrium Chemical Kinetics and Its Application). Moscow: Nauka, 1979.

[89] Polak, L. S., Ovsyannikov, A. A., Slovetskii, D. I., Vursel, F. B., *Teoreticheskaya i prikladnaya plazmokhimiya* (Theoretical and Applied Plasma Chemistry). Moscow: Nauka, 1979.

[90] *Kinetika i termodinamika khimicheskikh reaktsii v nizkotemperaturnoj plazme* (Kinetics and Thermodynamics of Chemical Reactions in Low-Temperature Plasma), Ed. Polak, L. S. Moscow: Nauka, 1979.

[91] Kompaniets, V. Z., Ovsyannikov, A. A., Polak, L. S. *Khimicheskie reaktsii v turbulentnykh potokakh gaza i plazmy* (Chemical Reactions in Gas and Plasma Turbulent Flows). Moscow: Nauka, 1979.

[92] Kompaniets, V. Z., Polak, L. S., Konstantinov, A. A. *et al. Matematicheskoe modelirovanie plazmokhimicheskikh protsessov s pomoshch'yu EVM, Khimicheskie reaktsii v nizko-temperaturnoj plasme* (Computer Simulation of Plasmochemical Processes. In *Chemical Reactions in Low-Temperature Plasma*. Moscow: Nauka, 1977, 6.

[93] Ivanov, Yu.A., Lebedev, Yu.A., Polak, L. S. *Metody kontaktnoj diagnostiki v neravnovesnoj plazmokhimii* (Contact Diagnostics Techniques in Nonequilibrium Plasma Chemistry). Moscow: Nauka, 1981.

[94] Benson, S. W., *The Foundations of Chemical Kinetics*. McGraw-Hill Series in Advanced Chemistry: XYII, N.Y., 1960.

[95] Rusanov, V. D., Fridman, A. A. *Fizika khimicheski aktivoj plazmy* (The Physics of Chemically Active Plasmas). Moscow: Nauka, 1984.

[96] Polak, L. S., Goldenberg, M.Ya., Levitskii, A. A., *Vychislitelnye metody v khimicheskoj kinetike* (Computational Methods in Chemical Kinetics). Moscow: Nauka, 1984.

[97] Prigogine, I., *Non-Equilibrium Statistical Mechanics*, Interscience, New York, 1962.

[98] Tikhonov, A. N., Arseniya, V.Ya., *Metody resheniya nekorrektnykh zadach* (Methods of Solution of Incorrect Problems). Moscow: Nauka, 1974.

[99] Tolman, H. C., *J.Am.Chem.Soc.* 1920, **42**, 2506.

[100] Moelwyn-Hughes, E. A., *The Kinetics of Reactions in Solutions* 2nd edn. Oxford: Clarendon Press, 1947.

[101] Hinshelwood, C. N. *The Kinetics of Chemical Change in Gaseous Systems*. 2nd Ed. Oxford: Clarendon Press, 1929.

[102] Polak, L. S., *Ocnovnye polozheniya obobshchennoj neravnovesnoj khimicheskoj kinetiki* (The Principles of Generalized Nonequilibrium Chemical Kinetics). Moscow: Preprint INKhs An SSSR, 1972, N.1.

[103] Rice, O. K., *Statistical Mechanics and Kinetics*. N.Y.: Wiley, 1967.

[104] Alexayev, B. V., *Matematicheskaya kinetika reagiruyushchikh gazov* (Mathematical Kinetics of Reacting Gases). Moscow: Nauka, 1982.

[105] Klimontovich, Yu.L., *Statisticheskaya fizika* (Statistical Physics). Moscow: Nauka, 1982.

[106] *Nobel-Symposium on Fast Reactions and Primary Processes in Chemical Kinetics*. N.Y.: Wiley, 1967.

[107] Slater, N. B. *Theory of Unimolecular Reactions*. N.Y.: Wiley, 1959.

[108] Johnston, H. S. *Gas Phase Reaction Rate Theory*. N.Y.: Ronald Press, 1966.

[109] Cotrell, T. L., McCorbey, J. C. *Molecular Energy Transfer in Gases* L.: Butterworths, 1961.

[110] Levin, R. D. *Quantum Mechanics of Molecular rate Processes*. Oxford: Pergamon Press, 1969.

[111] Kybo, R., *Statistical Mechanics* (Russian Ed.). Moscow, Mir, 1967.

[112] Glasstone, S. Laidler, K. J., Eyring, H., *The Theory of Rate Processes*. N.Y.-L.: 1941.

[113] Zwanzig, R., *Phys.Rev.* 1961, **124**, 983.

[114] Eliason, M., Hirschfelder, J., *J.Chem.Phys.* 1959, **30**, 1426.

[115] Shuler, K., *Chemischer Elementar Processes*. Berlin: Springer Verlag, 1968, 1.

[116] Serauskas, R. V., Schlag, E. F., *J.Chem.Phys.* 1965, **42**, 3009.

[117] Jangel, R., *Foundations of Classical and Quantum Mechanics*. L: Oxford Press, 1970.

[118] Huang Kerson. *Statistical Mechanics* (Russian Ed.). Moscow: Mir, 1966.

[119] Balescu, R. *Equilibrium and Nonequilibrium Statistical Mechanics*. N.Y.: Wiley, 1975.

[120] Polak, L. S., Khachojan, A.V., *Khim.Vys.Energ.* 1981, **15**, 26.

[121] Polak, L. S., Khachojan, A.V., *Khim.Vys.Energ.* 1982, **15**, 391.

[122] Karplus, M., Raff, L. M., *J.Chem.Phys.* 1964, **41**, 1266.

[123] Green, E. F. Moursund, A. L., Ross, J., *Adv.Chem.Phys.* 1966, **10**, 135.

[124] Schluter, D., *J.Quant.Spect.Rad.Transfer.* 1965, **5**, 87.

[125] Polak, L. S., *Reactions under Plasma Conditions*. N.Y.: Interscience-Wiley, 1971, **2**, 13.

[126] *Ocherki fiziki i khimii nizkotemperaturnoj plazmy* (Essays on Physics and Chemistry of Low-Temperature Plasma), Ed. Polak, L. S., Moscow: Nauka, 1974.

[127] *Modelirovanie i metody rascheta fiziko-khimicheskikh protsessov v plazme* (Modelling and Calculation Methods for Physicochemical Processes in Plasma). Ed. Polak, L. S. Moscow: Nauka, 1974.

[128] Arnold, L., Synergetics. *Dynamics of Synergetic Systems*. Berlin, Springer Verlag, 1980, **6**, 107.

[129] Zubarev, D. N., *Neravnovesnaya statisticheskaya termodinamika* (Non-equilibrium Statistical Thermodynamics). Moscow: Nauka, 1971.

[130] Michelson, M. L., *Chem.Eng.Sci.* 1977, **32**, 454.
[131] Oppenheim, J., Shuler, K., *Phys.Rev.* 1965, **138**, B, 1007.
[132] Kupperman, A., White, J. M., *J.Chem.Phys.* 1966, **44**, 4352.
[133] Corvin, K. K., Corrigan, S. J., *J. Chem.Phys.* 1969, **50**, 2570.
[134] Blomberg, C., *J.Stat.Phys.* 1981, **25**, 73.
[135] Kurz, T. G., *J.Appl.Probab.* 1977, **7**, 344.
[136] Van Kampen, N. G., *Adv.Chem.Phys.* 1976, **34**, 245.
[137] Gardiner, C. W., Chaturverdi, S., *J.Stat.Phys.* 1977, **17**, 429.
[138] Ortoleva, P., *Select Topics from The Theory of Nonlinear Physico-chemical Phenomena. Theoretical Chemistry. Periodicities in Chemistry and Biology.* N.Y.: Wiley-Interscience, 1978, **4**, 235.
[139] Polak, L. S., Mikhailov, A. S., *Samoorganizatsiya v neravnovesnykh fiziko-khimicheskikh sistemakh* (Self-Organization in Nonequilibrium Physicochemical Systems). Moscow: Nauka, 1983.
[140] Polak, L. S., Lebedev, Yu.A., Levitskii, A. A., *et al.*, *Novye problemy plazmokhimicheskoj kinetiki* (New Problems in Plasmochemical Kinetics). Preprint INKhS AN SSSR, 1981.
[141] Zwanzig, R., *J.Chem.Phys.* 1960, **55**, 1338.
[142] Greem, M. S., *J.Chem.Phys.* 1952, **47**, 67.
[143] Pauli, M. *Probleme der modernen Physik.* Zum A. Sommerfeld 60 Geburtstage. Leipzig: Springer Verlag, 1928, 30.
[144] Polak, L. S., *Pure and Appl.Chem.* 1974, **39**, 307.
[145] Salpeter, E. E., *Phys.Rev.* 1961, **122**, 1663.
[146] Kac, M., *Probability and Related Topics in Physical Sciences.* N.Y.: Wiley-Interscience, 1959.
[147] Van Hove, L., *Physica.* 1955, **21**, 517.
[148] Kawasaki, K., *Ann.Phys.* 1970, **61**, 1; *J.Phys.* 1973, **A6**, 1289.
[149] Polak, L. S., *Pure and Appl.Chem.* 1966, **13**, 345.
[150] Polak, L. S., *10th Intern.Conf. Gases. Invited Papers*, Oxford: Oxford Press, 1971, 113.
[151] Montroll, E. W., *Fundamental Problems in Statistical Mechanics.* Amsterdam: Wiley-Interscience, 1962, 230.
[152] Nakajama, E., *Prog.Theor.Phys.* 1958, **20**, 948.
[153] Oppenheim, J., *Adv.Chem.Phys.* 1958, **1**, 1.
[154] Prigogine, I., Resibois, P., *Physica.* 1961, **27**, 629.
[155] Resibois, P. *Physica.* 1963, **29**, 721.
[156] Swenson, R. J., *J.Math.Phys.* 1962, **3**, 1017.
[157] Vallander, S. V., Yegorov, I. A., Rydalevskaya, M. A. *Aerodinamika razrezhennykh gazov.* Leningrad: Izd.LGU, 1965, **2**, 14.
[158] Shuler, K., *5th Intern.Symp. on Combustion.* N.Y.: Reinhold Publishing, 1955, 56.
[159] Treanor, Ch.E., Marrone, P. V., *Phys.Fluids.* 1962, **5**, 1022.
[160] Zwolinski, B. J., Eyring, H., *J.Am.Chem.Soc.* 1947, **69**, 2702.

378   References

[161] Widom, B., *J.Chem.Phys.* 1961, **34**, 2050.
[162] Cercignani, C. *Theory and Practice of the Boltzmann Equation.* Scot. Academic: Edinburgh, 1975.
[163] Wachman, M., Hamel, B. B., *Rarefied Gas Dynamics.* Washington: Academic Press, 1967, **1**, 591.
[164] Shuler, K. E., Weiss, G. H., *J.Chem.Phys.* 1963, **38**, 505.
[165] Suider, R. F., *J.Chem.Phys.* 1964, **41**, 591.
[166] Masur, J., Rubin, R. J., *J.Chem.Phys.* 1959, **31**, 1351.
[167] Nitzan, A., Ortoleva, P., Deutch, J., Ross, J., *J.Chem.Phys.* 1974, **61**, 1056.
[168] Grossman, S., Schernner, R., *Z.Phys.* 1978, **B30**, 325.
[169] Ebeling, W., Malchow, H., *Ann.Phys.N.Folge.* 1979, **36**, 121.
[170] Gardiner, C. W., Chaturverdi, S., Wallis, D. F., *Ann. N.Y. Acad.Sci.* 1979, **316**, 453.
[171] Bertrand, G., *Synergetics, Far from Equilibrium*, Berlin, N.Y.: Springer Verlag, 1979, 147.
[172] *Primenenie vychislitelnoj matematiki v fizicheskoj i khimicheskoj kinetike* (Application of Computational Mathematics in Physical and Chemical Kinetics). Ed. Polak, L. S., Moscow: Nauka, 1969.
[173] Bugaenko, L. T., Kabakchi, S. A., *Metod statsionarnykh kontsentratsii v radiatsionnoj khimii* (Steady State Concentration Method in Radiation Chemistry). Moscow: Izd.Mosk.Univ., 1974.
[174] Porter, G., West, M., *Methods of Investigation of Fast Reactions* (Russian Ed.). Moscow, Mir, 1977.
[175] Kuzmin, M. G., Sadovskii, N. A., *Khim.Vys.Energ.* 1975, **9**, 291.
[176] Ware, W. R., *Creation and Detection of the Excited State.* Ed. Lamola, A. N.Y.: Marcell Dekker, 1971.
[177] Pikaev, A. K., Kabakchi, S. A., Makarov, I. E., Ershov, B. G., *Impuls'nyi radioliz i ego primenenie* (Pulse Radiolysis and Its Application). Moscow: Atomizdat, 1980.
[178] Tabata, Y., *Practical Applications of Pulse Radiolysis Techniqe.*, *Proc. Intern. Symp. Appl. Technol. Ion. Radiat.* 1982, **2**, 837.
[179] Blom, C. E., Moeller, K., Filgueira, R. R., *Chem.Phys.Lett.* 1987, **140**, 489.
[180] Van Brundt, R. T., Leep, D., *J.Appl.Phys.* 1981, **52**, 6588.
[181] Kuzmin, M. G., *Khimicheskaya i biologicheskaya kinetika* (in: *Chemical and Biological Kinetics*). Ed. Emanuel, N. M. Moscow: Izd.Mosk.Univ., 1983, 47.
[182] Mauser, H. *Formale Kinetik.* Dusseldorf: Bertelsmann Universitats Verlag, 1974.
[183] *Investigation of Rates and Mechanisms of Reactions. Techniques of Chemistry.* **VI.** Ed. Bernasconi, C. F., N.Y.: Wiley, 1986.

[184] Demas, J. N. *Excited State Lifetime Measurements.* N.Y.: Academic Press, 1983.

[185] O'Connor, D. V., Phillips, D., *Time-Correlated Single Photon Counting.* L.: Academic Press, 1984.

[186] Birks, J., Munro, I., *Usp.Fiz.Nauk.* 1967, **105**, 251.

[187] Demas, J. N., Adamson, A. W., *J.Phys.Chem.* 1971, **75**, 2464.

[188] Kuzmin, M. G., Guseva, L. N., *Zh.Prikl.Spektr.* 1972, **17**, 1015.

[189] Lakowicz, J. R., *Principles of Fluorescence Spectroscopy.* N.Y.-L.: Plenum Press, 1983.

[190] Noyes, R. M., *Progr.React.Kinet.* 1961, **1**, 131.

[191] Ware, W. R., Nemzek, T. L., *Chem.Phys.Lett.* 1973, **23**, 557.

[192] Ware, W. R., Lee, S. K., Brant, G. J., Chow, P. P., *J.Chem.Phys.* 1971, **54**, 4729.

[193] Agranovich, V. M., *Teoriya eksitonov* (Exciton Theory). Moscow: Nauka, 1968.

[194] Sadovskii, N. A., Kuzmin, M. G., *Dokl.Akad.Nauk SSSR.* 1975, **222**, 1380.

[195] *Modelling of Chemical Reaction System.* Berlin–New York: Springer Verlag, 1981.

[196] Lambert, J. D., *Computational Methods in Ordinary Differential Equations.* N.Y.: Wiley, 1973.

[197] Hartman, Ph., *Ordinary Differential Equations.* New York–London–Sydney: Wiley, 1964.

[198] Denn, M. M., *Stability of Reaction and Transport Process.* N.Y.: Cliffs, 1975.

[199] Slovetskii, D. I., *Mekhanizmy khimicheskikh reaktsii v neravnovesnoj plazme* (Mechanisms of Chemical Reactions in Nonequilibrium Plasma). Moscow: Nauka, 1980.

[200] Gagarin, S. G., Kolbanovskii, Yu.A., Polak, L. S., *Primenenie vychislitelnoj matematiki v khimicheskoj i fizicheskoj kinetike* (in: *Application of Computational Mathematics in Physical and Chemical Kinetics*), Moscow: Nauka, 1969, 82.

[201] Pavlov, B. V., Brin, E. F., *Khim.Fizika.* 1982, **1**, 509.

[202] Dmitriev, V. I. *Prostaya kinetika* (Simple Kinetics). Novosibirsk: Nauka, 1982, 382.

[203] Pisarenko, V. I., Pogorelov, A. G., Kononov, M. F., *Dokl.Akad.Nauk SSSR.* 1966, **167**, 859.

[204] Spivak, S. I., Gorskii, V. G., *Khim.Fizika.* 1982, **1**, 237.

[205] Porter, R. N., *Ann.Rev.Phys.Chem.* 1064, **26**, 317.

[206] Porter, R. N., Karplus, M., *J.Chem.Phys.* 1964, **40**, 1105.

[207] Banker, D. In: *Computational Techniques in Physics of Atomic and Molecular Collisions* (Russian Ed.). Moscow: Mir, 1974, 277.

[208]   Denisik, S. A., Malama, Yu.G., Lebedev, S. N., Polak, L. S., *Primenenie Vychislitelnoj matematiki v khimicheskoj i fizicheskoj kinetike* (in: *Application of Computational Mathematics in Chemical and Physical Kinetics*), Ed. Polak, L. S. Moscow: Nauka, 1969, 277.

[209]   Grinchak, M. B., Levitskii, A. A., Polak, L. S., Umanskii, S.Ya., *Khim. Fizika.* 1982, **1**, 331.

[210]   Grinchak, M. B., Levitskii, A. A., *Eksperientalnye i teoreticheskie issledovaniya plazmmokhimicheskikh protsessov* (in: *Experimental and Theoretical Studies on Plasmochemical Processes*), Moscow: Nauka, 1983, 127.

[211]   Robinson, P. J., Holbrook, K. A., *Unimolecular Reactions.* L.-N.Y.–Sydney–Toronto: Wiley-Interscience, 1972.

[212]   Kuznetsov, N. M., *Kinetika monomoleculyarnukh reaktsii* (The Kinetics of Unimolecular Reactions). Moscow: Nauka, 1982.

[213]   Luss, D., *Steady-State Multiplicity of Chemically Reacting Systems*. In: *Modelling of Chemical Reaction Systems*. Berlin–N.Y.: Springer Verlag, 1981.

[214]   Valter, B. V., Salnikov, I. E., *Ustojchivoct' rezhimov raboty khimicheskikh reaktorov* (The Stability of Operation Conditions of Chemical Reactors). Moscow: Khimiya, 1981.

[215]   Noyes, R. M., *Ber.Bunsenges.phys.Chem.* 1980, **84**, 295.

[216]   Perlmutter, D. D., *Stability of Chemical Reactors*. N.Y.: Cliffs, 1972.

[217]   Aris, R., *The Mathematical Theory of Diffusion and Reaction in Permeable Catalysts*. Oxford: Clarendon Press, 1975.

[218]   Varma, A., Aris, R., *Stirred Pots and Empty Tubes*. In: *Chemical Reactor Theory*. N.Y.: Prentice, 1977.

[219]   Luss, D., *Steady State Multiplicity and Criteria for Chemically Reacting Systems*. In: *Dynamics and Modelling of Reactive Systems*. N.Y.: Academic Press, 1980.

[220]   Zhivotov, V. K., Levitskii, A. A., Macheret, S. O., Polak, L. S., Rusanov, V. D., Fridman, A. A., *High Energ.Chem.*, 1985, **19**, 425.

[221]   Abramenkova, I. A., Bugaenko, L. T., Kalyazin, E. P., *Matematicheskoe modelirovanie kinetiki radiatsionno-khimicheskikh protsessov, Modelirovanie na EVM radiatsionnykh defektov v kristallakh.* (Mathematical Modelling of Kinetics of Radiation-Induced Chemical Processes. In: *Computer Simulation of Radiation Defects in Crystals*). Leningrad: RTP LIYaF, 1983, 157.

[222]   Batyuk, S. A., Mekhanizm radioliza v shirokom intervale doz. Diss.kand. khim.nauk (The Mechanism of Radiolysis in a Wide Dose Range. PhD Thesis in Chem.). Moscow. MGU, 1983.

[223]   Rosinger, E. L. J., Dixon, R. S., *Atomic Energy of Canada Ltd. Report* **AECL-5958**, 1977.

[224]   Boyd, A. W., Carver, M. B., Dixon, R. S., *Radiat.Phys.Chem.* 1980, **15**, 177.

[225]   Kummler, R., Leffert, C., Im, K., Piccirelli, R., Kevan, L., *J.Phys.Chem.* 1977, **81**, 2451.

[226]   Cremer H. Knocke, K. F., Rortgen, H., *Z.AMM.* 1973, **53**, 299.

[227]   Farrow, L. A., Edelson, *Int.J.Chem.Kinet.* 1974, **6**, 787.

[228]   Abramenkov, A. G., Abramenkova, I. A., *Khim.Vys.Energ.* 1979, **13**, 557, *Deposited Papers VINITI* No. 1943-79.

[229]   Bugaenko, V. L., Grishkin, V. L., *Paket programm dlya modelirovaniya protsessov khimicheskoj kinetiki. Opisanie metoda* (Software Package for Simulation of Chemical Kinetic Processes. Description of the Method). Moscow: Preprint ITEF, 1980. No. 50.

[230]   Bugaenko, V. L., Grishkin, V. L. *Paket programm dlya modelirovaniya protsessov khimicheskoj kinetiki. Opisanie vvoda dannykh* (Software Package for Simulation of Chemical Kinetic Processes. Description of Data Entry) Moscow: Preprint ITEF, 1981, No. 54.

[231]   Giria, G., Gopinathan, C., *Radiat.Phys.Chem.* 1980, **16**, 245.

[232]   Jonah, C. D., Matheson, M. S., Miller, J. R., Hart, E. J., *J.Phys.Chem.* 1976, **80**, 1276.

[233]   Kuppermann, A. In: *Action chimiques et biologique des radiations.* Ed. Haissinsky, M. Paris: Masson et Cie, 1961. **5**, 85.

[234]   Byakov, V. M., *Mechanism radioliza vody* (In: The mechanism of Water Radiolysis). Ed. Bugaenko, L. T. and Byakov, V. M. Moscow: Izd.Mosk. Univ., 1970, S.5.

[235]   Hayon, E., *Trans.Farad.Soc.* 1965, **61**, 723.

[236]   Mahlman, H. A., *J.Chem.Phys.* 1961, **35**, 936.

[237]   Tretyakova, N. I., Byakov, V. M., *Mekhanizm obrazovaniya i priroda predshestvennikov radioliticheskoj perekisi vodoroda* (Formation Mechanism and the Nature of Precursors of Radiolytic Hydrogen Peroxide). Moscow: Preprint ITEF, 1982, No. 115.

[238]   Byakov, V. M., Grishkin, V. L., *Rekombinatsiya ion-elektronnykh par v prisutstvii aktseptora* (Ion-Electron Pair Recombination in Presence of Scavenger). Moscow: Preprint ITEF, 1978, No. 4.

[239]   Klein, W., Schuler, R. H., *J.Phys.Chem.* 1973, **77**, 978.

[240]   Robinson, R. G., Freeman, G. R. *J.Chem.Phys.* 1968, **48**, 982.

[241]   Iha, K. N., Freeman, G. R., *J.Chem.Phys.* 1968, **48**, 5480.

[242]   Kuzmin, M. G., *Fizicheskaya Khimiya. Sovremennye Problemy*, Ed. Kolotyrkin Ya. M. Moscow: Khimya, 1985, 31.

[243]   *Einfurung in die Photochemie*, Unter Leitung von Becker H.G.O. Berlin: VEB Deutscher Verlag der Wissenschaften, 1983.

[244]   *Reactivity of the Photoexcited Molecule.* Interscience Publishers, N.Y. 1967.

[245]   Barltrop, J. A., Coyle, J. D., *Excited States in Organic Chemistry*. Cambridge: Wiley, 1975.

[246]   *Organic Photochemistry*. N.Y.: Marcell Dekker, 1973.

[247]   *Rearrangement in Ground and Excited States*. Ed. De Mayo P. N.Y.: Academic Press, 1980.

248]   Grabowski, Z. R., Rotkiewicz, K., Siemiasczuk, A., *Nouveau J. de chimie*. 1970, **3**, 443.

[249]   Suzuki, H. *Electronic Absorption Spectra and Geometry of Organic Molecules. An Application of Molecular Orbital Theory*. N.Y.-L.: Academic Press, 1967.

[250]   *Excited State*. **1** and **2**. Ed. Lim, E. C. N.Y.: Academic Press, 1974.

[251]   Grabowski, Z. R., *et al., J.Luminecence*. 1969, **18–19**, 420.

[252]   Bakhshiev, N. G., Knyazhanskii, M. I., Minkin, V. I., Osipov, O. A., Saidov, G. V., *Russ.Chem.Rev.* 1969, **38**, 740.

[253]   Neporent, B. S., *Izv.Akad Nauk SSSR*. Ser.Fiz. 1973, **37**, 236.

[254]   Porter, G. Balzani, V., Moggi, L., *Adv.Photochem.* 1974, **4**, 147.

[255]   *Spektroskopiya fotoprevrashchenii v molekulakh* (Spectroscopy of Phototransformations in Molecules). Leningrad: Nauka, 1977.

[256]   *Vozbuzhdennye molekuly. Kinetika prevrashchenii* (Excited Molecules. The Kinetics of Transformation). Leningrad: Nauka, 1982.

[257]   Medvedev, E. S., Osherov, V. I., *Teoriya bezyzluchatelnykh perekhodov v mnogoatomnykh moleculakh* (The Theory of Radiationless Transitions in Polyatomic Molecules). Moscow: Nauka, 1983.

[258]   Kuzmin, M. G., *Problem biophotokhimii. Trudy Moskovskogo obshch. ispytatelei prirody,* Moscow: Nauka, 1973, **49**, 79.

[259]   Kuzmin, M. G., Guseva, L. N., *Dokl.Akad.Nauk SSSR*. 1971, **200**, 375.

[260]   Nikitin, E. E., *Russ.Chem.Rev.* 1968, **37**, 716.

[261]   Letokhov, V. S., *Nelineinye selektivnye fotoprotsessy v atomakh i molekulakh* (Nonlinear Selective Processes in Atoms and Molecules). Moscow: Nauka, 1983.

[262]   Michl, J., *Topics Current Chem.* 1974, **46**, 1: *Pure and Appl. Chem.* 1975, **41**, 507.

[263]   Berstein, R. B., *Molecular Reactions Dynamics*. Oxford: Clarendon Press, 1974.

[264]   Turro, N. J., McVey, J., Ramamurthy, P., Lechtken, P., *Angew.Chem.* 1979, **91**, 597.

[265]   Kryukov, A. I., Sherstyuk, V. P., Dilung, I. I., *Fotoperenos elektrona i ego prikladnye aspekty* (Photoinduced Electron Transfer and Aspects of Its Applications). Kiev: Naukova dumka, 1982.

[266]   Birks, J. B., *Photochemistry of Aromatic Molecules*. N.Y.: Wiley, 1970.

[267]   *Organic Molecular Photophysics*. **1** and **2**. Ed. Birks, J. B., N.Y.: Wiley, 1975.

[268]   *The Exiplex*. Ed. Gordon, M. and Ware, W. R. N.Y.: Academic Press, 1975.

[269] Mulliken, R. S., Pearson, W. B., *Molecular Complexes*. N.Y.: Wiley, 1969.
[270] McGlynn, S. P., *Chem. Rev.* 1958, **58**, 1113.
[271] Andrews, L. J., Keefer, R. M., *Molecular Complexes in Organic Chemistry* (Russian Ed.). Moscow: Mir, 1967.
[272] Kampar, V. E., Neilands O.Ya., *Russ.Chem.Rev.* 1977, **46**, 503.
[273] Christodoulas, N., McGlynn, S. P., *J.Chem.Phys.* 1964, **40**, 43.
[274] Masuhara, H., Mataga, N., *Bull.Chem.Soc. Japan*, 1972, **45**, 43.
[275] Marcus, R. A., *J.Chem.Phys.* 1965, **43**, 679.
[276] Khairutdinov, R. F., Sadovskii, I. A., Parmon, V. N., Kuzmin, M. G., Zamaraev, K. I., *Dokl.Akad.Nauk SSSR.* 1975, **220**, 888.
[277] Rehm, W., Weller, A. *Israel J.Chem.* 1970, **8**, 259.
[278] Chibisov, A. K., *Russ.Chem.Rev.* 1981, **50**, 615.
[279] Rehm, W., Weller, A., *Ber.Bunsenges.phys.Chem.* 1969, **73**, 834.
[280] Plotnikov, V. G., Ovchnnikov, A. A., *Russ.Chem.Rev.* 1978, **47**, 247.
[281] Cornelisse, J., Lodder, G., Havinga, E., *Rev.Chem.Intermediates.* 1979, **2**, 231.
[282] Cornelisse, J., Havinga, E., *Chem.Rev.* 1975, **75**, 353.
[283] Rossi, R. A., de Rossi, R. H., *Aromatic Substitution by the SRN1 Mechanism*, Washington, D.C.: American Chemical Society, 1983.
[284] Kuzmin, M. G., Guseva, L. N., *Khim.Vys.Energ.* 1970, **4**, 24.
[285] Tomkiewicz, M., Klein, M., *Proc.Nat.Acad. Sci. USA.* 1973, **70**, 143.
[286] Dauben, W. G., Salem, L., Turro, N. J., *Acc.Chem.Res.* 1975, **8**, 41.
[287] Woodward, R. B., Hoffman, R., *The Conservation of Orbital Symmetry*. Weinheim, 1970.
[288] Husain, D., Donovan, R. J., *Adv.Photochem.* 1971, **8**, 1.
[289] Klopffer, W., *Adv.Photochem.* 1977, **10**, 311.
[290] Uzhinov, B. M., Martynov, I.Yu., Kuzmin, M. G., *Zh.Prikl.Spektr.* 1974, **20**, 495.
[291] Krasheninnikov, A. A., Shablya, A. V., *Optika i Spektr.* 1973, **34**, 1214.
[292] Eigen, M., Krise, W., Maass, G., De Mayer, L., *Prog. React. Kinet.* 1961, **1**, 286.
[293] Demyashkevich, A. B., Zaitsev, N. K., Kuzmin, M. G., *Khim.Vys.Energ.* 1982, **16**, 60.
[294] Marcus, R. A., *J.Phys.Chem.* 1968, **72**, 4249.
[295] Levich, V. G., German, E. D., Dogonadze, R. R., Kuznetsov, A. M., Kharkats, Yu.I., *Teor.Eksper.Khim.* 1970, **6**, 455.
[296] Shapovalov, V. L., Demyashkevich, A. B., Kuzmin, M. G., *Khim.Vys. Energ.* 1982, **16**, 433.
[297] *Molekulyarnaya fotonika* (Molecular Photonics). Moscow: Nauka, 1970.
[298] Zaitsev, N. K., Demyashkevich, A. B., Ivanov, V. L., Kuzmin, M. G., *Zh.Prikl.Spektr.* 1978, **29**, 496.
[299] Grigoryeva, T. M., Kuzmin, M. G., *Usp.Nauchn.Fotogr.* 1978, **19**, 177.

384 **References**

[300] Klein, U. K. A., Hafner, F. W., *Chem.Phys.Lett.* 1976, **43**, 141.
[301] Grigoryeva, T. M., Ivanov, V. L., Kuzmin, M. G., *Dokl.Akad.Nauk SSSR.* 1978, **238**, 603.
[302] Grigoryeva, T. M., Ivanov, V. L., Kuzmin, M. G., *Khim.Vys.Energ.* 1979, **13**, 325.
[303] Srinivasan, R., *Adv.Photochem.* 1963, **1**, 83.
[304] Turro, N. J., Dalton, J. C., Dawes, K., Farrington, G., Hautala, R., Morton, D. *et al. Acc.Chem.Res.* 1972, **5**, 92.
[305] Dauben, W. G., Salem, L., Turro, N. J., *Acc.Chem.Res.* 1975, **8** 41.
[306] Wagner, P. J., *Topics Current Chem.* 1976, **66**, 1.
[307] Scaiano, J. C., *J.Photochem.* 1973–1974, **2**, 81.
[308] Closs, G., Doubleday, C., *J.Am.Chem.Soc.* 1973, **95**, 2735.
[309] Walling, C., Gibson, M., *J.Am.Chem.Soc.* 1965, **87**, 3361.
[310] *Chemical Reactivity and Reaction Paths.* Ed. Klopman, G. New York–London–Sydney–Toronto: Wiley-Interscience, 1974.
[311] Goddard, W. A., *J.Am.Chem.Soc.* 1972, **94**, 793.
[312] *Mechanism of Molecular Migration.* Ed. Thyagarajan, B. S. L.: Interscience, 1969, **2**, 117.
[313] Longuet-Higgins, H. C., Abrahamson, E. W., *J.Am.Chem.Soc.* 1965, **87**, 2045.
[314] Dewar, M. J. S., *The Molecular Orbital Theory of Organic Chemistry.* McGraw-Hill, 1969.
[315] Van der Lugt, W.Th. A. M., Oosterhoff, L. J., *Chem.Comm.* 1968, 1235.
[316] Fischer, E., *Ber.Bunsenges.phys.Chem.* 1967, **73**, 739.
[317] Steinmetz, R., *Fortschr.chem.Fortsch.* 1967, **7**, 445.
[318] Van Tamelen, E. E., *Acc.Chem.Res.* 1972, **5**, 186.
[319] Fischer, E., *Fortchr.chem.Fortch.* 1967, **7**, 605.
[320] Dahne, S., *Z.Wiss.Photogr.Photophys.Photochem.* 1968, **62**, 183.
[321] Kropp, P. J., In: *Organic Photochemistry.* Ed. Chapman, O. L. N.Y.: Marcell Dekker, 1967, **1**, 1.
[322] Bellus, J. D., *Adv.Photochem.* 1971, **8**, 109.
[323] Balzani, V., Moggi, L., Manfrin, M. F., Bolletta, F., Laurence, G. S., *Coord.Chem.Rev.* 1975, **15**, 321.
[324] *Photochemistry of Coordination Compounds.* Eds. Balzani V. and Carassiti, V. N.Y.-L: Academic Press, 1970.
[325] Molin, Yu.N., Panfilov, V. N., Petrov, A. K., *Infrakrasnaya fotokhimiya* (Infrared Photochemistry). Novosibirsk: Nauka. 1985.
[326] Harley, H., Ponder, A. O., Bowen, B. J., Merton, T. R., *Phil.Mag.* 1922, **43**, 430.
[327] Kuhn, W., Martin, H., *Naturwiss*, 1932, **20**, 772; *Z.phys.Chem.* 1933, **21**, 93.
[328] Mrozowski, S., *Z.Phys.Chem.* 1932, **78**, 824.

[329] Zuber, K., *Nature*, 1935, **136**, 796; *Helv.Phys.Acta.* 1935, **8**, 487; *ibid.* 1936, **9**, 285.

[330] Levy, R., Janes, G. S., *Apparatus and method for the separation of Isotopes.* US Patent. 1973. No. 3772519. *Chem.Abstr.* 1974, **80**, 33090j.

[331] Nebenzahl, I., Levin, N., *Separation of uranium-235.* West German Patent. 1973, No. 2312194. *Chem.Abstr.* 1973, **79**, 152148y.

[332] Ivanov, L. N., Letokhov, V. S., *Kvantovaya elektronika.* **2**, (Quantum Electronics. 2), 1975.

[333] Tuccio, S. A., *et al.*, *IEEE J.Quant.Electr.* 1974, **QE-10**. 790.

[334] Ambartsumian, R. V., Letokhov, V. S., Makarov, G. N., Puretskii, A. A., *Pis'ma ZhTEF.* 1972, **15**, 709; *ibid.* 1973, **17**, 91.

[335] Houston, P. L., Nowak, A. V., Steinfeld, J. I., *J.Chem.Phys.* 1973, **58**, 58.

[336] *Photoselective Chemistry.* Parts 1 and 2, *Adv.Chem.Phys.* 1981, **47**.

[337] Lind, S. C. *The Chemical Effects of Alpha-particles and Electrons.* Chem. Catalog, New York, 1928.

[338] De Boer, K. *A Practical Guide to Spline Method* (Russian Ed.) Moscow: Radio i svyaz', 1985.

[339] Pikaev, A. K., *Khim.Vys.Energ.* 1985, **19**, 196.

[340] Raitsimring, A. M., Tsvetkov, Yu.D., *Khim.Vys.Energ.* 1980, **14**, 229.

[341] Pshezhetskii S.Ya., Dmitriev, M. T., *Radiatsionno-khimicheskie prevrash-cheniya v vozdushnoj srede* (Radiation Induced Chemical Transformations in Air Medium). Moscow: Atomizdat, 1978.

[342] Dzantiev, B. G., Ermakov, A. N., Zhitomirskii, B. M., Popov, V. N., *Khim.Vys.Energ.* 1982, **16**, 97.

[343] Machi, S., Tokunaga, D., Nishimura, K., Hashimoto, S., Kawakami, W., Washino, M. *et al.* *Radiat.Phys.Chem.* 1977, **9**, 371.

[344] Willis, C., Boyd, A. W., *Int.J.Radiat.Phys.Chem.* 1976, **8**, 71.

[345] Wilson, D. E., Armstrong, D. A., *Int.J.Radiat.Phys.Chem.* 1970, **2**, 297.

[346] Dixon, R. S., *Radiat.Res.Rev.* 1970, **2**, 237.

[347] Pikaev, A. K., *Sovremennaya radiatsionnaya khimia. Radioliz gazov i zhidkostei* (Modern Radiation Chemistry. Radiolyses of Gases and Liquids). Moscow: Nauka, 1986.

[348] Peterson, D. B. *The Radiation Chemistry of Gaseous Ammonia*, NSRDS-NBS. 1974, No. 44.

[349] Johnson, G. R., *Radiation Chemistry of Nitrous Oxide Gas. Primary Processes. Elementary Reactions and Yields*, NSRDS-NBS. 1973, No. 45.

[350] Pshezhetskii, S.Ya., *Mekhanism radiatsionno-khimicheskikh reaktsii. 2 izd.* (The Mechanism of Reactions in Radiation Chemistry. 2nd Ed.). Moscow: Khimiya, 1968.

[351] Bugaenko, L. T., Byakov, V. M., Kabakchi, S. A., *Khim.Vys.Energ.* 1985, **19**, 291.

[352] Allen, A. O., *The Radiation Chemistry of Water and Aqueous Solutions*, G. Van Nostrand Co. Inc., New Jersey, 1961.

[353] Bugaenko, L. T., *Vestn.Mosk.Univ. Ser. 2. Khim.* 1980, **23**, 523.

[354] Saraeva, V. V., Kalyazin, E. P., *Khim.Vys.Energ.* 1985, **19**, 218.

[355] Pichuzhkin, V. I., Saraeva, V. V., Bach, N. A., *Khim.Vys.Energ.* 1968, **2**, 151, 155.

[356] Topchiev, A. V., Polak, L. S., Chernyak, N.Ya., Glushnev, V. E., Vereshchinskii, I. V., Glazunov, P.Ya., *Dokl.Akad.Nauk SSSR.* 1960, **130**, 789.

[357] Saraeva, V. V., *Radiatsionno-khimicheskoe okislenie uglevodorov v zhidkoj faze., Sovremennye Problemy fizicheskoj khimii.* (Radiation Induced Oxidation of Hydrocarbons in Liquid Phase. In: Current Topics of Physical Chemistry). Ed. Topchieva, K. V. Moscow: Izd.Mosk Univ., 1975, **8**, 367.

[358] Tushurashvili, R. G., Nanobashvili, H. M., Basiladse, Ts.M., *Radiochem. Radioanal.Lett.* 1979, **38**, 99.

[359] *Praktikum po radiatsionnoj khimii.* (Laboratory Works on Radiation Chemistry). Ed. Saraeva, V. V. Moscow: Izd.Mosk.Univ. 1982.

[360] Tsetlin, B. L., Babkin, I.Yu., Kabanov, V.Ya., Ponomarev, A. N., *Khim. Vys.Energ.* 1985, **19**, 303.

[361] Ivanov, V. S., *Radiatsionnaya polimerizatsiya* (Radiation Polymerization). Leningrad: Nauka, 1967.

[362] Goldanskii, V. I., *Russ.Chem.Rev.* 1975, **44**, 1019.

[363] Lukhovitskii, V. I., Polikarpov, V. V., *Tekhnologiya radiatsionnoj emulsionnoj polimerizatsii* (Technology of Radiation Induced Emulsion Polimerization). Moscow: Atomizdat, 1980.

[364] Braginskii, R. P., Finkel, E. E., Leshchenko, S. S., *Stabilizatsiya radiatsionno-modifitsirovannykh poliolefinov* (Stabilization of Polyolefins Treated by Radiation). Moscow: Khimiya, 1973.

[365] P'yankov, G. N., Meleshevich, A. P., *et al.*, *Radiatsionnaya modifikatsiya polimernykh materialov* (Radiation Treatment of Polymeric Materials). Kiev: Tekhnika, 1969.

[366] Makhlis, F. A., *Radiatsionnaya fizika i khimiya polimerov* (Radiation Physics and Chemistry of Polymers). Moscow: Atomizdat, 1972.

[367] Shalaev, A. M., Adamenko, A. A., *Radiatsionno-stimulirovannoe izmenenie elektronnoj struktury* (Radiation Induced Changes in Electronic Structure). Moscow: Atomizdat, 1977.

[368] Vaviliov, B. S., Ukhin, N. A., *Radiatsionnye effekty v poluprovodnikakh i poluprovodnikovykh priborakh* (Radiation Induced Effect in Semiconductors and Semiconductor Base Devices). Moscow: Atomizdat, 1969.

[369] Vinetskii, V. L., Kholodar', G. A., *Radiatsionnaya fizika poluprovodnikov* (Radiation Physics of Semiconductors). Kiev: Naukova dumka, 1979.

[370] Byurganovskaya, G. V., Vargin, V. V., Leko, N. A., Orlov, N. F., *Dejstvie izluchenii na neorganicheskie stekla* (The Action of Radiations on Inorganic Glasses). Moscow: Atomizdat, 1968.

[371] Brekhovskikh, S. M., Viktorova, Yu.N., Landa, L. M., *Radiatsionnye defekty v steklakh* (Radiation Induced Defects in Glasses). Moscow: Energoatomizdat, 1982.

[372] Lushchik, Ch.B., Vitol, I. K., Elango, M. A., *Usp.Fiz.Nauk.* 1977, **122**, 223.

[373] Aluker, E. D., Lusis, D.Yu., Chernov, S. A., *Elektronnye vozbuzdeniya i radioluminestsentsiya shchelochno-galoidnykh kristallov* (Electronic Excitations and Radioluminescence in Alkali Metal Halide Crystals). Riga: Znanie, 1979.

[374] Pshezhetskii, S.Ya., Trakhtenberg, L. I., *Fizicheskaya khimiya. Sovremennye problemy*, Ed. Kolotyrkin Ya.M. Moscow: Khimiya, 1984, 175.

[375] Zakharov, Yu.A., Nevostruev, V. A., Ryabykh, S. M., Safonov, Yu.N., *Khim.Vys.Energ.* 1985, **19**, 398.

[376] Pshezhetskii, S.Ya., Kotov, A. G., Milinchuk, V. K., *et al.*, *EPR svobodnykh radikalov v radiatsionnoj khimii* (ESR of Free Radicals in Radiation Chemistry). Moscow: Kimiya, 1972.

[377] Lebedev, Ya.S., *Fizicheskaya khimiya. Sovremennye problemy*, Ed. Kolotyrkin, Ya.M. Moscow: Khimiya, 1985, 220.

[378] Milinchuk, V. K., *Khim.Vys.Energ.* 1985, **19**, 326.

[379] Rozno, A. G., Gromov, V. V., *Khim.Vys.Energ.* 1983, **17**, 223.

[380] Gromov, V. V., *Elektricheskii zaryad v obluchennykh materialakh* (Electric Charge in Irradiated Materials). Moscow: Energoatomizdat, 1982.

[381] Tyutnev, A. P., Vannikov, A. V., Saenko, V. S., *Khim.Vys.Energ.* 1983, **17**, 3.

[382] Salikhov, K. M., Medvinskii, A. A., Boldyrev, V. V., *Khim.Vys.Energ.* 1967, **1**, 381.

[383] Supe, A. A., Zubarev, V. E., Bugaenko, L. T., *Khim.Fizika.* 1986, **5**, 1626.

[384] Gromov, V. V., Kotov, A. G., *Khim.Vys.Energ.* 1985, **19**, 312.

[385] Kotov, A. G., Gromov, V. V., *Radiatsionnaya fizikokhimiya geterogennykh sistem* (Physical Radiation Chemistry of Heterogeneous Systems), Moscow: Energoatomizdat, 1988.

[386] Polak, L. S., Sergeev, P. A., Slovetskii, D. I., *Khim.Vys.Energ.* 1973, **7**, 387.

[387] Uhlenbeck, G. E., Ford G. *Lectures in Statistical Physics*. Providence R.I.: American Mathematical Society, 1963.

[388] Hintze, I. O., *Turbulence* (Russian Ed.). Moscow: Fizmatgiz, 1963.

[389] Lefever, R., Hortschemke, W. In: *Synergetics. Far from Equilibrium*. Berlin–N.Y.: Springer Verlag, 1979, 57.

[390] Schloegl, F., *Z.Phys.* 1972, **253**, 147.

[391] Lefever, R. In: *Fluctuations, Instabilities and Phase Transitions.* N.Y.: Plenum Press, 1975, 358.

[392] Kuznetsov, V. R., Lebedev, A. B., Sekundov, A. N., Smirnova, I. P., *Vliyanie pulsatsii kontsentratsii na diffuzionnoe gorenie, Khimicheskaya fizika goreniya i vzryva.* (The Effect of Concentration Pulsations on Diffusive Combustion. In: Chemical Physics of Combustion and Explosion). Chernogolovka: IKhF AN SSSR, 1977, 57.

[393] Kompaniets, V. Z., Ovsyannnikov, A. A., Polak, L. S., Epstein, I. L., *Plazmokhimia-79. Tezisy dokladov III Vsesoyuzn.Simp. po plasmokhimii.* Moscow: Nauka, 1979, 88.

[394] Kompaniets, V. Z., Polak, L. S., *Neravnovesnaya khimicheskaya kinetika i ee primenenie* (In: *Nonequilibrium Chemical Kinetics and Its Application*). Moscow: Nauka, 1979, 227.

[395] Kompaniets, V. Z., Polak, L. S., Konstantinov, A. A., Epstein, I. L., *Khimicheskie reaktsii v neravnovesnoj plazme* (In: *Chemical Reactions in Nonequilibrium Plasma*). Moscow: Nauka, 1977, 6.

[396] Grinchak, M. V., Kesei Ch., Kompaniets, V. Z. *et al.*, *Sintez v nizko-temperaturnoj plazme* (In: *Syntheses in Low-temperature Plasmas*). Moscow: Nauka, 1980, 88.

[397] Glansdorff, P., Prigogine, I., *Thermodynamic Theory of Structure, Stability and Fluctuations.* N.Y.: Wily-Interscience, 1971.

[398] Kompaniets, V. Z., Polak, L. S., Epstein, I. L., *Plazmokhimicheskie reaktory i protsessy* (In: *Plasmochemical Reactors and Processes*). Moscow: Nauka, 1979, 135.

[399] Chung, P. M., *Combus.Sci.Technol.* 1976, **13**, 123.

[400] Chung, P. M., *Phys.Fluids.* 1970, **13**, 1153.

[401] Chung, P. M., *Phys.Fluids.* 1973, **16**, 980.

[402] Donaldson, C., Varma, A. K., *Combust.Sci.Technol.* 1976, **13**, 5.

[403] Reynolds, W., *Ann.Rev.Fluid Mech.* 1976, **8**, 183.

[404] Lauder, B. E., Spalding, D. B., *Mathematical Models of Turbulence.* L.: Academic Press, 1972.

[405] Gosman, A. D., Puw, W. M., Runchal, A. K. *et al.*, *Heat and Mass Tranfer in Recirculating Flows.* N.Y.: Academic Press, 1969.

[406] Chung, P. M., *Phys.Fluids.* 1972, **15**, 1375.

[407] Resibois, P. De Leener, M., *Classical Kinetic Theory of Fluids.* New York–London–Sydney–Toronto: Wiley-Interscience, 1977.

[408] Griem, H. R., *Plasma Spectroscopy.* New York–San Fransisco–Toronto–London: McGraw Hill, 1964.

[409] Hill, R. A., *J.Quant.Spectr.Rad.Transfer,* 1967, **7**, 401.

[410] Volfkovich, S. I., Egorov, A. P., Rogovin, Z.Zh. *et al.*, *Obshchaya khimicheskaya tekhologiya.* **1** (General Industrial Chemistry. Vol. 1). Moscow: GNTI khim.lit., 1959.

[411] *Metody i protsessy khimicheskoj tekhnologii.* **1** (Methods and Processes of Industrial Chemistry). Moscow–Leningrad: Izd.AN SSSR, 1955.

[412] *Problemy khimii i khimicheskoj tekhnologii* (Problems of Chemistry and Industrial Chemistry). Moscow: Nauka, 1977.

[413] Breger, A.Kh., *Radiatsionno-khimicheskaya tekhnologia. Ee zadachi i metody* (Industrial Radiation Chemistry. Its Tasks and Methods). Moscow: Atomizdat, 1978.

[414] Rudoj, V. A., Putilov, A. V., *Radiatsionnaya tekhnologia za rubezhom* (Radiation Processing in Foreign Countries). Moscow: Energoatomizdat, 1983.

[415] *Transactions of the Fourth Int. Meeting on Rad. Processing, Radiat.Phys. Chem.* 1983, **22**, 1. No. 1–2., 2 No. 3–5.

[416] *Transactions of the Fifth Int. Meeting on Rad. Processing, Radiat.Phys. Chem.* 1985, **26**, No. 1–5.

[417] Chepel, L. V., *Primenenie uskoritelej elektronov v radiatsionnoj khimii* (Use of Electron Accelerators in Radiation Chemistry). Moscow: Atomizdat, 1975.

[418] Goldin, V. A., *Metody i ustrojstva dlya radiatsionnoteckhnologicheskikh issledovanii s izotopnymi istochnikami izluchenii* (Methods and Devices for Radiation Processing Studies with Isotope Sources of Radiation). Moscow: Energoatomizdat, 1982.

[419] Abramyan, E. A., *Promyshlennye uskoriteli elektronov* (Industrial Electron Accelerators). Moscow: Energoatomizdat, 1986.

[420] Syrkus, N. P., Starizny, E. S., Rudoj, V. A., Putilov, A. V., *Radiatsionnaya tekhnologiya i kompleksnye energokhimicheskie proizvodstva* (Radiation Processing and Combined Energochemical Production Technologies). Moscow: Atomizdat, 1980.

[421] Goldin, V. A., Chistov, E. D., *Ustanovki i apparaty radiatsionnoj tekhnologii* (Apparatuses and Setups for Radiation Processing) Moscow: Energoatomizdat, 1985.

[422] Bugaenko, L. T., Kalyazin, E. P., *Radiatsionno-khimicheskaya tekhnologiya, Problemy khimii i khimicheskoj technologii* (Radiation Processing. In: *Problems of Chemistry and Industrial Chemistry*). Moscow: Nauka, 1977, 35.

[423] Kalyuzhny, S. V., Sklya, V. I., Kovalev, G. V., *et al.*, *Tekhnicheskaya bioenergetika, Tezisy II Vsesoyuzn.Soveshch. Saratov.* September 15–19, 1985. Moscow: Nauka, 1985, 24.

[424] Goldin, V. A., Stepanov, G. D., *Khim.Vys.Energ.* 1985, **19**, 260.

[425] Pikaev, A. K., *Sovremennaya radiatsionnaya khimiya. Osnovnye polozheniya. Eksperimentalnaya tekhnika i metody* (Modern Radiation Chemistry. Basic Principles. Methods and Instruments). Moscow: Nauka, 1985.

[426] *Uskoriteli dlya promyshlennykh radiatsionno-technologicheskikh*

processov (Industrial Accelerators for Radiation Processing). Moscow: Tekhsnabeksport, 1979.

[427]   Generalova, V. V., Gurskii, M. N., *Khim.Vys.Energ.* 1985, **19**, 272.

[428]   Ivanov, V. I., *Kurs dozimetrii* (A Treatise on Dosimetry). Moscow: Atomizdat, 1978.

[429]   Generalova, V. V., Gurskii, M. N., *Dozimetriya v radiatsionnoj tekhnologii* (Dosimetry in Radiation Processing). Moscow: Izd.standartov, 1981.

[430]   Gochaliev, G. Z., *Tekhnologicheskaya dozimetriya* (Industrial Dosimetry). Moscow: Energoatomizdat, 1984.

[431]   Chertov, A. G., *Edinitsy fizicheskikh velichin* (Units of Physical Quantities). Moscow: Vysshaya shkola, 1977.

[432]   Bukhman, F. A., Melamed, V. G., Polak, L. S., Khait, Yu.L., *Dokl. Akad. Nauk SSSR.* 1967, **177**, 876.

[433]   Bukhman, F. A., Melamed, V. G., Polak, L. S., Chait, Y. L., *Primenenie vychislitel'noj matematiki v khimicheskoj i fizicheskoj kinetike* (In: Application of Computational Mathematics in Chemical and Physical Kinetics). Moscow: Nauka, 1969.

[434]   Vursel, F. B., Lysov, G. V., Polak, L. S., Blinov, L. M., *Khim.Vys.Energ.* 1971, **5**, 105.

[435]   *Plazmokhimicheskie reaktsii i protsessy* (Plasmochemical Reactions and Processes). Ed. Polak, L. S. Moscow: Nauka, 1981.

[436]   *Zh.Vsesoyuzn.Khim.Obshch. im D.I. Mendeleeva.* 1973, **18**, 242–332.

[437]   *Uspol'zovanie atomnoj energii v khimicheskoj tekhnologii* (Utilization of Atomic Energy in Industrial Chemical Processes). Moscow: NIITEKhim. 1983.

[438]   Ershov, B. G., Petryaev, E. P., *Khim.Vys.Energ.* 1985, **19**, 483.

[439]   Dzhagatspanyan, R. V., Kosorotov, V. I., Filippov, M. T., *Vvedenie v raditsionnuyu teknologiyu* (Introductory Radiation Processing). Moscow: Atomizdat, 1979.

[440]   Aver'yanov, S. V., Goldin, V. A., *Tekhnologiya radiatsionnoj vulkanizatsii termostojkikh samoslipayushchikhsya elektroizolyatsionnykh materialov* (The Technology of Radiation Curing of Thermostable Self-Adhesive Insulating Materials). Moscow: Atomizdat, 1980.

[441]   Kuzminskii, A. S., Fedoseeva, T. S., Kaplunov, M.Ya., *Tekhnologiya radiatsionnoj vulkanizatsii i modifitsirovaniya elastomerov* (The Technology of Radiation Cure and Modification of Elastomers). Moscow: Energiya, 1982.

[442]   Finkel, E. E., Karpov, V. L., Berlyant, S. M., *Tekhnologiya raditsionnoj modifikatsii polimerov* (Radiation Processing of Polymers). Moscow: Energiya, 1983.

[443]   Bach, N. A., Vannikov, A. V., Grishina, A. D., *Electroprovodnost' i*

paramagnetizm polimernykh poluprovodnikov (Conductivity and Paramagnitism of Polymeric Semiconductors). Moscow: Nauka, 1971.

[444] Shiryaeva, G. V., Kozlov, Yu.D., *Tekhnologiya radiatsionnogo otverzhdeniya pokrytii* (The Technology of Radiation Cure of Coatings). Moscow: Atomizdat., 1980.

[445] Chutny, B., Kucera, J., *Radiat.Res.Rev.* 1975, **5**, 1.

[446] Vereshchinskii, I. V., *Khim.Vys.Energ.* 1970, **4**, 483.

[447] Zagorets, P. A., Poluektov, V. A., Shostenko, A. G., *Khim.Vys.Energ.* 1985, **19**, 393.

[448] Dzhagatspanyan, R. V., Yemel'yanov, V. I., *Tekhnologiya radiatsionnokhimicheskikh proizvodstv v gazovoj faze* (Gas Phase Radiation Processing Production Units). Moscow: Energiya, 1982.

[449] Hutson, T., Jr., Logan, R. S., *Chem.Eng.Progr.* 1972, **68**, 76.

[450] Governal, L. J., Clarke, J. T., *Chem.Eng. Progr.* 1956, **52**, 281.

[451] Hyring, H. G., Knol, H. W., *Chem.Process Eng.* 1964, **45**, 560, 619, 690, ibid. **46**, 38.

[452] *Ullmans Encyklopadie der technischen Chemie.* 4. Weinheim: Aufl. Verlag Chemie, 1975, **9**, 525.

[453] Broich, F., *Fette, Seifen, Anstrichm.* 1970, **72**, 17.

[454] Joschek, H. I., *Chem.Ztg.* 1969, **93**, 655.

[455] Asinger, F., *Ber.Deutsch Chem.Ges.* 1944, **77**, 191.

[456] Orthner, L., *Angew.Chem.* 1950, **62**, 302.

[457] Graf, R., *Justus Liebigs Ann.Chem.* 1952, **578**, 50.

[458] Beerman, C., *Eur.Chem.News. Normal Paraffin Supplement.* 1966, 2 Dec., 36.

[459] Hartig, H., *Chem.Ztg.* 1975, **99**, 179.

[460] Turner, P., *Inf.Chem.* 1970, **9**, 51.

[461] Sanders, G. M., Pot, J., Havinga, E., *Fortschr.Chem.Org.Natur.* 1969, **27**, 131.

[462] Arnold, D. P., Demayo, P. D., *Chem.Technol.* 1971, **1**, 615.

[463] Fischer, M., Wierdorf, W.-W., Nurrenbach, A., et al., *Photochemical Isomerisation of cis-isomers of vitamin A compounds and their derivatives.* West German Patent. 1974. 2210800 BASF AG. *Chem.Abstr.* 1974, **80**, 11623h.

[464] Fischer, M., *Angew.Chem.* 1978, **90**, 17.

[465] *Organicheskie fotokromy* (Organic Photochroms). Ed. Yeltsov, A. V., Leningrad: Khimiya, 1982.

[466] Barachevskii, V. A., Lashkov, G. I., Tsekhomskii, V. A., *Fotokhromizm i ego primeneiya* (Photochromism and Its Applications). Moscow: Khimiya, 1977.

[467] Kuzmin, M. G., *Dokl.Akad.Nauk SSSR.* 1963, **151**, 1371.

[468] Brederlow, G., Fill, E., Witte, K. J., *The High-Power Iodine Laser.* Berlin: Springer Verlag, 1983.

[469] Haydon, S. G., *Spestrosc.Lett.* 1975, **8**, 815.

[470] *Excimer Lasers*, Ed. Phodes, Ch.K., Berlin: Springer Verlag, 1979.

[471] Ablesov, V. K., Denisov, Yu.I., *Protochnye khimicheskie lazery* (Chemical Flow Lasers). Moscow: Energoatomizdat, 1987.

Symbols for quantities

$a, b, \ldots$	reaction rate order
A	absorbance, atomic mass, pre-exponential factor
A_r	work of rearrangement of medium
c	velocity of light, concentration
c_{Ac}	concentration of solute (scavenger)
c_i	concentration of i-th component
D	diffusion coefficient
e	charge of electron
e_{qu}^-	quasi-free electron
e_s^-	solvated electron
e_{tr}^-	trapped electron
E	kinetic energy, mean energy lost by electron
E'	kinetic energy of expelled electron
E^*	excitation energy
E_0	initial energy
E°	standard electrode potential
E_a	activation energy of chemical reaction
E_e	energy of electron
E_i	energy of ion beam
E_n	energy of neutron
E_{th}	threshold energy of chemical reaction
E_b	binding energy of electron in atom or of nucleon in nucleus
E_j	energy of electronic transition
f_i	oscillator strength
f	vibrational wave function, distribution function
ΔF	change in Helmholtz free energy
$F(r)$	distance distribution function
G	energetic field, Gibbs free energy
G_o	initial energetic yield

G_{fi}	free ion yield
$\mathbf{H_o}$	Hamiltonian
$\mathbf{H}$	perturbation operator
ΔH	enthalpy
h	Planck's constant
I	ionization potential, intensity, transmitted radiation flux
I_i	ionization potential of i-th level
k	Boltzmann's constant, reaction rate constant
k_D, k_{-D}	rate constants for diffusion approach and diffusion separation
k_d	radical disproportionation rate constant
k_c	radical combination rate constant
k_{in}	rate constant for reaction in 'cage'
K	equilibrium constant
l	optical pathlength. Thickness of layer of substance, target thickness
l_t	thermalization pathlength
m_e	mass of electron
m_i	mass of fast ion
m_n	mass of neutron
m_{nu}	mass of nucleus
m_0	rest mass of electron
n, N	number of particles in unit volume
N_A	Avogadro's number
P	probability
p	momentum
R	range, radius
Ry	Rydberg constant
r_o	Onsager radius
R_{12}	transition moment
s	spin-wave function
S	stopping power, entropy
t	time
T	temperature
T_e	electron temperature
T_n	neutral atom temperature
T_p	temperature of plasma
u	odd-wave function
U	potential well depth
v	linear velocity
v_e	velocity of electron
W	energy, average energy of ion pair formation
x	thickness, pathlength
Z	charge of nucleus
z	charge of ion

g	even-wave function
β	ratio of velocity of particle to velocity of light
ε	molar extinction coefficient, fraction of energy transferred, static permittivity
ε_0	absolute permittivity *in vacuo*
ε_∞	permittivity at infinite frequency
θ	angle
λ_t	transport pathlength
μ	linear coefficient of interaction, transition operator, mobility of charge particle in electric field, chemical potential
$1/\mu$	free pathlength
μ/ρ	mass coefficient of interaction
ν	frequency, frequency factor
ν_{λ_j}	stoichiometric coefficient
$\tilde{\nu}$	wave number
ρ	density
σ	cross-section of process, reaction cross-section, substituent constant in Taft or Hammett equations
$\sigma(E)$	energy-dependent cross-section of process
τ_1	dielectric relaxation time of medium
τ_s	solvation time
φ_e	electronic wave function
φ	angle, wave function, quantum yield
Ψ	wave function

Index